PLANT SCIENCE RESEARCH AND PRACTICES

PHOTOSYNTHESIS

FUNCTIONAL GENOMICS, PHYSIOLOGICAL PROCESSES AND ENVIRONMENTAL ISSUES

Plant Science Research and Practices

Additional books in this series can be found on Nova's website
under the Series tab.

Additional e-books in this series can be found on Nova's website
under the e-book tab.

PLANT SCIENCE RESEARCH AND PRACTICES

PHOTOSYNTHESIS

FUNCTIONAL GENOMICS, PHYSIOLOGICAL PROCESSES AND ENVIRONMENTAL ISSUES

NAFEES KHAN
EDITOR

nova publishers
New York

Library of Congress Cataloging-in-Publication Data

Photosynthesis : functional genomics, physiological processes and environmental issues / editor: Nafees Khan (Department of Botany, Aligarh Muslim University, India).
pages cm. -- (Plant science research and practices)
ISBN 978-1-63463-304-8 (hardcover)
1. Photosynthesis. 2. Photosynthesis--Genetic aspects. 3. Photosynthesis--Environmental aspects. I. Khan, Nafees A., editor. II. Series: Plant science research and practices.
QK882.P383 2014
572'.46--dc23

2014039711

Published by Nova Science Publishers, Inc. † New York

CONTENTS

PREFACE

Photosynthesis is a well-regulated process utilizing solar energy for storage as chemical energy to meet the plant's metabolic needs. It is the key phenomenon that augments plant growth and development under all conditions and is considered as the ultimate source of life for all plant and human life. Although the basic processes of photosynthesis and their importance to the agricultural system for food security have been recognized about two centuries ago, there is an increasing need for intense research considering the population explosion. As the demand for food is expected to increase twofold by 2030, the efforts by plant scientists are necessary to improve the yield of food crops for sustainable development. The improvements in yield of crops can be attained by the protection of photosynthetic apparatus in the current changing environment, where degradation of soil, air and water quality has posed challenges to photosynthetic potential of plants. It has been estimated that the increasing salinization of land will result in the loss of 50% cultivated land by 2050. The situation will further aggravate with the contamination of soil and water by heavy metals. The concentration of CO2 is expected to rise to as much as 500-1000 ppm by 2100 which can lead to increases in global mean temperature by approximately 1°C to 3°C above the present value by 2025 and 2100, respectively. Photosynthesis is one the first processes inhibited under the changing environment leading to reduction of growth and productivity of plants. The research focus on targeting mechanisms that could help in the survival of plants under the changing environment is essential for modulating photosynthetic processes to suit the climate changes. There are several processes which can be manipulated to maintain photosynthetic potential of plants. These include up-regulation of photosynthesis-related enzymes and controlling genes, enhancing source-sink potential and inhibiting the feed control. Functional genomics can also be a significant approach to understand the photosynthetic gene function under the influences of climate changes. This book provides information on the functional and structural genomics for better understanding and analysis of photosynthesis under changing climatic conditions. The central theme of the present book is to unravel the mechanisms of photosynthesis under optimal and stressful environments and gain insight into the processes which can lead to a better understanding of the photosynthetic inhibition due to various constraints and provide measures to improve photosynthetic mechanisms under these conditions.

Chapter 1 - Various growth and development phenomenon were described earlier in terms of physiology. With the advent of technology, these have been described in terms of biochemical aspects. With further advancement in technology, molecular approach is being utilized for the study of growth and development in organisms. In the last two decades, molecular regulation, through genes controlling growth and development, has been studied extensively. Likewise, photosynthesis (a metabolic process) has also been influenced by various factors and the genes responsible has been studied. Photosynthesis is vital to all aspects of plant biology as the provider of energy and assimilator for growth and reproduction, but how it is regulated by abiotic stresses and biotic stresses and their molecular regulatory mechanisms is still not very clear. Photosynthesis is central to the performance of autotrophic plants, not in isolation but rather in combination with the processes that determine growth and development as part of the whole organism's function, reproductive performance and survival. Different environmental factors – abiotic such as temperature, water, and light and biotic such as insect, pests, or pathogens; together with the competition from other organisms – vary unpredictably. Since the mechanism of photosynthesis involves various components, including photosynthetic pigments and photosystems, the electron transport system, and CO_2 reduction pathways, any damage at any level caused by stress may reduce the overall photosynthetic capacity of a green plant. In this chapter, the molecular regulation of photosynthetic response of green plants to environmental stress factors, like extremes of light, water, and temperature together with biotic factors are discussed. Also, chloroplasts are shown as sensor of stress.

Chapter 2 - Elevated atmospheric CO_2 has been shown to influence many key physiological processes of plants. Leaf photosynthesis is typically stimulated during growth with double the present atmospheric CO_2 concentration ([CO_2]). It is not unusual for CO_2 in the rhizosphere to be up to more than 10-fold higher than the atmospheric [CO_2] due to the activities of microorganisms. The authors often question whether CO_2 level in the root-zone (RZ) could regulate plant growth. By using aeroponically-grown lettuce plants, the authors carried out a series of experiments to study the effects of elevated RZ [CO_2] on the photosynthesis of vegetable crops at different air and RZ temperatures in an environmentally controlled room and fluctuating greenhouse. Under two different temperature regimes of 28°C/22°C and 36°C/30°C (day/night) in the environmentally controlled room, it was found that elevated RZ [CO_2] stimulated photosynthetic CO_2 assimilation, with greater increase at higher temperatures. Elevated RZ [CO_2] decreased stomatal conductance at both temperatures, reducing transpirational water loss. When whole lettuce plants were grown under ambient temperatures ranging from 20 to 36°C under a maximal photosynthetic photon flux density (PPFD) of 2100 μmol m^{-2} s^{-1} inside the greenhouse, the results indicated that elevated RZ [CO_2] protected lettuce plants from dynamic photoinhibition of photosystem II. It was supported by the fact that midday depression of chlorophyll fluorescence F_v/F_m ratios was alleviated in elevated RZ [CO_2] treated plants. Elevated RZ [CO_2] was also found to reduce the negative impacts of high air temperature on photosynthesis, N metabolism, and growth of lettuce plants, resulting in enhanced productivity. The effects of elevated RZ [CO_2] on photosynthesis, productivity, nitrate, total reduced nitrogen, total leaf soluble and Rubisco proteins under different root-zone temperature (RZTs) was also studied in a tropical greenhouse. The results demonstrated that increased photosynthesis and productivity under elevated RZ [CO_2] at different RZTs was partially due to the alleviation of midday water loss, both dynamic and chronic photoinhibition, as well as, a higher turnover of Calvin Cycle with higher ribulose-1,5-bisphosphate carboxylase/oxygenase (Rubisco) proteins. Effects of RZ [CO_2] at different air and RZT temperatures on different aspects of photosynthesis and growth were discussed.

Chapter 3 - Leaf senescence is the last step of leaf development and can be induced by many genetic and environmental factors. The chloroplast and many leaf pigments undergo a progressive degeneration. The leaf functionality is slowed down and several physiological processes are altered. The chlorophyll *a* fluorescence and its derivate indexes can be used to monitor leaf senescence. The loss of leaf functionality is tightly correlated with the photosynthetic activity. The chlorophyll *a* fluorescence and its derived parameters can be suggested as tool to monitor the leaf senescence. Natural and induced senescence such as dark incubation, detachment, salinity stress, ethylene exposure, heavy metal stress etc. will be taken in consideration. The physiological, biochemical and molecular basis connected with the photosynthesis and chlorophyll fluorescence will be described and discussed.

Chapter 4 - The phenomenon of carbon isotope heterogeneity in a photosynthetic biomass became known at the beginning of 1960s, after the pioneering works of Abelson and Hoering (1960) and Park and Epstein (1960, 1961), but the real mechanism of carbon isotope fractionation leading to uneven carbon isotope distribution in a biomass and the reasons for the isotope patterns of the biomolecules remain vague. In this chapter, it is shown that ^{13}C distribution is tightly bound to the oscillatory character of metabolic processes and to the real biosynthetic pathways of metabolites. It was found that due to oscillations, photosynthesizing cells function in regime of filling/depleting metabolic pools. In the phase of depletion the isotope effect-generating reactions are followed by a Raleigh effect, resulting in the dependence of the isotopic patterns of the molecules on the sequence of their syntheses, i.e., in the strict temporal organization of cell processes. It is shown that oscillations in photosynthesis consist of a phase of CO_2 assimilation and a phase of photorespiration. Each phase is followed by carbon isotope fractionation having the effects of opposite signs. As a result, carbon isotope heterogeneity emerges. Post-photosynthetic metabolism, mainly presented as the metabolism of the glycolytic chain, is also presented through oscillations that have two phases: glycolysis and gluconeogenesis. In glycolysis, carbon isotope fractionation occurs. This is one more reason for the isotope heterogeneity of a biomass and works in combination with biosynthetic pathways to determine the isotopic patterns of post-photosynthetic metabolites. Knowledge of the nature of carbon isotope heterogeneity provides a new delicate tool for studies on living organisms. Some possible fields for carbon-isotope data applications with the presented model are considered.

Chapter 5 - There is ample evidence in existing scientific literature that photorespiration acts towards reducing photoinhibition in plants. The internal generation of CO_2, resulting from the oxidation of ribulose bisphosphate by the enzyme RUBISCO acting as an oxygenase, is often sufficient to prevent photoinhibition partly or even completely, whereas photoinhibition is observed, when photorespiration is prevented. To prevent photorespiration in terrestrial plants it is, however, necessary to create a highly artificial composition of air that is replete in CO_2 and/or deplete in oxygen. The use of highly artificial and manipulated atmospheres makes it difficult to assess the environmental significance of photorespiratory alleviation of photoinhibition.

In contrast, aquatic plants naturally grow in an environment where carbon availability as well as ambient oxygen concentration varies from location to location as a result of varying carbonate content in the underlying soil as well as diurnally in the same location as a result of biological processes (photosynthesis and respiration). Probably as a result of the availability of free, dissolved, CO_2, often limiting for photosynthesis due to the low diffusion rate of CO_2 in water, aquatic plants have evolved a number of different physiological carbon accumulation strategies, including bicarbonate use, inducible C_4-type photosynthesis and CAM photosynthesis. This makes aquatic plants an obvious tool for the study of photorespiration/photoinhibition interactions.

In this chapter, the author am presenting data from a study on photoinhibition in three aquatic plants with different carbon uptake strategies: *Callitriche cophocarpa*, a plant reportedly relying entirely on dissolved CO_2 in the water for photosynthesis, *Elodea canadensis*, a bicarbonate user, and *Crassula helmsii*, an aquatic CAM-plant, taking up CO_2 during the night, storing it as malate for photosynthetic use the following day. The plants were grown under varying carbon-availability, and were compared based on their photosynthetic oxygen evolution, degree of photoinhibition (evaluated using PAM-fluorometry), and their activity of antioxidative enzymes.

Photosynthetic oxygen evolution decreased with decreasing carbon availability in *Elodea* and *Callitriche*, but not in the aquatic CAM plant *Crassula*, where the internal malate pool partly compensates for low ambient carbon availability. Photoinhibition increased with increasing carbon availability (decreasing photorespiration) in both *Callitriche* and *Elodea*, but not in *Crassula*, where the internal malate pool may prevent any significant photorespiration as well as any significant photoinhibition. Photoinhibition was higher in the bicarbonate user *Elodea*, compared to *Callitriche*, as bicarbonate uptake functions as a carbon concentration mechanism, reducing photorespiration. The antioxidative enzyme ascorbate peroxidase evinced higher activities with lower carbon availability, especially in *Callitriche*, to a lesser degree in *Elodea* and not at all in *Crassula*.

The results confirm the previously described interaction between carbon uptake physiology and photoinhibition. Interestingly, CAM-photosynthesis may prevent both photorespiration and photoinhibition, but further studies are needed.

Chapter 6 - Geological records indicate that plant life on the land has aquatic origins, started approximately 460 million years ago. However, nearly all of the land-dwelling plants are very sensitive to the inundation. Nonetheless, the most basic attributes of flooding tolerance seemingly involved an adaptive *return* to aquatic competence by terrestrial plants. The negative impact of flooding on terrestrial plant life is a consequence of oxygen deprivation. Flooding often disrupts plant water relations, gas exchange and stomatal conductance. Under the diffusional limitation of carbon uptake imposed by stomatal closure, the photosynthetic capacity of flood-sensitive plants would be significantly declined and can exhibit non-stomatal inhibition of photosynthesis with alterations in plant hormone concentrations induced by tissue anoxia. A tight regulation of the plant water status is thus required to maintain photosynthesis, nutrient uptake and growth in flooded environments. Here, the authors review the importance of stomatal attributes and photosynthesis for the survival of flooded land-dwelling plants.

Chapter 7 - Photosynthetic organisms sustain life by converting light energy into chemical energy. Photosynthesis is a key phenomenon contributing substantially to plant growth and development. It is a well regulated and multistep process to produce assimilatory power NADPH and ATP to synthesize carbohydrate. The changing environmental conditions have forced plant scientists to better understand photosynthetic process more importantly for agricultural sustainability. The abiotic stresses such as salt stress, heavy metal stress, water stress, heat stress, chilling stress, nutrient stress etc. adversely affect photosynthesis because of induced oxidative stress. Various mechanisms may provide protection to the photosynthetic apparatus under abiotic stress, but phytohormones have potential to modify biochemical, physiological processes and molecular mechanisms in plants to optimize photosynthesis under optimal and stressful environments. They are a strong tool to protect and alleviate stress by modulating metabolic processes in plants. The present chapter gives an insight into the role of phytohormones in modifying biochemical, physiological and molecular responses to adapt plants in stressful environments.

Chapter 8 - Photosynthesis is undoubtedly a vital biological process which contributes substantially towards plant's growth and development. The enzyme ribulose 1, 5-bisphosphate carboxylase (Rubisco) is the most abundant enzyme on planet and plays an important role in autotrophic carbon fixation. It regulates the pool size of important photosynthetic intermediates. It is the key enzyme of the Calvin cycle which catalyzes, the carboxylation and oxygenation of ribulose 1, 5-bisphosphate (RuBP) and is the main enzyme fixing the vast majority of atmospheric carbon, acting as the natural sink for CO_2. Abiotic stress hampers the process of photosynthesis in most of the plants by modifying the ultrastructure of the organelles, concentration of various pigments and metabolites including enzymes involved in this process. It also influences stomatal regulation. Salinity is currently the most important abiotic stress that unfavorably affects the plant growth process and production potential through its impact on various physiological and biochemical process, including the imperative process of photosynthesis. One of the most prominent effects of stresses is the stomata closure. It results in lowering the concentration of intercellular CO_2, which in turn causes deactivation of Rubisco. Understanding the role of Rubisco in the biosphere is very important in today's context, with increased rate of fossil carbon liberated by human activities and flaring the problem of global warming. Further, it is important to enhance our understanding on the expression patterns of Rubisco genes under salt stress, involved in photosynthesis, in order to develop transgenic lines of different crops with enhanced photosynthetic capacity under stressful conditions. Understanding the details of the adverse effect of salinity stress on photosynthesis, focusing on Rubisco, can lead to the design of new salt-tolerant genotypes with better abitity to protect photosynthetic potential under salinity stress. The present study focuses on structural and functional aspects of the Rubisco in photosynthesis under optimal and salt stress conditions.

Chapter 9 - Oxidative stress is an imbalance between reactive oxygen species (ROS) generation and scavenging of the reactive intermediates leading to damage and redox disturbances in living cells. It leads to disruption and impairment of cellular sugars, proteins, lipids and nucleic acids. Plants under stress tend to generate and accumulate ROS hindering signal cascades at organelle and cellular levels. Soluble sugars such as fructans, raffinose family oligosaccharides (RFOs), trehalose and polyols in plants have long been characterized to activate antioxidant pathways or themselves act as ROS scavengers under abiotic stress. Herein, the authors discuss their antioxidant functioning during environmental stress in cell organelles such as chloroplasts, mitochondria and peroxisomes. This physiological sugar-antioxidant interplay exerts a control over various genetic and molecular approaches facilitating favourable strategies in developing plants of agronomic and commercial importance.

Chapter 10 - Plants are constantly exposed to pathogens infection and mechanisms of defense must be triggered to avoid their deleterious effect on host physiology. Alterations on host secondary metabolism as a result of pathogens infection have been investigated by many researchers; however, studies reporting their effects on primary metabolism are yet poorly characterized. Defense induction generates energetic costs for their hosts, which require, for instance, modifications on photosynthetic and respiratory performance besides changes on source-sink patterns, leading to reductions in plant fitness then translated in production and profit losses. It is invaluable the importance of photosynthesis in resistance against pathogens, so the development of strategies that favors plant physiology and enhance the activation of plant defense responses can provide a powerful tool for disease control, delaying or reducing the use of fungicides or pesticides since the demand for sustainable plant-protection strategies is urgent. The application of silicon as a broad plant-protection strategy has been currently evidenced; despite virtually nothing is known about the molecular basis and regulation of silicon-mediated disease resistance. The aim of this chapter is to introduce to the readers some aspects of silicon involvement on plant physiology and thus encourage the discussion and investigation on this intriguing subject.

Chapter 11 - The search of new inhibitors of the photosynthetic process through the screening of natural products is described in this chapter. Additionally, synthetic endeavors toward the development of more effective analogues are also analyzed.

Chapter 12 - According to established knowledge, only 0.1% of the total light energy (visible and invisible) that reaches the surface of our planet earth is transformed into free chemical energy through the chlorophyll molecule. If only the human knew he had the ability to dissociate the molecule of water,and do it irreversibly. With this small fraction the food chain is initiated. Explaining the existing biomass based only on photosynthesis in plants requires the acceptance of dogmas such as the source of human energy is glucose or the ability of the eukaryotic cell to amplify the ATP; concepts that although they are widely accepted cannot be explained rationally. Hence the author's discovery of the intrinsic capacity of the melanin molecule to dissociate and re-form the water molecule involves a conceptual revolution of biblical proportions. It explains and demonstrates that photosynthesis exists in humans and therefore in mammals, and in each and every living entity where melanin is present. Its primary function is the production of free chemical energy from photonic energy, a behavior unique in nature.

In: Photosynthesis
Editor: Nafees Khan

ISBN: 978-1-63463-304-8

Chapter 1

PHOTOSYNTHESIS: MOLECULAR REGULATION THROUGH BIOTIC AND ABIOTIC FACTORS

Saumya Srivastava ***** ***and Yogesh Kumar Sharma*** **†**
Plant Physiology Laboratory, Department of Botany,
University of Lucknow, Lucknow, India

ABSTRACT

Various growth and development phenomenon were described earlier in terms of physiology. With the advent of technology, these have been described in terms of biochemical aspects. With further advancement in technology, molecular approach is being utilized for the study of growth and development in organisms. In the last two decades, molecular regulation, through genes controlling growth and development, has been studied extensively. Likewise, photosynthesis (a metabolic process) has also been influenced by various factors and the genes responsible has been studied. Photosynthesis is vital to all aspects of plant biology as the provider of energy and assimilator for growth and reproduction, but how it is regulated by abiotic stresses and biotic stresses and their molecular regulatory mechanisms is still not very clear. Photosynthesis is central to the performance of autotrophic plants, not in isolation but rather in combination with the processes that determine growth and development as part of the whole organism's function, reproductive performance and survival. Different environmental factors – abiotic such as temperature, water, and light and biotic such as insect, pests, or pathogens; together with the competition from other organisms – vary unpredictably. Since the mechanism of photosynthesis involves various components, including photosynthetic pigments and photosystems, the electron transport system, and CO_2 reduction pathways, any damage at any level caused by stress may reduce the overall photosynthetic capacity of a green plant. In this chapter, the molecular regulation of photosynthetic response of green plants to environmental stress factors, like extremes of light, water, and temperature together with biotic factors are discussed. Also, chloroplasts are shown as sensor of stress.

---

* Saumya Srivastava e-mail: sonata906@gmail.com.
† Yogesh Kumar Sharma e-mail id: yogesh_s26@yahoo.com.

Keywords: Photosynthesis, Abiotic, Biotic, Molecular regulation

INTRODUCTION

Abiotic factors, such as supply of water, temperature and salinity, and biotic factors like insect, pests or pathogens strongly affect plant growth and development. Extremes of these conditions adversely affect many different mechanisms associated with plant responses and adaptation to stress. This may result in altered morphology and physiology, inhibition of cell division, and more importantly changes in metabolism, including photosynthesis. Life on earth ultimately depends on energy derived from the sun. Photosynthesis is the only process of biological importance that can harvest this energy. It is generally regarded as the major mechanism of energy input into the living world.

The term *photosynthesis* means "synthesis using light." Photosynthetic organisms use solar energy to synthesize carbon compounds that cannot be formed without the input of energy. More specifically, light energy drives the synthesis of carbohydrates from carbon dioxide and water with the generation of oxygen. Energy stored in these molecules can be used later to power cellular processes in the plant and can serve as the energy source for all forms of life. The present chapter attempts to discuss the effect of various biotic and abiotic stresses on photosynthetic mechanism, together with molecular regulation of stress responsive genes in photosynthesis.

GENERAL IMPACT OF STRESS ON PLANT GROWTH AND DEVELOPMENT

Stress is considered as an alteration in any environmental factor that has an impact on the plant by affecting its biochemical and physiological response to such changes, and may on occasions lead to damage or injury. In general, the stressful situations cause a series of physiological changes in the plant to offset these situations that are intended to maintain the plant's vital functions. Stress can be caused by both abiotic and biotic factors. Abiotic factors mainly include extreme temperatures, water stress, light and nutrition stress, presence of heavy metals, salinity etc. Biotic factors generally include damage done to plants by living organisms, such as bacteria, viruses, fungi, parasites, insects, weeds, etc.

Among stresses, the most important are those caused by low temperatures (stress due to cold or frost) or high temperatures (stress caused by excessive heat). Plants suffer severe damage that limits their optimum development, above or below a certain temperature, depending on the plant. Exposure to the cold paralyses enzymatic activity and induces a fall in the fluidity of the cellular membranes, which means that the passage of water and nutrients through them may be affected, and the plant may stop producing. If the temperature falls sharply, the plant runs a risk of freezing, with the ensuing formation of ice crystals inside the cells, leading to considerable cellular dehydration. On the contrary, prolonged exposure to extremely high temperatures causes damage to the cells due to the inactivation of enzymes and de-naturation of proteins, leading to an increase in the fluidity of the cellular membranes, i.e., the permeability of solutes through them becomes altered.

The heat also intensifies plant respiration in relation to photosynthesis which may lead to reduction of growth. Next important abiotic factor is the water stress. It may include a shortage of water (drought), and also an excess of this element (root asphyxia). Drought occurs when plant transpiration is greater than its water absorption. One of the first signs is a loss of cellular turgidity, which leads to the dehydration of the cell, closing of the stomata and a reduction in the photosynthetic capacity. Also, due to poor water management, the salinity of the soil is fast becoming a great problem in international agriculture. Stress due to excess salinity causes toxicity problems in plants due to the high saline concentration, whilst at the same time, producing an osmotic effect that prevents the plant from absorbing the water and nutrients it needs. Besides above abiotic factors, deficiency of mineral elements and presence of heavy metals in toxic concentrations generally leads to severe hampering of plant growth and development.

Among biotic factors, most harmful is the insects group, as it can cause great damage to crops. Insects can be classified depending on the way in which they cause harm to the plant: sap-sucking insects, insects that gnaw shoots, tunnelling insects, vectors of pathogenic agents, etc. Infestation by nematodes is increasingly important, due to the practical disappearance of conventional chemical disinfectants. These insects which live in the ground usually attack the roots, leading to deformities and gallnuts, but the symptoms are extremely visible in the top part of the plant, in the form of withering, yellowing, loss of vitality and rot. Infection due to fungi, bacteria or viruses is also important. The virulence of these infectious agents depends to a great extent on the resistance of the plant, and the conditions in which it grows.

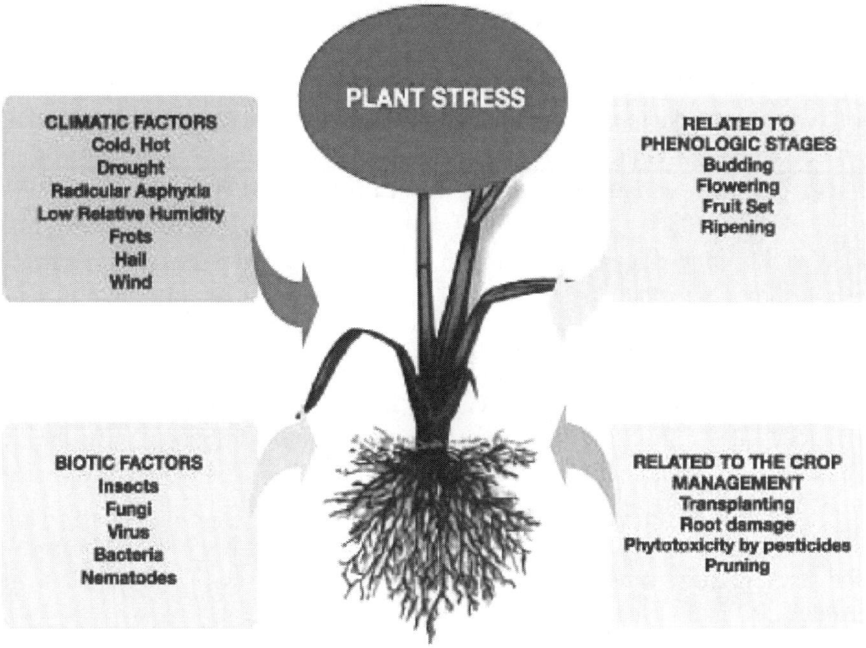

Source: http://plantstress.bioiberica.com/Training/What_is/Plant_stress.html.

Figure 1. Different factors contributing to plant stress.

The symptoms triggered in a plant due to infection by a pathogenic organism are quite varied: necrosis and deformities, rot and exudation, or alterations in colour, such as the foliar mosaic pathology caused by some viruses. Depending on the state or severity of the infection, losses in production can be quite considerable.

PHOTOSYNTHESIS: A BRIEF OVERVIEW

The most active photosynthetic tissue in higher plants is the mesophyll of leaves. Mesophyll cells have many chloroplasts, which contain the specialized light-absorbing green pigments, the chlorophylls. In the chloroplast, light energy is converted into chemical energy by two different functional units called photosystems. The absorbed light energy is used to power the transfer of electrons through a series of compounds that act as electron donors and electron acceptors. The majority of electrons ultimately reduce $NADP^+$ to NADPH and oxidize H_2O to O_2. Light energy is also used to generate a proton motive force across the thylakoid membrane, which is used to synthesize ATP. Plants and some photosynthetic prokaryotes have two reaction centers, photosystem I and photosystem II, that function in series. The two photosystems are spatially separated: PSI is found exclusively in the non-stacked stroma membranes, PSII largely in the stacked grana membranes. The reaction center chlorophylls of PSI absorb maximally at 700 nm, and those of PSII at 680 nm. Photosystems II and I carry out noncyclic electron transport, oxidize water to molecular oxygen, and reduce $NADP^+$ to NADPH. The photooxidation of water is modeled by the five-step S state mechanism. A tyrosine residue of the D1 protein of the PSII reaction center functions as an electron carrier between the oxygen evolving complex and P680. The electron flow ends with the reduction of $NADP^+$ to NADPH by a membrane-bound, ferrodoxin–NADP reductase. A portion of the energy of photons is also initially stored as chemical-potential energy, largely in the form of a pH difference across the thylakoid membrane. This energy is quickly converted into chemical energy during ATP formation by action of an enzyme complex known as the ATP synthase. The photophosphorylation of ADP by the ATP synthase is driven by a chemiosmotic mechanism. The reduction of CO_2 to carbohydrate via the carbon-linked reactions of photosynthesis is coupled to the consumption of NADPH and ATP synthesized by the light reactions of thylakoid membranes. Photosynthetic eukaryotes reduce CO_2 via the Calvin cycle that takes place in the stroma, of chloroplasts. Here, CO_2 and water are combined with ribulose-1,5-bisphosphate to form two molecules of 3- phosphoglycerate, which are reduced and converted to carbohydrate. The continued operation of the cycle is ensured by the regeneration of ribulose-1,5-bisphosphate. The carbohydrates synthesized by the Calvin cycle are converted into storage forms of energy and carbon: sucrose and starch. Sucrose, the transportable form of carbon and energy in most plants, is synthesized in the cytosol, and its synthesis is regulated by phosphorylation of sucrose phosphate synthase. Starch is synthesized in the chloroplast.

Photosynthetic activity in the intact leaf is an integral process that depends on many biochemical reactions. Different environmental factors can limit photosynthetic rates. Reports are available on the effects of stress factors like high light (Demmig-Adams and Adams 2000; Biswal et al. 2003), UV radiation (Biswal et al. 2003; Vass 1997; Broschi and Strid 2003), water deficit (Biswal et al. 2003), and extreme temperatures (Biswal and Biswal 1999;

Carpentier 1999; Khan et al. 2013a) on photosynthetic activities. Leaf anatomy is highly specialized for light absorption, and the properties of palisade and mesophyll cells ensure uniform light absorption throughout the leaf. In addition to the anatomic features of the leaf, chloroplast movements within cells and solar tracking by the leaf blade help maximize light absorption. At high photon flux, photosynthesis in most plants is CO_2 limited, but the limitation is substantially lower in C_4 and CAM plants because of their CO_2-concentrating mechanisms. Diffusion of CO_2 into the leaf is constrained by a series of different points of resistance. The largest resistance is usually that imposed by the stomata, so modulation of stomatal apertures provides the plant with an effective means of controlling water loss and CO_2 uptake. Both stomatal and nonstomatal factors affect CO_2 limitations on photosynthesis.

Temperature responses of photosynthesis reflect the temperature sensitivity of the biochemical reactions of photosynthesis and are most pronounced at high CO_2 concentrations. Because of the role of photorespiration, the quantum yield is strongly dependent on temperature in C_3 plants but is nearly independent of temperature in C_4 plants. Functional changes in the photosynthetic apparatus in response to prevailing temperatures in their environment have an important effect on the capacity of plants to live in diverse habitats.

CHLOROPLAST AS SENSOR OF STRESS

Chloroplasts are essential organelles that are the site of photosynthesis in green plants. Chloroplasts were formed when a eukaryotic cell swallowed an ancient cyanobacterium. Chloroplasts are now at the heart of plant energy metabolism and are particularly sensitive to changes in the environment and the specific nutrient requirements of the plant[1]. The chloroplast is a semiautonomous cell organelle, with its own DNA and a complete protein synthesis apparatus. Many of the proteins that make up the photosynthetic apparatus, as well as all the chlorophylls and lipids, are synthesized in the chloroplast. Other proteins are imported from the cytoplasm and are encoded by nuclear genes. Chloroplastic signaling pathways may play a key role in driving stress adaption and survival in plants. The organelle has potential to absorb light, split water molecules to liberate oxygen, and initiate electron transfer reactions, resulting in production of ATP and NADPH, which are used for fixation of carbon dioxide. Under normal conditions, there is a perfect coordination between light-induced electron transfer reactions associated with the thylakoid membrane and carbon dioxide fixation by the Calvin cycle in stroma. But during stress, a loss of coordination may cause leakage of electron to oxygen, which may produce toxic oxygen-free radicals.

Therefore, light absorbed by the pigments and oxygen liberated by chloroplast in this condition become harmful to plants. Second, the excited reaction center of photosystem II (PS II) results in the production of a strong oxidant (P680+) required for the liberation of oxygen. But under stress condition, the strong oxidant becomes long-lived and can oxidize lipids, pigments, and proteins of the membrane.

Thus, the potential of chloroplast for photoexcitation leading to liberation of oxygen, primarily responsible for survival of higher organisms in this planet, makes the leaves a major target of stress in green plants[2].

[1] Source: http://www.lgbp.univ-mrs.fr/rubrique.php3?id_rubrique=33.
[2] Source: Biswal (2005).

Among all the components of the photosynthetic apparatus that are known as the targets of stress, PS II and ribulose- 1,5-bisphosphate carboxylase oxygenase (Rubisco) are considered as the major stress sensors (Biswal et al. 2003; Biswal 2005). In natural environment, most of the stresses are sensed by photosynthetic organisms through these components. PS I is known to be relatively less sensitive, than PS II, to most stresses, but it is highly sensitive to stresses such as freezing (Biswal et al. 2003). The modifications of the chloroplast in response to various environmental stresses have been widely studied and, thus the literature in the area is vast. The stress is sensed at the levels of pigment composition, structural organization, primary photochemistry and the CO_2 fixation (Biswal et al. 2003; Biswal 2005). Extensive reviews are available on the photosynthetic response of plants to various environmental stresses (Pareek et al. 2010; Biswal 2005; Li et al. 2009; Sage and Kubien 2007; Allakhverdiev et al. 2008; Shinozaki and Yamaguchi- Shinozaki 2007; Lawlor 2009; Lawler and Tezara 2009; Pfannschmidth et al. 2009; Baena- Gonzalez and Sheen 2008; Biswal et al. 2008; Khan et al. 2012; Khan et al. 2013b).

PHOTOSYNTHETIC RESPONSE TO VARIOUS BIOTIC AND ABIOTIC STRESSES

Studies on effects of various abiotic factors on photosynthesis have been very extensively carried out in the past, however, literature available on response of photosynthesis towards biotic factors is still very limited.

Effect of Biotic Stresses on Photosynthesis

Plants are subject to a innumerable biotic stresses inflicted by various living organisms from viruses to mammals, many of which affect photosynthesis, either by altering its underlying metabolism or by reducing photosynthetic leaf area[3]. Pathogens (fungal, bacteria or viral agents) and animal pests causes on average, a 15% and 18% reduction in crop yield, respectively (Oerke and Dehne 2004).

Reports suggest that photosynthetic activity can be strongly influenced by herbivore feeding, depending on the duration of feeding, the herbivore species and the plant species and also the capacity of the attacked plant to cope with the herbivore. Direct effects of herbivore feeding are due to removal of leaf area or disruption of photosynthetically active tissue[4]. Besides this, herbivory can have indirect impacts on the photosynthetic activity of the remaining leaf tissue via (Nabity et al. 2009) - severed vasculature (leading to disruption of the leaf veins and subsequent water loss by transpiration, which ultimately leads to increased stomatal closure and reduced photosynthetic CO_2 assimilation), altered sink demand of the plant, resulting in increased photosynthesis of the remaining leaves, autotoxicity, as many defensive secondary compounds (e.g., nicotine, furanocumarins) are toxic to the plant itself-down-regulation of photosynthesis-related genes, for example due to the signalling substance jasmonic acid.

[3] Source: http://www.life.illinois.edu/delucia/Publications/chapter%2022%20web%20copy.pdf.
[4] Source: Schmidt (2009).

Most insect–plant gas exchange studies have used leaf-mass consumers (defoliators) as the stressor. A highly variable photosynthetic responses was observed in the past in response to defoliation injury. Studies have revealed a continuum of responses, from decreases to increases in photosynthetic rates of remaining leaves, plants, and plant canopies (Welter 1989).

There have been many reports on pests and insects feeding on plants and eventually causing disturbance in their photosynthetic machinery. For example, it has been reported that aphid feeding on cotton for 18 days, led to reduced CO_2 assimilation and increased transpiration (Shannag et al. 1998; Gomez et al. 2006). Similarly, infestation with silverleaf whitefly for 60 days significantly reduced the photosynthetic rate of cotton plants, however, there were no effects on stomatal conductances, intercellular CO_2 concentration or leaf chlorophyll content (Lin et al. 1999). Evidence suggests that the cell content-feeding thrips (Lei and Wilson 2004) or spider mites (Bondada et al. 1995; Reddall et al. 2004, 2007) reduce net photosynthesis in the oldest leaves of cotton after a long infestation time. Not only this, long-term infestation with spider mites have strongly decreases chlorophyll concentration, stomatal conductances, and consequently transpiration, of cotton planted even in the un-damaged leaf portions (Bondada et al. 1995; Reddall et al. 2004). This is attributed to the grana disintegration of the chloroplast which is probably caused by toxins in the mites´ saliva (Bondada et al. 1995). Caterpillars, also severely affect photosynthetic activity of the host plant, as they remove large amounts of plant tissue. Herbivory damages the leaf tissue, and may facilitate the invasion of pathogens via wounds. It is reported that pathogen attack can rapidly decrease photosynthetic activity of the plant, while causing an increase in respiration and photorespiration (Berger et al. 2007).

Viruses are also known to alter photosynthetic machinery in plants. Photosynthetic reductions by viral infection are generally accompanied by chlorosis and structural changes in photosynthetic organelles (Goodman et al. 1986), reduced mesophyll conductance to CO_2 (Sampol et al. 2003), altered photochemistry and primary photosynthetic reactions (Balachandran et al. 1997; Rahoutei et al. 1999; 2000; Bertamini et al. 2004; Sajnani et al. 2007), inhibition of Rubisco and other photosynthetic enzymes (Balachandran et al. 1997; Sampol et al. 2003; Bertamini et al. 2004) etc. Gene-expression studies show that virus infection induces down regulation of numerous genes involved in photon capture and thylakoid processes, CO_2 uptake and photosynthesis metabolism (Golem and Culver 2003; Whitham et al. 2003; Kokkinos et al. 2006; Espinoza et al. 2007). Viruses also induce changes in chloroplast ultrastructure and starch accumulation[5].

In contrast to viruses, the effects of fungi on host photosynthesis are more varied. Decreased photosynthetic rates together with chloroplast degeneration have been reported in fungal infected plants. Fungi (mostly leaf disease fungi) causes reduction in CO_2 assimilation, down regulation of Calvin cycle genes, disturbances in PSII functions, PSI inhibition and also stomatal closure etc (Baron et al. 2012). As far as bacteria are concerned, nature of photosynthetic limitations imposed by them is quite variable, although probably non-stomatal effects predominate. Infection with parasitic nematodes causes decreased transpiration and photosynthesis (Fatemy et al. 1985; Postuka et al. 1986; Melakeberhan et al. 1990; Schans and Arntzen 1991; Asmus and Ferreaz 2002), and associated decreases in leaf chlorophyll content (Siddiqui and Mahmood 1999).

[5] Source: http://www.life.illinois.edu/delucia/Publications/chapter%2022%20web%20copy.pdf.

Effect of Abiotic Stresses on Photosynthesis

In plants, green leaf is the major target of environmental stresses, both PS I and PS II, and Rubisco being the primary stress sensors in the many chloroplasts in the leaf (Figure 2). The stress-induced changes in the sensors result, in energy imbalance, changes in the redox status of electron transport components of thylakoids and alteration of cellular sugar level. These changes are suggested to initiate signalling cascades that lead to physical and molecular readjustments for stress adaptation. Short-term adaptation includes non-photochemical quenching (NPQ) of excess unused quanta, and modulation of toxic oxidative environment through several antioxidant systems. Expression of stress-responsive genes regulated by redox and sugar signalling is a long-term adaptive strategy of plants (Biswal et al. 2011).

Light

Sunlight is the ultimate source of energy for life on earth, but when this light is absorbed in excess, dramatically reduces the efficiency of photosynthetic process. This process, leading to downregulation of photosynthesis by excess of unutilized light, is known as photoinhibition, a high light stress syndrome. PS II is a major site of photoinhibition. D1 turnover in PS II is established to be the central event of photoinhibitory changes of the chloroplasts. In high-light conditions, D1 is rapidly degraded and synthesized and its level reflects the net balance between photodamage and the repair process. The nature of damage of electron transport components responsible for induction of D1 protein degradation is still unclear.

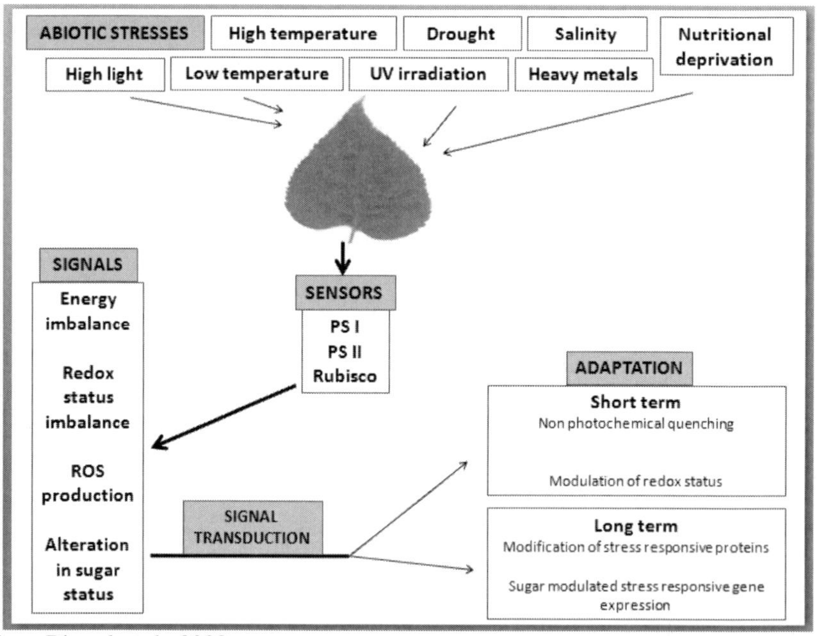

Modified from Biswal et al., 2008.

Figure 2. Green leaves as target of environmental stresses.

The free radical mediated induction of the degradation has been suggested by many authors (Biswal et al. 2003; Barber 1995; Krause 1994a).

Zeaxanthin, is considered as an effective xanthophyll for thermal dissipation of excess quanta during photoinhibition. The studies with mutants reveal that the thermal dissipation in photosynthetic organisms requires the synthesis of specific polypeptides in addition to a pH gradient across the membrane and formation of zeaxanthin from violaxanthin by de-epoxidase activity. Recently, Elrad et al. (2002) have shown a gene, namely, *Lhcbm1* that codes for a light harvesting polypeptide participating in the process of thermal dissipation. Besides, PSII photoinhibition, light stress also induces PSI photoinhibition. However, the nature of photoinhibition of PS I in many ways is different from that of PS II.

In the case of PS I, photoinhibition occurs at chilling stress and the inhibition takes place in the presence of oxygen at relatively low-light conditions. It is proposed that the temperature dependent loss of protective mechanism at the acceptor side of the photosystem results in the production of oxygen-free radicals, which attack the iron–sulfur centers of the complex and ultimately degrade the product of the *psaB* gene, one of the two major subunits of the PS I reaction center (Biswal 2005).

Water

Plants may experience stress both during drought (water deficit) and flooding (too much water). Water stress causes a significant reduction in the content of photosynthetic pigments, proteins, and lipids (Deo and Biswal 2001; Behera et al. 2003). Severe water stress resulting in desiccation of leaves causes not only a dramatic loss of the pigments but also disorganization of the thylakoid membranes (Deo and Biswal 2001). Photochemical reactions associated with PS II have been shown to be more susceptible to stress than those associated with PS I (Sundari and Raghavendra 1990). The loss of chloroplast capacity to fix carbon dioxide in green leaves under stress is attributed at least partly to stress induced closure of leaf stomata, a response exhibited by abscisic acid treatment of plants Graan and Boyer (1990). They have also proposed that one component of photosynthetic inhibition during dehydration is non stomatal. Kanechi et al. (1996) however, have shown high sensitivity of Rubisco activity to dehydration condition. At the molecular level, changes in the gene expression as a response of plants to water stress have been reported, and the possible functions of gene products induced by the stress have been critically reviewed (Bray et al. 2000; Bray 1997).

Temperature

Extreme temperatures lead to changes in the quantity of pigments, organization of thylakoids, primary photochemistry, and carbon dioxide fixing ability of chloroplasts (Biswal and Biswal 1999; Carpentier 1999). Chilling temperature leads to the loss of thylakoid photochemical reactions, and the reactions associated with PS II inactivation are suggested to be the main target of the stress (Krause 1994b). Low temperature also results in loss of efficiency of the Calvin cycle (Grafflage and Krause 1993).

High-temperature stress is demonstrated to cause a decline in electron transport efficiency of PS I, PS II, and the whole chain between PS I and PS II (Carpentier 1999).

UV Irradiation

UV radiation is known to significantly modify the structure and function of the photosynthetic apparatus (Vass 1997; Broschi and Strid 2003). Molecular events associated with the response of green plants to the radiation have been recently reviewed by Biswal et al. (2003) and Broschi and Strid (2003).

While UV-B and UV-C are demonstrated to be highly detrimental, UV-A brings about mild, but significant, damaging and non damaging effects on plant growth and photosynthesis. The PS II complex of thylakoids, have been demonstrated to be the main target of UV radiation (Kulandaivelu and Nedunchezhian 1989; Joshi et al. 1991; 1994; Wilson and Greenberg 1993).

Although the specific target molecules or components of PS II complex as the primary site of radiation-induced damage are not identified, the damage of reaction-center II including degradation (Wilson and Greenberg 1993; Friso et al. 1994) and structural modification (Nayak et al. 2003; Joshi et al. 1994; 1997) of D1 protein has been suggested.

Other Stresses

When plants are exposed to environmental stresses and the availability of CO_2 within the leaf is restricted and/or the synthesis of ATP is impaired, the activity of the Calvin cycle is reduced, but PS II remains active.

In these conditions, the concentration of the final electron acceptor $NADP^+$ is generally very low. This leads to an excess of excitation energy in the photosystems. If not dissipated, electrons accumulate in the electron transport chain and are transferred to oxygen (Mehler reaction): such processes generate ROS, e.g., superoxide (as in case of heavy metal stress and nutrition deficiency). Because of their high reactive potential, ROS react with and damage many cellular components (e.g., proteins, DNA and lipids), constituting oxidative stress, so they must then be detoxified by scavenging mechanisms such as superoxide dismutase (SOD) and the glutathione–ascorbate cycle (Noctor and Foyer 1998).

ROS also inactivate the photochemical reaction centre of PSII, causing photoinhibition. Recently, it has been proposed that most environmental stresses inactivate PSII by inhibiting the mechanisms for repairing photodamage rather than by directly attacking it (Murata et al. 2007). ROS generated by different stresses inhibit PSII repair by suppressing the transcription and translation of *psbA* genes encoding D1, one of the proteins of the PSII complex (Nishiyama et al. 2001; 2004; Allakhverdiev et al. 2002).

In the case of multistress factors, the stress effects are shown to be additive (Nayak et al. 2003). When plants experience different stresses in a sequence, their exposure to one stress develops provision of tolerance to another stress (Behera et al. 2003).

Molecular Regulation of Stress Responsive Genes in Photosynthesis

Earlier, the studies of plant responses to environmental stress were limited to physiological and biochemical parameters. However, owing to the enormous improvement in the area of plant molecular biology in the last few years, the stress response is now being examined at molecular level. Attempts have been made to identify and characterize the stress responsive genes, their expression and regulation during stress treatments.

In order to minimize the damaging effect of abiotic and biotic stress, the genes code for certain structural proteins and enzymes that are required for readjustment of cellular components of chloroplast in plants.

Surprisingly, the stress itself induces either overexpression or limited expression of the genes of these proteins in a highly regulated manner for the readjustment. High light induced enhancement of the production of transcripts for the synthesis of D1 protein replacing the photodamaged protein is one of the examples of these proteins (Shapira et al. 1997). This suggests that a stress can induce either reduction or enhancement of the accumulation of transcripts in order to produce the desired amount of certain specific proteins for adaptation under adverse environmental conditions.

Another category of proteins synthesized in response to stress is stress-specific and the proteins are not necessarily structural components of chloroplasts, but are believed to play a role in modulating the stress effect. Among these proteins, heat shock protein (HSP) and early light inducible protein (ELIP) are extensively studied with particular reference to chloroplast. Also, expression of osmotins in response to water stress is known. The plants including green algae are reported to produce HSP in response to heat stress (Vierling 1991). HSPs are known to be the members of multigene subfamilies.

Although heat is a major stress responsible for inducing the synthesis of these proteins, other stress factors may also cause their synthesis. ELIP is also a class of stress responsive proteins. The formation of transcripts of ELIP has been reported to be significantly enhanced with increase in the intensity of light (Potter and Kloppstech 1993). These proteins remain stable as long as the leaves are exposed to high irradiation but start degrading when the intensity of light is lowered (Potter and Kloppstech 1993). During light stress, the breakdown of pigment protein complexes may release free chlorophylls, which on absorption of light may produce oxygen free radicals, toxic to chloroplast. The ELIPs may absorb these free pigments, providing the proper surface and thus may protect the organelle against stress.

Protective role of late embryogenesis abundant (LEA) proteins against water stress in green plants are also well studied (Bray et al. 2000). Overexpression of LEA proteins in transgenic plants has been shown to develop resistance in plants to drought (Bray et al. 2000).

Several environmental and biotic stress factors often lead to development of oxidative stress in plants. Stress eventually leads to formation of oxygen free radicals which have the potential to damage or kill cellular components including chloroplasts (Figure 3).

The formation of oxygen-free radicals is known to induce expression of genes for the synthesis of antioxidant enzymes and other nonenzymatic antioxidants that constitute the antioxidant defense system against oxidative stress. Hydrogen peroxide is suggested to act as a stress signaling molecule in the signal transduction pathway (Foyer et al. 1997). The participation of protein kinases is also suggested (Mahalingam and Federoff 2003).

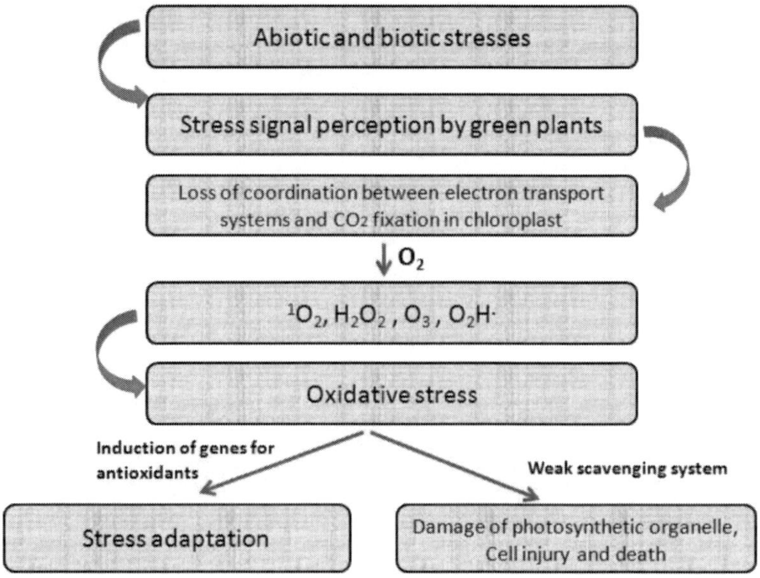

Modified from Biswal, 2005.

Figure 3. Signal perception and their transmission during oxidative stress.

Sugar, in addition to its role as a major source of energy, acts as a signalling molecule in cellular and whole-plant metabolic network. Sugar production through photosynthesis and changes in its level at different environmental conditions determine the growth, development and stress response (Rolland et al. 2002). Stress-induced changes in cellular status of sugar integrate carbon metabolism of photosynthesis, respiration and nitrogen in a well-regulated and coordinated manner (Wilson et al. 2006). Sugar-linked signal transduction pathway is interconnected with other cellular signalling pathways. Extensive literature is available on the interconnection between sugar signalling and the signaling pathways associated with ethylene, abscisic acid, cytokinins, auxin, G-protein and calcium (Rolland et al. 2002; 2006; Kudla et al. 2010). Under stressful environment, the decline in photosynthetic efficiency may result in sugar-limited conditions in plants. Such a condition generates a signalling cascade that finally makes the energy available from other sources primarily by upregulating some specific catabolic processes (Fujiki et al. 2005; Mohapatra et al. 2010) and thus protecting cells against nutrient stress. Several dark inducible genes responsible for enhancing catabolic pathways have been identified (Fujiki et al. 2001). These genes are expressed in dark-induced sugar starvation conditions. The role of one of the genes, *din2*, which codes for the cell wall-bound β -glucosidase is well known (Fujiki et al. 2005). These cell wall-bound enzymes cause the breakdown of cell-wall polysaccharides to sugars, and thus provide an alternative source of energy for the survival of plants under sugar-limiting conditions. The expression of the genes for these processes is suggested to be mediated by SnRK1 (sucrose non-fermenting 1-related kinase 1), the master regulator of transcription (Baena-Gonzalez and Sheen 2008). SnRK1 integrates all diverse sugar signalling and triggers catabolic processes necessary to sustain respiration and metabolic activity of energy-depriving cells.

The molecular mechanism of SnRK1 response to stress has been studied in *A. thaliana* (Baena-Gonzalez et al. 2007).

In response to sugar deficit, SnRK1 in *Arabidopsis* is observed to repress many genes associated with cellular biosynthetic processes, including the synthesis of lipids, proteins, sucrose, starch, amino acids and nucleotides (Rolland et al. 2002). Downregulation of biosynthetic pathways is a strategy of the cell to conserve energy under stress conditions.

Fortunately, due to gene manipulation, it is now possible to have plants with the capacity of stress tolerance. Attempts have been made to transfer these stress adaptive genes by effective transfer techniques and the subsequent regulated transcription in stress sensitive plants to make them stress resistant. Murata et al. (1992) produced plants by genetic manipulation, which can effectively resist chilling stress. Since the level of unsaturated fatty acids in membranes determines their sensitivity to chilling stress, they have been successful in manipulating the level of unsaturated fatty acids in chloroplast membranes by the transformation of plants with cDNA to plants with glycerol-3-phos-phate acetyl transferases from stress resistant species. Similarly, Hayashi et al. (1997) have demonstrated enhanced photosynthetic tolerance of *Arabidopsis* plants to salt and cold stress with transformation of plants with *CodA* gene for choline oxidase.

Transcription Factors (TFs) Involved in Regulation of Stress Response in Photosynthesis

Several studies have demonstrated that plants under stress conditions downregulate the expression of genes involved in photosynthesis and carbohydrate metabolism (Seki et al. 2002; Hannah et al. 2006; Wong et al. 2006). Many expression studies performed on plants subjected to different abiotic stresses have shown that a large number of genes are differentially expressed under stress, involving both stomatal and nonstomatal photosynthesis mechanisms. Thus, it is understandable that TFs modulating the expression of these genes play a critical role in plant response/adaptation to environmental cues. TFs are proteins with a DNA domain that binds to the cis-acting elements present in the promoter of a target gene.

They induce (activators) or repress (repressors) the activity of the RNA polymerase, thus regulating gene expression. The first indication that TFs could be involved in stomata regulation appeared when it was found that ABI3 (an ABI3/VP1 B3-type TF, also involved in controlling accumulation of chlorophyll and anthocyanins) suppressed the effect of the *abi1* mutation on stomatal regulation (Parcy and Giraudat 1997). While *abi1* plants do not regulate their stomata correctly (leading to increased transpiration) and are less sensitive to ABA, *abi1* plants ectopically expressing ABI3 showed a wild-type phenotype.

Meanwhile, it has been reported that modulation of transcription plays an important role in controlling guard cell activity. Recently two MYB-type TFs were identified as regulators of stomatal movements. *AtMYB60*, a *R2R3-MYB* gene of *Arabidopsis*, was shown to be specifically expressed in guard cells, and its expression regulated by light conditions, ABA and water stress (Cominelli et al. 2005).

A second member of the *Arabidopsis thaliana* family of R2R3-MYB TFs, *AtMYB61*, is also specifically expressed in guard cells in a manner consistent with its involvement in the regulation of stomatal aperture (Liang et al. 2005). However, it has been shown that *AtMYB60* and *AtMYB61* have distinct expression patterns and functions.

While *AtMYB60* gene expression is induced by light, leading to an increased stomatal aperture, *AtMYB61* transcription is repressed by light, although it has the same effect on stomatal movement. It was demonstrated that *AtMYB61* expression is both sufficient and necessary to cause reductions in stomatal aperture. Moreover, Liang et al. (2005) showed that AtMYB61 gene expression was not altered in plants treated with ABA, salt and drought, known to induce stomatal closure. Thus, AtMYB61 seems to act via a mechanism parallel to that responsible for closing stomata in response to water deficit (Saibo et al. 2009).

In addition to stomatal regulation, abiotic stress conditions reduce the photosynthetic rate by controlling the expression of genes involved in non-stomatal processes associated with photosynthesis and carbohydrate metabolism. Among the genes whose expression is controlled by the environmental cues are those that encode chlorophyll a/b-binding (CAB) proteins, PSI and PSII subunits, oxygen-evolving complex and Rubisco subunits. Most genes associated with photosynthesis are under the control of a transcriptional regulatory network evolved to control plant response to external stimuli. Transcriptional profiling studies have shown that although some are upregulated, many photosynthesis-related genes and genes for carbohydrate metabolism are downregulated under abiotic stress conditions. Those genes may be the target for TFs such as the Cys2/His2-type zinc-finger proteins STZ and AZF2.

These TFs were shown to function as transcription repressors under drought, cold and high salinity stress conditions (Sakamoto et al. 2004). The expression of both STZ and AZF2 genes is induced mainly in leaves under drought stress, supporting the hypothesis that they play a role regulating photosynthesis related genes.

Tolerance of plants to abiotic stresses is well known to be a multigenic trait. For that reason, plant improvement using genes that play a role in the abiotic stress response is frequently insufficient to improve stress tolerance significantly. To overcome this, TFs that regulate several stress-responsive genes (e.g., the AP2/EREBP family) have often been used to manipulate plants in order to have a broader response. The results obtained in terms of stress tolerance are often much better than using a single gene encoding a non-regulatory protein and the observed effects on photosynthesis efficiency or photosynthetic machinery are normally positive (Saibo et al. 2009). Both stomatal and non-stomatal responses to abiotic stress involve transcriptional regulation and consequently the involvement of many TFs. Some of these TFs can mediate the regulation of photosynthetic-related gene expression by the redox state. Several TFs have been shown to regulate stomatal aperture and so mediate responses to adverse environmental conditions. Expression of genes coding for components regulating many aspects of photosynthetic metabolism is affected by stress, and TFs have a substantial role in regulating them and thus carbon metabolism. Adjustment to stress conditions may be improved by modifying the TFs in plants.

CONCLUSION

Photosynthesis in plants is undoubtedly a major sensor of stress. Various stress factors have generally been studied separately in different contexts and with limited objectives. It is now imperative to study the effect of multiple stresses on photosynthetic efficiency simultaneously. Multiple stresses should not only include a combination of abiotic factors but should also incorporate biotic factors.

Plants in a natural environment experience several stress factors, which may operate simultaneously or in various conjunctions. This is unlike laboratory conditions where these factors are examined almost solely in isolation. The interacting stress factors also exhibit significant seasonal or diurnal variations with increase or decrease in the intensity of individual stress components. In the last few years, attempts have been made to simulate natural stress conditions in laboratories and examine the effects of multiple stresses on photosynthesis. However, there is definitely scope for further investigation in this area.

Significant progress has been made in the area of molecular biology of stress response in plants but many aspects still remain unclear and in need of further investigation. Stress responsive genes, both stress resistant and stress sensitive, have been identified in a few cases. However, the details of molecular and biochemical mechanisms of natural stress resistance have to be precisely understood before proper manipulation of gene transfer technology can be effectively applied in the production of stress resistant plants. Also, literature available on molecular response of photosynthetic elements towards biotic stresses is limited and there is definite scope for further research and expansion.

REFERENCES

Allakhverdiev, S. I., Nishiyama, Y., Miyairi, S., Yamamoto, H., Inagaki, N., Kanesaki, Y., Murata, N., (2002). Salt stress inhibits the repair of photodamaged photosystem II by suppressing the transcription and translation of *psbA* genes in *Synechocystis*. *Plant Physiol.* 130, 1443-1453.

Allakhverdiev, S. I., Kreslavski, V. D., Klimov, V. V., Los, D. A., Carpentier, R., Mohanty, P., (2008). Heat stress: an overview of molecular responses in photosynthesis. *Photosynth. Res.* 98, 541-550.

Asmus, G. L. and Ferraz, L. C. C. B., (2002). Effect of population densities of *Heterodera glycines* race 3 on leaf area, photosynthesis and yield of soybean. *Fitipatol. Bras.* 27, 273-278.

Baena-Gonzalez, E. and Sheen, J., (2008). Convergent energy and stress signalling. *Trends Plant Sci.* 13, 474-482.

Baena-Gonzalez, E., Rolland, F., Thevelein, J. M., Sheen, J., (2007). A central integrator of transcription networks in plant stress and energy signalling. *Nature*. 448, 938-943.

Balachandran, S., Hurry, V. M., Kelley, S. E., Osmond, C. B., Robinson, S. A., Rohozinski, J., Seaton, G. G. R., (1997). Concepts of plant biotic stress. Some insights into stress physiology of virus-infected plants, from the perspective of photosynthesis. *Physiol. Plant.* 100, 203-213.

Barber, J., (1995). Molecular basis of vulnerability of photosystem II to damage by light. *Aust. J. Plant Physiol.* 22, 201-208.

Baron, M., Flexas, J. and Delucia, E. H., (2012). *Photosynthetic responses to biotic stress in Terrestrial Photosynthesis in a changing environment: A molecular, physiological and ecological approach,* ed. J. Flexas, F. Loreto and H. Medrano. Cambridge University Press.

Behera, S. K., Nayak, L., Biswal, B. (2003). Senescing leaves possess potential for stress adaptation: the developing leaves acclimated to high light exhibit increased tolerance to osmotic stress during senescence. *J. Plant Physiol.* 160, 125-131.

Berger, S., Sinha, A. K., Roitsch, T., (2007). Plant physiology meets phytopathology: plant primary metabolism and plant-pathogen interactions. *Journal of Experimental Botany.* 58, 4019-4026.

Bertamini, M., Muthuchelian, K., Nedunchezhian, N., (2004). Effect of grapevine leafroll on the photosynthesis of field grown grapevine plants (*Vitis vinifera* L. Cv. Lagrein). *J. Phytopathol.* 152, 145-152.

Biswal, B. and Biswal, U. C., (1999). Photosynthesis under stress: stress signals and adaptive response of chloroplasts. In: Pessarakli, M., ed. *Handbook of Plant and Crop Stress.* New York: Marcel Dekker, Inc., 315-336.

Biswal, B., Joshi, P. N., Raval, M. K., and Biswal, U. C., (2011). Photosynthesis, a global sensor of environmental stress in green plants: stress signalling and adaptation. *Curr. Sci.* 101.

Biswal, B., (2005). Photosynthetic response of green plants to environmental stress: Inhibition of photosynthesis and adaptational mechanisms. In: Pessarakli, M., ed. *Handbook of Photosynthesis.* CRC Press, Florida, US, 2nd edn., 739-749.

Biswal, B., Raval, M. K., Biswal, U. C., and Joshi, P. N., (2008). Response of photosynthetic organelle to abiotic stresses: modulation by sulfur metabolism. In: Khan, N. A., et al. *Sulfur Assimilation and Abiotic Stress in Plants.* Springer Verlag, Berlin, 167-192.

Biswal, U. C., Biswal, B., Raval, M. K., (2003). *Chloroplast Biogenesis: From Proplastid to Gerontoplast.* Dordrecht, The Netherlands, Kluwer Academic Publishers.

Bondada, B. R., Oosterhuis, D. M., Tugwell, N. P., Kim, K. S., (1995). Physiological and cytological studies of two-spotted spider mite, *Tetranychus urticae* K (Acari, Tetranychidae) injury in cotton. *Southwestern Entomol.* 20, 171-180.

Bray, E. A., Bailey-Serres, J., Weretilynk, E., (2000). Responses to abiotic stresses. In: Buchanan, B., Gruessem, W., Jones, R., eds. *Biochemistry and Molecular Biology of Plants.* Rockville, MD: American Society of Plant Physiologists, 1158-1203.

Bray, E. A., (1997). Plant responses to water deficit. *Trends Plant Sci.* 2, 48–54.

Broschi, M., Strid, A. (2003). Molecular events following perception of ultraviolet-B radiation by plants. *Physiol. Plant.* 117, 1-10.

Carpentier, R., (1999). Effect of high-temperature stress on the photosynthetic apparatus. In: Pessarakli, M., ed. *Handbook of Plant and Crop Stress.* New York: Marcel Dekker, Inc., 337–348.

Cominelli, E., Galbiati, M., Vavasseur, A., Conti, L., Sala, T., Vuylsteke, M., Leonhardt, N., Dellaporta, S. L., Tonelli, C., (2005). A guard-cell-specific MYB transcription factor regulates stomatal movements and plant drought tolerance. *Curr. Biol.* 15, 1196-1200.

Demmig-Adams, B., Adams III, W. W., (2000). Harvesting sunlight safely. *Nature.* 403, 371-374.

Deo, P. M. and Biswal, B. (2001). Response of senescing cotyledons of clusterbean to water stress in moderate and low light: possible photoprotective role of β-carotene. *Physiol. Plant.* 112, 47-54.

Elrad, D., Niyogi, K. K., Grossman, A. R., (2002). A major light harvesting polypeptide of photosystem II functions in thermal dissipation. *Plant Cell* 14, 1801-1816.

Espinoza, C., Vega, A., Medina, C., Schlauch, K., Cramer, G., Arce-Johnson, P., (2007). Gene expression associated with compatible viral diseases in grapevine cultivars. *Funct. Integ. Geno.* 7, 95-110.

Fatemy, Y., Trinder, K. E., Wisgfield, J. N., Evans, K., (1985). Effects of *Globodera rostochiensis,* water stress and exogenous abscisic acid on stomatal function and water use of Cara and Pentland Dell potato plants Revue Nématol. *Revue Nématol.* 8, 249-255.

Foyer, C. H., Lopez-Delgado, H., Dat, J. F., Scott, I. M., (1997). Hydrogen peroxide and glutathione associated mechanisms of acclimatory stress tolerance and signaling. *Physiol. Plant.* 100, 241-254.

Friso, G., Spetea, C., Giocometti, G. M., Vass, I., Barbato, R., (1994). Degradation of photosystem II reaction center D1 protein induced by UV B radiation in isolated thylakoids. Identification and characterization of C- and Nterminal breakdown products. *Biochim. Biophys. Acta.* 1184, 78-84.

Fujiki, Y., Nakagawa, Y., Furumoto, T., Yoshida, S., Biswal, B., Ito, M., Watanabe, A., Nishida, I., (2005). The response to darkness of the late-responsive dark-inducible genes is positively regulated by leaf age and negatively by calmodulin-antagonist-sensitive signalling in *Arabidopsis thaliana. Plant Cell Physiol.* 46, 1741-1746.

Fujiki, Y., Yoshikawa, Y., Sato, T., Inada, N., Ito, M., Nishida, I., Watanabe, A., (2001). Dark inducible gene from *Arabidopsis thaliana* is associated with leaf senescence and repressed by sugar. *Physiol. Plant.* 111, 345-352.

Golem, S. and Culver, J. N., (2003). Tobacco mosaic virus induced alterations in the gene expression profile of *Arabidopsis thaliana. Mol. Plant Microbe Interact.* 16, 681-688.

Gomez, L. D., Baud, S., Gilday, A., Li, Y., Graham, I. A., (2006). Delayed embryo development in the *Arabidopsis trehalose-6-phosphate synthase 1* mutant is associated with altered cell wall structure, decreased cell division and starch accumulation. *Plant J.* 46, 69-84.

Goodman, R. N., Kira΄ly, Z., Wood, K. R., (1986). *Physiology of Plant Disease*. University of Missouri Press, Columbia.

Graan, T. and Boyer, J. S., (1990). Very high CO_2 partially restores photosynthesis in sunflower at low water potentials. *Planta* 181, 378-384.

Grafflage, S. and Krause, G. H. (1993). Alterations of properties of ribulose bisphosphate carboxylase related to cold acclimation. In: Lee, P. L., Christersson, L., eds. *Advances in Cold Hardiness.* Boca Raton, FL: CRC Press, 113-124.

Hannah, M. A., Wiese, D., Freund, S., Fiehn, O., Heyer, A. G., Hincha, D. K., (2006). Natural genetic variation of freezing tolerance in *Arabidopsis. Plant Physiol.* 142, 98-112.

Hayashi, H., Alia, Mustardy, L., Deshnium, P., Ida, M., Murata, N., (1997). Transformation of *Arabidopsis thaliana* with the *Cod A* gene for choline oxidase; accumulation of glycinebetaine and enhanced tolerance to salt and cold stress. *Plant J.* 12, 133-142.

Joshi, P. N., Biswal, B., Biswal, U. C., (1991). Effect of UV-A on aging of wheat leaves and role of phytochrome. *Environ. Exp. Bot.* 31, 267-276.

Joshi, P. N., Biswal, B., Kulandaivelu, G., Biswal, U. C., (1994). Response of senescing wheat leaves to ultraviolet-A light: changes in energy transfer efficiency and PS II photochemistry. *Radiat. Environ. Biophys.* 33, 167-176.

Joshi, P. N., Ramaswamy, N. K., Raval, M. K., Desai, T. S., Nair, P. M., Biswal, U. C., (1997). Response of senescing leaves of wheat seedlings to UV-A radiation; inhibition of PSII activity in light and darkness. *Environ. Exp. Bot.* 38, 237-242.

Kanechi, M., Uchida, N., Yasuda, T., Yamaguchi, T., (1996). Nonstomatal inhibition associated with inactivation of Rubisco in dehydrated coffee leaves under unshaded and shaded conditions. *Plant Cell Physiol.* 37, 455-460.

Khan, M. I. R., Syeed, S., Nazar, R., Anjum, N. A., (2012) An insight into the role of salicylic acid and jasmonic acid in salt stress tolerance. In: Khan, N., Nazar, R., Iqbal, N., Anjum, N., editors. *Phytohormones and abiotic stress tolerance in plants.* Springer Heidelberg Dordrecht: London, New York, 277-300.

Khan, M. I. R., Iqbal, N., Asim Masood, A., Tasir, S. Per., Khan, N. A., (2013a). Salicylic acid alleviates adverse effects of heat stress on photosynthesis through changes in proline production and ethylene formation. *Plant Signal. Behav.* 8:11, e26374.

Khan, M. I. R., Asgher, M., Khan, N. A., (2013b) Rising temperature in the changing environment: A serious threat to plants. *Climate Change Environ. Sustain.* 1, 25-36.

Kokkinos, C. D., Clark, C. A., Mcgregor, C. E., Labonte, D. R., (2006). The effect of sweet potato virus disease and its viral components on gene expression levels in sweet potato. *J. Am. Soc. Hort. Sci.* 131, 657-666.

Krause, G. H., (1994a). The role of oxygen in photoinhibiton of photosynthesis. In: Foyer, C. H., Mullineaux, P. M., eds. *Causes of Photooxidatative Stress and Amelioration of Defense Systems in Plants.* Boca Raton, Fl: CRC Press, 43-76.

Krause, G. H., (1994b). Photoinhibition induced by low temperatures. In: Baker, N. R., Bowyer, J. R., eds. *Photoinhibition of Photosynthesis. From Molecular Mechanism to the Field.* Oxford: Bios Scientific Publishers, 331-348.

Kudla, J., Batistic, O., Hashimoto, K., (2010). Calcium signals: the lead currency of plant information processing. *Plant Cell*, 22, 1-23.

Kulandaivelu, G. and Nedunchezhian, N., (1989). Effect of UV-B enhanced radiation on growth and photosynthetic activities on higher plants and their defense mechanisms. In: Biswal, U. C., Britton, G., eds. *Trends in Photosynthesis Research.* Bikaner, India: Agro Botanical Publishers, 215-229.

Lawlor, D. W., (2009). Musings about the effects of environment on photosynthesis. *Ann. Bot.*, 103, 543-549.

Lawor, D. W. and Tezara, W., (2009). Causes of decreased photosynthetic rate and metabolic capacity in water-deficient leaf cells: a critical evaluation of mechanisms and integration of processes. *Ann. Bot.* 103, 561-579.

Lei, T. T. and Wilson, L. J., (2004). Recovery of leaf area through accelerated shoot ontogeny in thrips-damaged cotton seedlings. *Ann. Bot.* 94, 179-186.

Li, Z., Wakao, S., Fischer, B. B., and Niyogi, K. K., (2009). Sensing and responding to excess light. *Annu. Rev. Plant Biol.* 60, 239-260.

Liang, Y. K., Dubos, C., Dodd, I. C., Holroyd, G. H., Hetherington, A. M., Campbell, M. M., (2005). *AtMYB61*, an R2R3-MYB transcription factor controlling stomatal aperture in *Arabidopsis thaliana. Curr. Biol.* 15, 1201-1206.

Lin, T. B., Schwartz, A., Saranga, Y., (1999). Photosynthesis and productivity of cotton under silverleaf whitefly stress. *Crop Sci.* 39, 174-184.

Mahalingam, R. and Fedoroff, N. (2003). Stress response, cell death and signaling: the many faces of reactive oxygen species. *Physiol. Plant.* 119, 56-68.

Melakeberhan, H., Ferris, H., Dias, J. M., (1990). Physiological responses of resistant and susceptible *Vitis venifera* cultivars to *Meloidogyne incognita*. *J. Nematol.* 22, 224-230.

Mohapatra, P. K., Patro, L., Raval, M. K., Ramaswamy, N. K., Biswal, U. C., Biswal, B., (2010). Senescence induced loss in photosynthesis enhance activity of cell wall β - glucosidase. *Physiol. Plant.* 138, 348-355.

Murata, N., Ishizaki-Nishizawa, O., Higashi, S., Hayashi, H., Tasaka, Y., Nishida, I., (1992). Genetically engineered alteration in the chilling sensitivity of plants. *Nature.* 356, 710-713.

Murata, N., Takahashi, S., Nishiyama, Y., Allakhverdiev, S. I., (2007). Photoinhibition of photosystem II under environmental stress. *Biochimi. Biophy. Acta.* 1767, 414-421.

Nabity, P. D., Zavala, J. A., DeLucia, E. H., (2009). Indirect suppression of photosynthesis on individual leaves by arthropod herbivory. *Ann. Bot.* 103, 655-663.

Nayak, L., Biswal, B., Ramaswamy, N. K., Iyer, R. K., Nair, J. S., Biswal, U. C., (2003). Ultraviolet-A induced changes in photosystem II of thylakoids: effects of senescence and high growth temperature. *J. Photochem. Photobiol. B.* 70, 59-65.

Nishiyama, Y., Allakhverdiev, S. I., Yamamoto, H., Hayashi, H., Murata, N., (2004). Singlet oxygen inhibits the repair of photosystem II by suppressing the translation elongation of the D1 protein in *Synechocystis* sp. PCC 6803. *Biochemistry* 43, 11321-11330.

Nishiyama, Y., Yamamoto, H., Allakhverdiev, S. I., Inaba, M., Yokota, A., Murata, N., (2001). Oxidative stress inhibits the repair of photodamage to the photosynthetic machinery. *EMBO J.* 20, 5587-5594.

Noctor, G. and Foyer, C. H., (1998). Ascorbate and glutathione: keeping active oxygen under control. *Ann. Rev. Plant Physiol. Plant Mol. Biol.* 49, 249-279.

Oerke, E. C. and Dehne, H. W., (2004). Safeguarding production—losses in major crops and the role of crop protection. *Crop Prot.* 23, 275-285.

Parcy, F. and Giraudat, J., (1997). Interactions between the ABI1 and the ectopically expressed ABI3 genes in controlling abscisic acid responses in *Arabidopsis* vegetative tissues. *Plant J.* 11, 693-702.

Pareek, A., Sopory, S. K., Bohnert, H. J., Govindjee, (2010). (eds.), *Abiotic Stress Adaptation in plants: Physiological, Molecular and Genomic Foundation*, Springer, Dordrecht, The Netherlands.

Pfannschmidt, T., Brautigam, K., Wagner, R., Dietzel, L., Schroter, Y., Steiner, S., Nykytenko, A., (2009). Potential regulation of gene expression in photosynthetic cells by redox and energy state: approaches towards better understanding. *Ann. Bot.* 103, 599-607.

Postuka, J. W., Dropkin, V. H., Nelson, C. J., (1986). Photosynthesis, photorespiration, and respiration of soybean after infection with root nematodes. *Photosyn.* 20, 405-410.

Potter, E. and Kloppstech, K., (1993). Effects of light stress on the expression of early light — inducible proteins in barley. *Eur. J. Biochem.* 214, 779–786.

Rahoutei, J., Barón, M., García-Luque, I., Droppa, M., Neményi, A., Horváth, G., (1999). Effect of tobamovirus infection on thermoluminescence characteristics of chloroplasts from infected plants. *Zeitschrift für Naturforschung*. 54, 634–639.

Rahoutei, J., García-Luque, I., Barón, M., (2000). Inhibition of photosynthesis by viral infection: effect on PSII structure and function. *Physiol. Plant.* 110, 286-292.

Reddall, A., Sadras, V. O., Wilson, L. J., Gregg, P. C., (2004). Physiological responses of cotton to two-spotted spider mite damage. *Crop Sci.* 44, 835-846.

Reddall, A. A., Wilson, L. J., Gregg, P. C., Sadras, V. O., (2007). Photosynthetic response of cotton to spider mite damage: Interaction with light and compensatory mechanisms. *Crop Sci.* 47, 2047-2057.

Rolland, F., Baena-Gonzalez, E., Sheen, J., (2006). Sugar sensing and signalling in plants: conserved and novel mechanisms. *Annu. Rev. Plant Biol.* 57, 675-709.

Rolland, F., Moore, B., Sheen, J., (2002). Sugar sensing and signaling in plants. *Plant Cell* (Suppl.). 14, S185-S205.

Sage, R. E. and Kubien, D. S., (2007). The temperature response of C_3 and C_4 photosynthesis. *Plant Cell Environ.* 30, 1086-1106.

Saibo, N. J. M., Lourenco, T., Oliveira, M. M., (2009). Transcription factors and regulation of photosynthetic and related metabolism under environmental stresses. *Ann. Bot.* 103, 609-623.

Sajnani, C., Zurita, J. Z., Roncel, M., Ortega, J. M., Barón, M., Ducruet, J. M., (2007). Changes in photosynthetic metabolism induced by tobamovirus infection in *Nicotiana enthamiana* studied *in vivo* by chlorophyll thermoluminescence. *New Phytol.* 175, 120-130.

Sakamoto, H., Maruyama, K., Sakuma, Y., Meshi, T., Iwabuchi, M., Shinozaki, K., Yamaguchi-Shinozaki, K., (2004). *Arabidopsis* Cys2/His2-type zinc-finger proteins function as transcription repressors under drought, cold, and high-salinity stress conditions. *Plant Physiol.* 136, 2734-2746.

Sampol, B., Bota, J., Riera, D., Medrano, H., Flexas, J., (2003). Analysis of the virus-induced inhibition of photosynthesis in malmsey grapevines. *New Phytol.* doi: 10.1046/j.1469-8137.2003.00882.x.

Schans, J. and Arntzen, F. K., (1991). Photosynthesis, transpiration and plant growth characters of different potato cultivars at various densities of *Globodera pallida. Netherlands J. Plant Pathol.* 97:297-310.

Schmidt, L., (2009). Impact of wounding and biotic stress on primary metabolism and growth processes of different plant species. (PhD Thesis). Institute of Chemistry and Dynamics of the Geosphere, Phytosphere (ICG-3). Forschungszentrum Julich GmbH, Germany.

Seki, M., Narusaka, M., Ishida, J., Nanjo, T., Fujita, M., Oono, Y., Kamiya, A., Nakajima, M., Enju, A., Sakurai, T., Satou, M., Akiyama, K., Taji, T., Yamaguchi-Shinozaki, K., Carninci, P., Kawai, J., Hayashizaki, Y., Shinozaki, K., (2002). Monitoring the expression profiles of 7000 *Arabidopsis* genes under drought, cold and high-salinity stresses using a full-length cDNA microarray. *Plant J.* 31, 279-292.

Shannag, H. K., Thorvilson, H., El-Shatnawi, M. K., (1998). Changes in photosynthetic and transpiration rates of cotton leaves infested with the cotton aphid, *Aphis gossypii*: Unrestricted infestation. *Ann. Appl. Biol.* 132, 13-18.

Shapira, M., Lers, A., Heifetz, P. B., Irihimovitz, V., Osmond, C. B., Gillham, N. W., Boynton, J. E. (1997). Differential regulation of chloroplast gene expression in *Chlamydomonas reinhardtii* during photoacclimation: light stress transiently suppresses synthesis of Rubisco LSU protein while enhancing synthesis of the PS II D1 protein. *Plant Mol. Biol.* 33, 1001-1011.

Shinozaki, K. and Yamaguchi-Shinozaki, K., (2007). Gene networks involved in drought stress response and tolerance. *J. Exp. Bot.* 58, 221-227.

Siddiqui, Z. A. and Mahmood, I., (1999). Effects of inoculations of *Heterodera cajani*, *Meloidogyne incognita* and *Fusarium udum* and *Bradyrhizobium japonicum* on the wilt disease complex of pigeon pea. *Indian Phytopathol.* 52, 66-70.

Sundari, D. S. and Raghavendra, A. S., (1990). Sensitivity of photosynthesis by spinach chloroplast membranes to osmotic stress *in vitro*: Rapid inhibition of O_2 evolution in presence of magnesium. *Photosynth. Res.* 23, 325–330.

Vass, I., (1997). Adverse effects of UV-B light on the structure and function of the photosynthetic apparatus. In: Pessarakli, M., ed. *Handbook of Photosynthesis.* New York: Marcel Dekker, Inc., 931–949.

Vierling, E., (1991). The role of heat shock proteins in plants. *Annu. Rev. Plant Physiol. Plant Mol. Biol.* 42, 579–620.

Welter, S. C., (1989). Arthropod impact and plant gas exchange, In: Bernays, E. A., ed. *Insect-plant Interactions,* Vol. 1, CRC Press, Boca Raton, 135.

Whitham, S. A., Quan, S., Chang, H. S., Cooper, B., Estes, B., Zhu, T., Wang, X., Hou, Y. M. (2003). Diverse RNA viruses elicit the expression of common sets of genes in susceptible *Arabidopsis thaliana* plants. *Plant J.* 33, 271-283.

Wilson, M. I. and Greenberg, B. M. (1993). Protection of the D1 photosystem II reaction center protein from degradation in ultraviolet radiation following adaptation of *Brassica napus* L. to growth in ultraviolet-B. *Photochem. Photobiol.* 57, 556–563.

Wilson, K. E., Ivanov, A. G., Oquist, G., Grodzinski, B., Sahan, F., Huner, N. P. A., (2006). Energy balance, organellar redox status and acclimation to environmental stress. *Can. J. Bot.* 84, 1355-1370.

Wong, C. E., Li, Y., Labbe, A., Guevara, D., Nuin, P., Whitty, B., Diaz, C., Golding, G. B., Gray, G. R., Weretilnyk, E. A., Griffith, M., Moffatt, B. A., (2006). Transcriptional profiling implicates novel interactions between abiotic stress and hormonal responses in *Thellungiella*, a close relative of *Arabidopsis*. *Plant Physiol.* 140, 1437-1450.

In: Photosynthesis
Editor: Nafees Khan

ISBN: 978-1-63463-304-8

Chapter 2

ELEVATED ROOT-ZONE CO_2 ON PHOTOSYNTHESIS OF VEGETABLE CROPS AT DIFFERENT AIR AND ROOT-ZONE TEMPERATURES

Jie He*

National Institute of Education, Nanyang Technological University, Singapore, Singapore

ABSTRACT

Elevated atmospheric CO_2 has been shown to influence many key physiological processes of plants. Leaf photosynthesis is typically stimulated during growth with double the present atmospheric CO_2 concentration ([CO_2]). It is not unusual for CO_2 in the rhizosphere to be up to more than 10-fold higher than the atmospheric [CO_2] due to the activities of microorganisms. We often question whether CO_2 level in the root-zone (RZ) could regulate plant growth. By using aeroponically-grown lettuce plants, we carried out a series of experiments to study the effects of elevated RZ [CO_2] on the photosynthesis of vegetable crops at different air and RZ temperatures in an environmentally controlled room and fluctuating greenhouse. Under two different temperature regimes of 28°C/22°C and 36°C/30°C (day/night) in the environmentally controlled room, it was found that elevated RZ [CO_2] stimulated photosynthetic CO_2 assimilation, with greater increase at higher temperatures. Elevated RZ [CO_2] decreased stomatal conductance at both temperatures, reducing transpirational water loss. When whole lettuce plants were grown under ambient temperatures ranging from 20 to 36°C under a maximal photosynthetic photon flux density (PPFD) of 2100 μmol m^{-2} s^{-1} inside the greenhouse, the results indicated that elevated RZ [CO_2] protected lettuce plants from dynamic photoinhibition of photosystem II. It was supported by the fact that midday depression of chlorophyll fluorescence F_v/F_m ratios was alleviated in elevated RZ [CO_2] treated plants. Elevated RZ [CO_2] was also found to reduce the negative impacts of high air temperature on photosynthesis, N metabolism, and growth of lettuce plants, resulting in enhanced productivity. The effects of elevated RZ [CO_2] on photosynthesis, productivity, nitrate, total reduced nitrogen, total leaf soluble and Rubisco proteins under

* Corresponding Author Address; email: jie.he@nie.edu.sg.

different root-zone temperature (RZTs) was also studied in a tropical greenhouse. The results demonstrated that increased photosynthesis and productivity under elevated RZ [CO_2] at different RZTs was partially due to the alleviation of midday water loss, both dynamic and chronic photoinhibition, as well as, a higher turnover of Calvin Cycle with higher ribulose-1,5-bisphosphate carboxylase/oxygenase (Rubisco) proteins. Effects of RZ [CO_2] at different air and RZT temperatures on different aspects of photosynthesis and growth were discussed.

Keywords: Air temperature, Chlorophyll fluorescence, Elevated root-zone [CO_2], NO_3^- uptake, Photosynthetic CO_2 assimilation, Relative water content, Rubisco, Root-zone temperature, Stomatal conductance, Total reduced N

INTRODUCTION

Most of the studies on plant responses to elevated CO_2 have focused on atmospheric CO_2. Elevated atmospheric CO_2 has been shown to influence many key physiological processes of plants (Drake et al. 1997; Curtis and Wang 1998; DeLucia et al. 1999; Hymus et al. 2001; Yin 2002; Ainsworth and Long 2005; Ainsworth and Rogers 2007; Taub et al. 2008; Kontunen-Soppela et al. 2010; Crous et al. 2012; Evans 2013; Ruiz-Vera 2013). Typically, leaf photosynthesis during growth has been stimulated as a result of the doubling of current [CO_2] (Drake et al. 1997; Curtis and Wang 1998; De Lucia et al. 1999; von Caemmerer and Evans 2010; Ruiz-Vera 2013). It is very common for [CO_2] in the rhizosphere to be more than 10-fold higher than that of atmospheric CO_2 concentration, due to soil microorganism activity and root respiration (De Jong and Schappert 1972; Norstadt and Porter 1984; Schlesinger 1997; Raich et al. 2002). Hence, elevated RZ [CO_2] should also be considered as one of the important environmental factors that may influence plant growth, development, and physiology in ways that are similar to elevated atmospheric [CO_2]. However, compared to the large amounts of information on plant responses to elevated atmospheric CO_2, little research has been conducted on the role of elevated RZ CO_2 in plant biology.

It has been reported that both bicarbonate (HCO_3^-) and dissolved CO_2 gas can diffuse into root cells (Chang and Roberts 1992; Carmer and Richards 1999; Teskey and McGuire 2002; Ford et al. 2007). For example, Chang and Roberts (1992) reported that the source of HCO_3^- in root cytosol was not only originating from respired CO_2 (mitochondrial) but also from sources outside of the root cell. Cramer and Richards (1999) found that tomato (*Lycopersicon esculentum*) roots grown at elevated RZ [CO_2] of 5000 ppm takes up 9 times more CO_2 than roots grown at 360 ppm. Dissolved CO_2 in the xylem sap of roots could be carried upward in the stem when trees are transpiring. CO_2 in xylem sap is a mixture of CO_2 transported from soil and released by respiration of roots and lower parts of stem (Teskey and McGuire 2002). Ford et al. (2007) also reported that the stems of pine seedlings assimilated carbon that was transported from the root via the root anaplerotic pathway.

Although some studies have demonstrated that elevated RZ [CO_2] increased plant growth (Viktorand Cramer 2003; Cramer et al. 2005; He et al. 2004; 2007; 2010; 2013), negative impacts on plant growth due to elevated RZ [CO_2] has also been reported (Cramer 2002; Boru et al. 2003). The impact of elevated RZ [CO_2] on plant is dependent on the species, pH of growth medium, mineral nutrition, RZ [CO_2] (Cramer and Lips 1995; Cramer and Titus 2001;

Viktor and Cramer 2003; Bialczyk et al. 2007), RZ O_2 concentration (Boru et al. 2003; Cramer et al. 2005), growth, irradiances (Cramer and Richards 1999), as well as, air and root-zone temperatures (He et al. 2004; 2007; 2010; 2013). According to these researchers, elevated RZ [CO_2] influences both below- and above-ground physiological processes, such as mineral nutrient absorption, photosynthetic CO_2 assimilation (*A*), photosynthetic utilization of radiant energy, stomatal conductance (g_s), and water use efficiency (WUE).

In this review paper, the main focus of discussion is that of the effects of elevated RZ [CO_2] on photosynthesis, plant water relations, and productivity of aeroponically grown vegetable crops at different air and RZ temperatures (RZT).

Elevated RZ [CO_2] Impacts on Photosynthetic Gas Exchange and Water Relations

The atmospheric [CO_2] has increased over 70 μmol mol^{-1} in the last 50 years and it was recorded at over 400 μmol mol^{-1} in May 2013 (http://www.esrl.noaa.gov/gmd/dv/iadv/). It has been demonstrated that increased [CO_2] enhanced photosynthesis for a wide range of C_3 species (Curtis and Wang 1998; Long et al. 2004; Nowak et al. 2004; Ainsworth and Long 2005; Ainsworth and Rogers 2007). Stomata, through which CO_2 and water vapor diffuse into and out of the leaf, play a central role for not only in the regulation and control of photosynthesis but also transpiration (Farquhar and Sharkey 1982). Stomata have also developed mechanisms to sense and respond to various environmental factors such as surrounding [CO_2], light intensity, and air humidity. Stomatal response to [CO_2] was first demonstrated more than 70 years ago by Freudenberger (1940). Since rhizosphere [CO_2] could be 10-fold higher than ambient [CO_2] (Norstadt and Porter 1984; Schlesinger 1997; Raich et al. 2002) due to the activities of soil microorganisms, we often question whether elevated RZ [CO_2] could regulate photosynthesis and water relations of soilless culture plants. Using an aeroponic system, we investigated the relationships among photosynthetic CO_2 assimilation, *A*, stomata conductance, g_s, transpiration, and midday relative water content (RWC) for crisphead-type lettuce (*Lactuca sativa* L. cv. 'Wintergreen') (He et al. 2004). In our study, the lettuce plants were grown under photosynthetic photon flux density (PPFD) of 650 μmol m^{-2} s^{-1} (12 h photoperiod) at 36°C/30°C and 28°C/22°C (day/night), with ambient RZ [CO_2] and three elevated RZ [CO_2] levels (2000, 10 000, and 50 000 ppm). In comparison to ambient RZ [CO_2], elevated RZ [CO_2] stimulated *A* at both temperatures, with a greater increase at the higher temperature under the same elevated RZ [CO_2] (He et al. 2004). It was also found that under both temperatures, g_s was lower in plants grown under all elevated RZ [CO_2] than ambient RZ [CO_2], reducing transpiration rates. Higher midday RWC was also observed in plants grown under elevated RZ [CO_2] compared to those grown under ambient RZ [CO_2]. The percentage changes of g_s, transpiration rates, and RWC under elevated RZ [CO_2] was greater at higher (36°C/30°C) than lower temperatures (28°C/22°C). Higher internal [CO_2] normally favours ribulose-1,5-bisphosphate carboxylase/oxygenase (Rubisco) carboxlyation over oxygenation (Peterson 1983). Under elevated atmospheric [CO_2], one of the major benefits of an increase in photosynthesis is the reduction of photorespiration (Long et al. 2004; Ainsworth and Rogers 2007). Reducing photorespiration may prove to be useful in a future atmosphere where elevated [CO_2] and high temperatures or limiting water have been predicted. This was based upon the assumption that increasing *A* under elevated RZ [CO_2] also resulted in the concomitant reduction in photorespiration, especially under higher temperatures (Peterhansel and Maurino 2011).

g_s has long been known to respond to atmospheric [CO_2], where reduced g_s could account for the observed reduction in leaf transpiration in plants grown under elevated atmospheric [CO_2] (Drake et al. 1997; Ainsworth and Long 2005; Lobell and Gourdji 2012) and RZ [CO_2] (He et al. 2004). However, Drake et al. (1997) reported that reduced g_s under elevated atmospheric [CO_2] did not limit photosynthesis. Despite the lower g_s observed in plants grown under elevated [CO_2], the means of internal [CO_2]/ambient [CO_2], were found to be similar for plants grown under both ambient and elevated [CO_2]. In our study with aeroponically grown plants, the lower g_s values at elevated RZ [CO_2] (He et al. 2004) indicated that similar physiological changes that occur in plants grown at elevated atmospheric [CO_2] (Drake et al. 1997) may have also taken place in plants grown under elevated RZ [CO_2] (He et al. 2004). Increased *A* at elevated RZ [CO_2] could reflect higher internal [CO_2] transport from roots, as *A* was measured at ambient [CO_2] (He et al. 2004). In our aeroponic system, though there was a lack of interaction between plant roots and microorganisms, the roots should still be capable of incorporating HCO_3^- into organic acids through phosphoenolpyruvate carboxylase (Chang et al. 1992; Vuorinen et al. 1992; Cramer et al. 1993; Cramer and Lips 1995; Edwards 1998; Cramer et al. 2005; Ford et al. 2007; Berveiller and Damesin 2008; Dimou et al. 2009). It had been reported that the enzyme carbonic anhydrase, which rapidly catalysed the conversion of dissolved CO_2 to H_2CO_3 and HCO_3^-, is found in root cell cytosol (Tetu 2007; Dimou et al. 2009). Assimilation of HCO_3^- from the nutrient solution by the roots had been widely observed (Vapaavuori and Pelkonen 1985; Vuorinen et al. 1992; Cramer et al. 1993; Cramer and Richards 1999; Cramer et al. 2005). In our study (He et al. 2004), as pH was about 6.5 inorganic carbon could gain entry into the roots as CO_2 and HCO_3^-. Hence, enhanced photosynthesis, despite lower g_s, was still possible due to the contribution of internal CO_2 from dissolved inorganic carbon (DIC), transported from RZ to shoot (Cramer and Richards 1999).

Using the same crisphead-type lettuce (*L. sativa* L. cv. 'Wintergreen'), grown in the same environmentally controlled rooms (He et al. 2004) to study light response curves of *A* and g_s (He et al. 2010), we found that there were no significant differences in *A* and g_s under a low PPFD of 300 μmol m^{-2} s^{-1} among the different RZ [CO_2] treatments. However, substantial differences in *A* and g_s were observed when the plants were grown under different elevated RZ [CO_2] at PPFD of ≥ 600 μmol m^{-2} s^{-1} (Figure 1). Further, higher *A* was observed when PPFD was increased from 300 to 1200 μmol m^{-2} s^{-1}. Plants had significantly higher *A* at all elevated than ambient RZ [CO_2] at both temperature regimes (Figures 1A, 1B). Though *no significant difference was observed in A* between plants grown at RZ [CO_2] of 10 000 and 50 000 ppm, these plants demonstrated a much higher *A* than those grown under ambient and elevated RZ [CO_2] of 2000 ppm. We can deduce that the impact of RZ [CO_2] on *A* was 300 ppm < 2000 ppm < 10 000 ppm ≈ 50 000 ppm, suggesting that elevated RZ [CO_2] of 10 000 ppm was sufficient to achieve the maximal photosynthetic rate. It was also found that plants grown at 28°C/22°C (Figure 1A) had higher *A* than those grown at 36°C/30°C (Figure 1B) under the same PPFD. Further, g_s also increased with increasing PPFD and the magnitudes of increase were much lower at 36°C/30°C (Figure 1D) than 28°C/22°C (Figure 1C). However, under a low PPFD of 300 μmol m^{-2} s^{-1}, the g_s values at 36°C/30°C (Figure 1D) were significantly higher than that at 28°C/22°C (Figure 1C) under all RZ [CO_2].

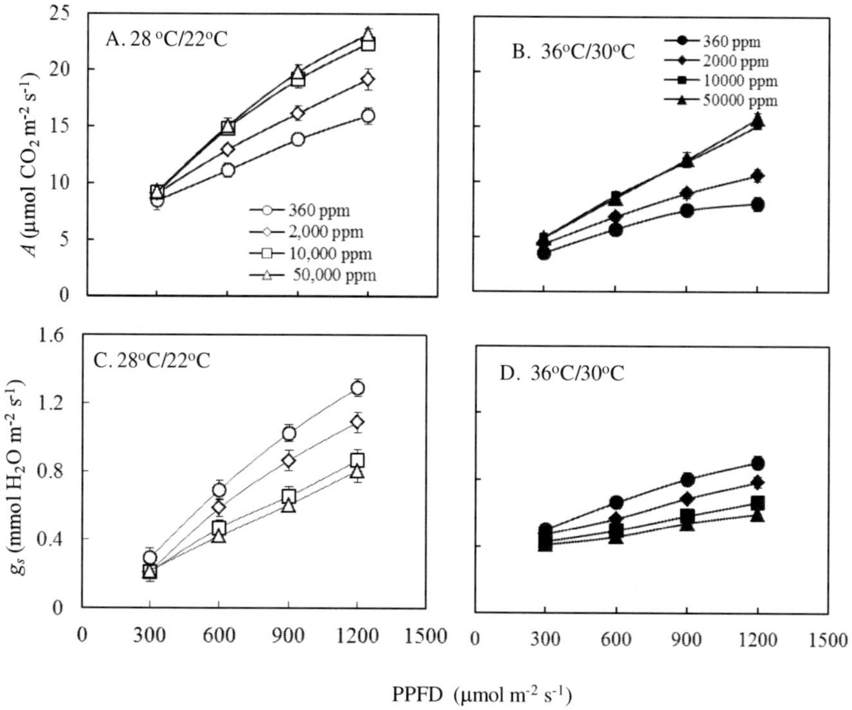

Figure 1. Light response curve of A (A, B) and g_s (C, D) of lettuce plants grown under different levels of elevated RZ [CO_2] in two different environmentally controlled rooms at 28°C/22°C and 36°C/30°C (day/night) at a PPFD of 650 µmol m^{-2} s^{-1} (12h photoperiod) for 2 weeks. Each point is the mean of 4 measurements of 4 different leaves from two different bins. Vertical bars represent the standard errors (redrawn from He et al. 2010).

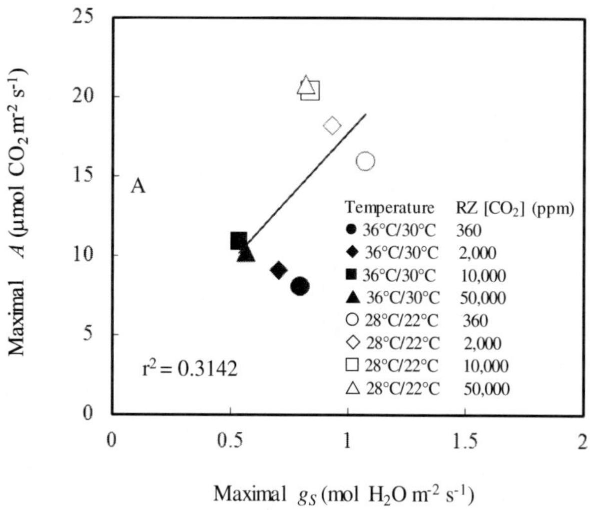

Figure 2. Correlations between maximal A and maximal g_s (data derived from Figure 1) of lettuce plants grown under different levels of elevated RZ [CO_2] at 28°C/22°C and 36°C/30°C (day/night ± 0.5°C) at a PPFD of 650 µmol m^{-2} s^{-1} (12h photoperiod) for three weeks (modified and redrawn from He et al. 2010).

In theory, the only growth-limiting factors in our aeroponic system were that of light availability and CO_2, since water stress had been reduced due to the continuous misting of nutrient solution to the roots of the plants. However, the amount of water absorbed by a plant is determined by volume of nutrient solution with which its roots are in contact with horizontally and vertically (Qin et al. 2007; He et al. 2009). As such, poor root development in plants grown at high air (He et al. 2004; 2010) and RZTs (He et al. 2013) could be the main factor affecting the ability of plants to uptake water, under elevated RZ [CO_2] (He et al. 2001). Furthermore, increased transpiration rates at higher temperatures (He et al. 2004) could account for the significantly higher g_s values at high (Figure 1D) than low air temperature regimes (Figure 1C), under all RZ [CO_2] and low PPFD level of 300 μmol m^{-2} s^{-1} (He et al. 2010). The magnitudes of increase in g_s from 300 to 1200 μmol m^{-2} s^{-1} were much lower at high than low temperatures (Figures 1C, 1D), further implying that under high PPFD and temperatures, elevated RZ [CO_2] resulted in higher midday leaf RWC while lower g_s reduced the loss of water and, thus, exhibited high WUE (Farquhar et al. 1989).

Based on the above discussion, it was clear that higher *A* under elevated RZ [CO_2] was not due to increased g_s as there was no correlation between maximal *A* and maximal g_s (Figure 2). These findings further confirmed that there was more internal CO_2 available to plants grown under elevated RZ [CO_2] reported previously by our research team (He et al. 2004; 2007). Inorganic carbon could enter roots either as CO_2 or HCO_3^- and be transported to the shoots via the transpiration stream (Vapaavuori and Pelkonen 1985; Bialczyk et al. 1996; Cramer and Lips 1995; Cramer and Richards 1999). Dissolved CO_2 had also been reported to be carried upwards in the xylem sap of stem, when trees are transpiring (Teskey and McGuire 2002; 2005). Thus, xylem-transported CO_2 could also be used as carbon source and fixed in photosynthetic tissues (Hibberd and Quick 2002; McGuire et al. 2009; Bloemen et al. 2013).

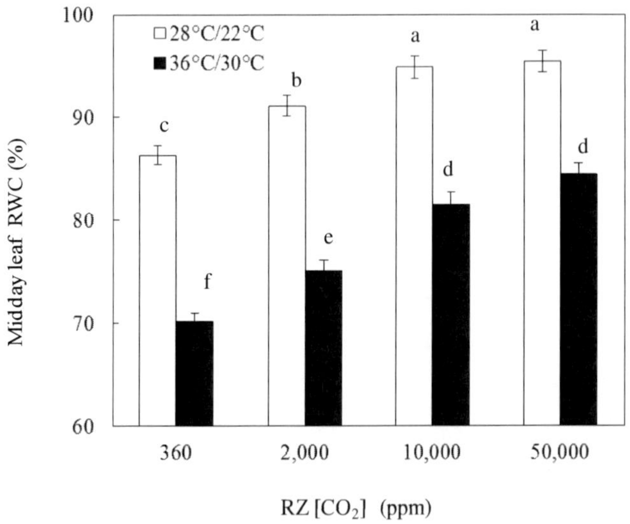

Figure 3. Midday leaf RWC of lettuce plants grown under different elevated RZ [CO_2] in two different environmentally controlled rooms at 28°C/22°C and 36°C/30°C (day/night) at a PPFD of 650 μmol m^{-2} s^{-1} (12h photoperiod) for 2 weeks. Each point is the mean of 5 measurements of 5 different leaves from two different bins. Vertical bars represent the standard errors. Means with different letters above the columns are statistically different (p <0.001) as determined by Tukey's multiple comparison test. (redrawn from He et al.. 2010).

Leaf transpiration decreased when stomata were partially closed allowing plants to have a higher WUE. Plant water status generally improved under elevated ambient CO_2 (Clifford et al. 1993; Tezara et al. 2002) partially due to reduced transpiration demand. Tezara et al. (2002) observed lowered g_s and reduced transpiration, but increased *A* and **WUE** by 230% and 380%, respectively, between sunflower plants (*Helianthus annuus* L. cv. Avante) that were well-watered and those suffering from severe **water** deficit under elevated ambient [CO_2]. In our study (He et al. 2010), higher leaf midday RWC (Figure 3) was observed at elevated RZ [CO_2] and the increase of RWC was greater at 36°C/30°C than at 28°C/22°C. For example, compared to plants grown at ambient RZ [CO_2], midday leaf RWC increased by 18% and 7% in lettuce plants grown at 36°C/30°C and 28°C/22°C, respectively (Figure 3).

In the greenhouse of Massey University, New Zealand, where the same crisphead-type lettuce (*L. sativa* L. cv. 'Wintergreen') were grown aeroponically, all plants were exposed to fluctuating ambient temperatures and light intensity. Maximal ambient temperatures and PPFD during midday, on sunny days, inside the greenhouse were 36°C and 2100 $\mu mol\ m^{-2}\ s^{-1}$, respectively. Treatments of different RZ [CO_2] started three weeks after transplanting and diurnal changes in *A* and g_s were measured six days after treatment (Figure 4). No significant increases in *A* (Figure 4C) and g_s (Figure 4D) were observed during the first 2 to 3 hours following sunrise. The highest *A* recorded in all plants was at about 1100h when PPFD (Figure 4A) and ambient temperature (Figure 4B) were about 1200 $\mu mol\ m^{-2}\ s^{-1}$ and 30°C, respectively. The *A* of plants grown in elevated RZ [CO_2] were significantly enhanced compared to that at ambient RZ [CO_2]. Similar to the experiments carried out in the environmentally controlled room (He et al. 2004; 2010), the values of *A* were very similar between plants grown at RZ [CO_2] of 10 000 and 50 000 ppm, but they were significantly higher than RZ [CO_2] at ambient and 2000 ppm. During midday at about 1300h, *A* decreased significantly. However, this midday depression of *A* was alleviated in all plants under elevated RZ [CO_2]. When ambient temperature decreased from 1400 to 1600h, *A* increased again, but it further declined with the decreases in PPFD in the later afternoon. It was interesting to note that elevated RZ [CO_2] significantly decreased g_s between 1100 and 1400 h (Figure 4D). Plants grown at high levels of RZ [CO_2] of 10 000 and 50 000 ppm had the lowest g_s and those at ambient RZ [CO_2] had highest g_s. Table 1 summarizes the midday depression of *A*, g_s, A/g_s and RWC. Magnitude of midday depression in *A* was highest in plants grown under ambient RZ [CO_2] while the lowest g_s was found in plants under elevated RZ [CO_2] of 10 000 and 50 000 ppm accompanied by the highest observed midday RWC. Thus far, only a few studies had been carried out to investigate the possible impacts of root carbon uptake on *A*, g_s, and WUE of plants grown in the greenhouse. Our results showed that plants grown under elevated RZ [CO_2] had higher WUE supported by higher A/g_s (Table 1). However, effects of RZ [CO_2] on gas exchange and water relations could vary with growth conditions. Cramer and Richards (1999) found that tomato plants (*L. esculentum*) grown under low light intensities and high 80% RH had lower g_s and transpiration rates, and thus, higher WUE if they were grown with elevated rhizosphere DIC of 5000 ppm. However, plants grown at 40% RH did not exhibit the same results. Our study carried out in both the greenhouse and environmentally controlled room suggested that the impacts of RZ [CO_2] on *A* and g_s varied with levels of light (He et al. 2007) and ambient temperatures (He et al. 2004, 2007; 2010).

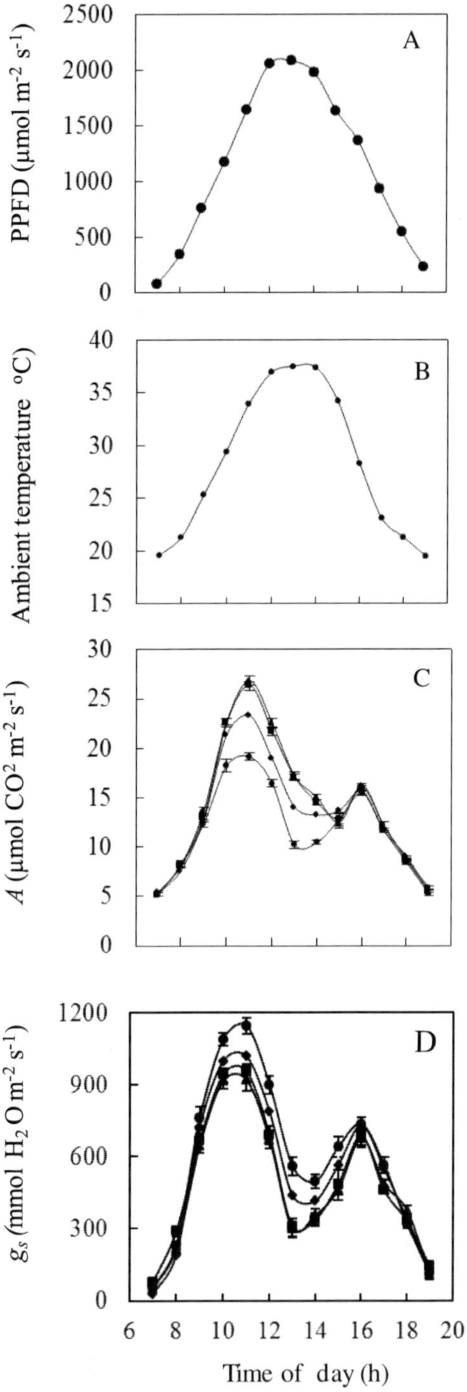

Figure 4. Diurnal changes of PPFDs (A) and ambient temperature (B), *A* (C), and g_s (D) of lettuce plants grown under different levels of elevated RZ [CO_2] (●, ambient RZ CO_2; ◆, 2000 ppm RZ CO_2; ■, 10,000 ppm RZ CO_2 and ▲, 50,000 ppm RZ CO_2) on a sunny day (6 days after different elevated RZ [CO_2] treatments) in a greenhouse. Each point is the mean of 5 measurements from 5 different leaves. Vertical bars represent the standard errors (redrawn from He et al. 2007).

Table 1. *A*, g_s and RWC during midday (at 1300h) in lettuce plants grown in a greenhouse on a sunny day

RZ [CO_2]. (ppm)	*A* (μmol CO_2 m^{-2} s^{-1})	g_s(mmol H_2O m^{-2} s^{-1})	A/g_s	Midday RWC
360 (ambient)	10.21 ± 0.39[a]	496 ± 16[a]	0.021	70.52 ± 0.28[a]
2000	14.07 ± 0.32 (137%)[b]	399 ± 24 (79%)[b]	0.035	77.21 ± 0.31 (109%)[b]
10000	17.09 ± 0.29 (167%)[c]	308 ± 17 (62%)[c]	0.055	85.65 ± 0.22 (121%)[c]
50000	17.22 ± 0.39 (169%)[c]	301 ± 23 (60%)[c]	0.057	85.11 ± 0.43 (120%)[c]

Measurements were made on day 6 after transferring to different RZ [CO2]. Each point is the mean of 5 measurements on the same age of leaves (the 4th leaf from the base) from 5 different plants. Percentages (parentheses) are relative to ambient RZ [CO2]. Means with different letters are statistically different ($p < 0.001$) as determined by Tukey's multiple comparison test (modified from He et al. 2007).

In Singapore, the temperate vegetable crops, such as lettuce, have been successfully grown in the tropics by cooling only the RZ (He and Lee 1998a; b; He et al. 2001). We found that high ambient-RZT (A-RZT) and light reduced *A* and caused shoot water deficit of lettuce plants by altering the balance between water uptake by the root system and water loss via the shoot (He et al. 2001; He 2009; 2010). Cooling the RZ of lettuce plants not only alleviated stomatal limitations of photosynthesis measured by *A*, g_s, and RWC but also mitigated non-stomatal limitations of photosynthesis, such as photosystem II (PSII) functionality, and Rubisco protein and activity, even though their aerial parts were exposed to ambient temperatures from 24 to 38°C (He and Lee 2001; 2004). Using the same lettuce plants (*L. sativa* L. cv. 'Wintergreen') in a tropical greenhouse in Singapore, we recently studied the effects of elevated RZ [CO_2] and RZT on photosynthetic gas exchange and water relations (He et al. 2013). In our study, the aerial parts of plants were subjected to the fluctuations of ambient temperatures ranging from 24 to 38°C. The maximum PPFD at plant canopy on sunny days was about 1200 μmol m^{-2} s^{-1} and RH was between 65 to 95%, whilst roots were grown in either 20°C-RZT or A-RZT. Different RZ [CO_2] was imposed on plants at each of the two RZT three weeks after transplanting. Light-saturated photosynthetic CO_2 assimilation (A_{sat}) and stomatal conductance ($g_{s\ sat}$) were determined from attached leaves two weeks after different treatments in the greenhouse (Figure 5). At both RZT, plants had higher A_{sat} at each of the three different elevated RZ [CO_2] compared to ambient RZ [CO_2] of 360 ppm (Figure 5A). However, the values of $g_{s\ sat}$ were significantly lower in lettuce plants grown at elevated than ambient RZ [CO_2] of 360 ppm (Figure 5B). A_{sat} increased with increasing RZ [CO_2], where RZ [CO_2] of 10 000 ppm was sufficient to achieve highest A_{sat} and maximum productivity (to be discussed next) (Figure 5A). However, $g_{s\ sat}$ was significantly lowered at higher than ambient RZ [CO_2] (Figure 5B) and therefore, increased A_{sat} under elevated RZ [CO_2] was not due to increased $g_{s\ sat}$ (Figure 6). These findings further confirmed that under elevated RZ [CO_2], dissolved CO_2 from nutrient solution could be transported upward via the xylem sap of the stem with the transpiration stream (Teskey and McGuire 2005) and fixed in leaves (McGuire et al. 2009). Changes of midday RWC (Table 2) under different RZ [CO_2] at both RZTs were similar to those of A_{sat} (Figure 5A). In all, the responses of A_{sat}, $g_{s\ sat}$, and midday leaf RWC to different RZ [CO_2] were similar at both RZTs.

Table 2. Midday leaf RWC (%) of lettuce plants grown under different levels of elevated RZ [CO_2] at 20°C-RZT and A-RZT for 2 weeks

RZ [CO_2]. (ppm)	A-RZT	20°C-RZT
360 (ambient)	65.2 ± 0.89[f]	82.2 ± 1.21[c]
2000	68.2 ± 1.03 (105%)[e]	87.2 ± 0.70 (1065)[b]
10000	75.9± 0.76 (116%)[d]	94.± 0.88 (114%)[a]
50000	76.3 ±0.87 (117%)[d]	94.6 ± 0.61 (115%)[a]

Each value is the mean of 5 measurements of 5 different leaves (n = 5). Percentages (parentheses) are relative to ambient RZ [CO2]. Means with different letters are statistically different ($p < 0.001$) as determined by Tukey's multiple comparison test (modified from He et al. 2013).

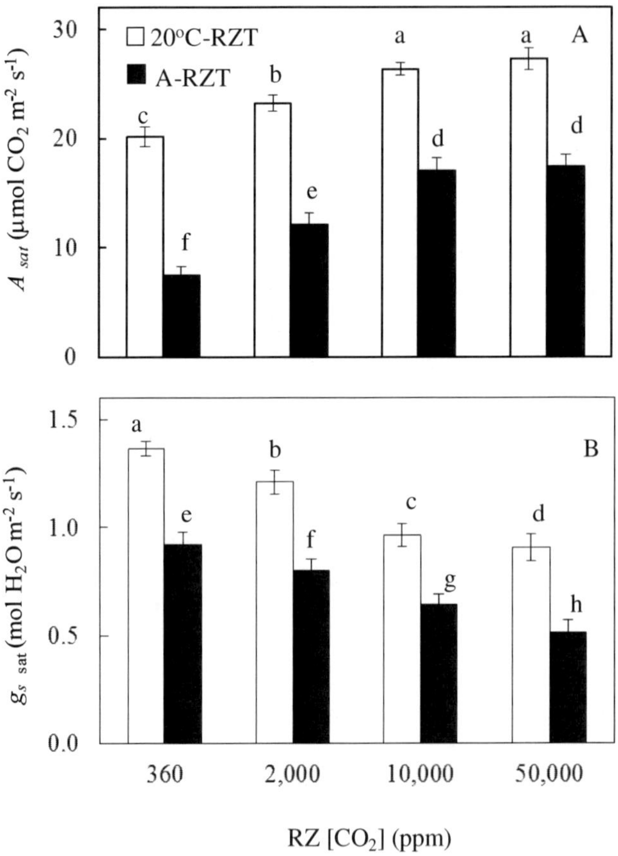

Figure 5. A_{sat} (A) and $g_{s\ sat}$ (B) of lettuce plants grown under different elevated RZ [CO_2] at 20°C- RZT (open bars) and A-RZT (closed bars) for 2 weeks. Each value is the mean of 5 measurements of 5 different leaves (n = 5). Vertical bars represent the standard errors. Means with different letters above the bars are statistically different (p < 0.001) as determined by Tukey's multiple comparison test (redrawn from He et al. 2013).

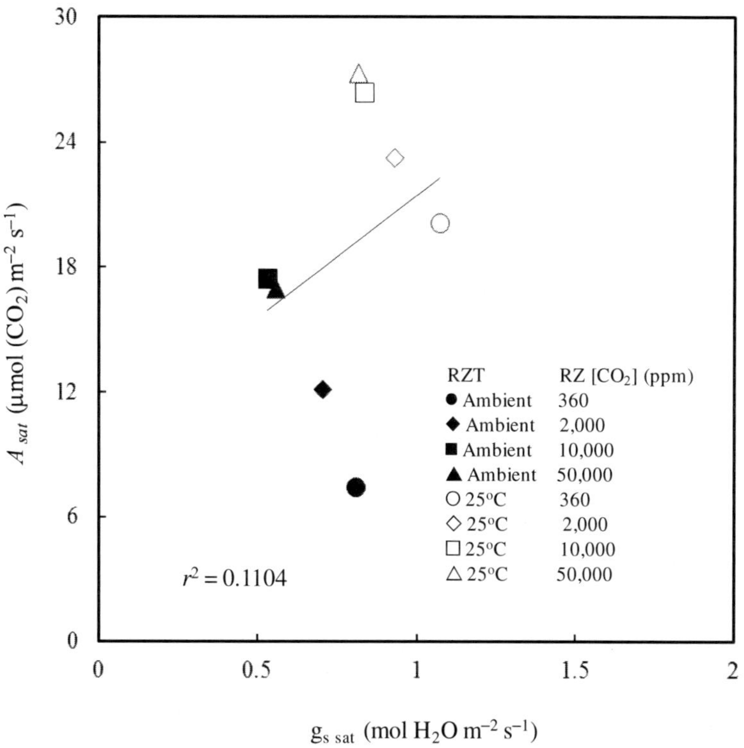

Figure 6. Correlation between *A sat* and $g_{s\ sat}$ (data derived from Figure 5) of lettuce plants grown under different levels of elevated RZ [CO_2] at 20°C-RZT and A-RZT for three weeks (re-drawn from He et al. 2013).

Using detached branches of *Populus deltoides*, Bloemen et al. (2013) allowed them to take up $^{13}CO_2$-labelled solution at either high or low $^{13}CO_2$ concentrations. They reported that there was ^{13}C assimilated in photosynthetic tissues. The ratio of $^{13}CO_2$ assimilated from the internal source to from the atmosphere was highest for the branches under the high light (HL) and high vapour pressure deficit (VPD) treatment corresponding to higher transpiration, than under low VPD. They concluded that the relative contribution of internal CO_2 recycling to tree carbon gain is strongly dependent on factors controlling transpiration, respiration, and photosynthesis. In the our study, although we were unable to provide the amount of elevated RZ [CO_2] needed to obtain similar effects on *A*, the results of higher *A* observed in plants treated with higher elevated RZ [CO_2] supported our hypothesis that elevated RZ [CO_2] enhanced photosynthesis which was due to higher [CO_2] supplied to the RZ. Our previous (He et al. 2001) and present (Table 2) studies showed that temperate lettuce grown in tropical greenhouse at A-RZT had midday leaf RWC below 80-70%. These plants had lower $g_{s\ sat}$ and showed symptoms of wilting during midday. However, stomata of these plants were still partially open and all plants were still transpiring. Although DIC was not measured directly, the amount in the nutrient solution might still be higher in the transpiration stream (despite the lower g_s and transpiration rate) when plants were supplied with higher RZ [CO_2] of 2000, 10 000 and 50 000 ppm, as compared to ambient RZ [CO_2] of 360 ppm.

It was reported by Retallack (2001) and Lake et al. (2001) that stomatal density of leaves varied inversely with [CO_2] in air around leaves. In all our studies carried out in both environmentally controlled room and greenhouse, newly emerged leaves which developed after the plants were treated with different levels of elevated RZ [CO_2] were selected for *A* and g_s measurements. Our results led us to postulate that elevated RZ [CO_2] could cause lower stomatal density and hence lower g_s. In our studies, when elevated RZ [CO_2] was supplied, inorganic carbon could enter roots either as CO_2 or HCO_3^- and be transported to the shoot via transpiration stream (Vapaavuori and Pelkonen 1985; Bialczyk et al. 1996; Cramer and Lips 1995) and hence regulated the development of stomata in the newly developing leaves. Consequently, a higher *A* with lower g_s from plants grown at elevated RZ [CO_2] could partially result from using inorganic carbon as CO_2 or HCO_3^- via transpiration stream from roots. Therefore, effects of elevated RZ [CO_2] on stomatal development merit our further studies.

ELEVATED RZ [CO_2] PROTECTS PLANTS FROM MIDDAY DEPRESSION OF PHOTOSYNTHESIS

Photosynthesis is directly and dramatically influenced by the amount of light striking on leaves. Depression of photosynthesis, such as photoinhibition, can occur when plants are grown under excessive sunlight coupled with other environmental stresses (Björkman and Powles 1984; Demmig and Björkman 1987; Osmond 1994; He et al. 1996; 2000). Midday depression of photosynthesis of plants grown in the greenhouse and natural conditions under high light and temperatures have been reported by ourselves (He et al. 1996; 2001; He and Lee 2004) and others (Long et al. 1994; Horton et al. 1996; Pons and Welschen 2003), where such midday depression of photosynthetic CO_2 assimilation was associated with stomatal closure (Pereira and Chaves 1993; He et al. 2001). We have reported that temperate lettuce grown in the aeroponic systems in the tropical greenhouse showed substantial midday depression of *A* and g_s on sunny days, though RZ cooling could reduce the midday depression of *A*, mainly due to the closure of stomata (He et al. 2001). In our further study, we have also found that hot A-RZT reduced photosynthetic utilization of radiant energy in temperate lettuce grown in the greenhouse especially during midday (He and Lee 2004).

It has been reported that the higher atmospheric CO_2 may decrease photoinhibition and thus enhance plant growth (Hymus et al. 1999; 2001; Roden et al. 1999; Kitao et al. 2007). Working with tomato (*L. esculentum*) seedlings, Cramer and Richards (1999) reported that increased biomass was observed in both control and salinity-stressed plants grown in enriched rhizosphere DIC, at high temperatures (daily maximum of 37°C) and light (PPFD of 1500 $\mu mol\ m^{-2}\ s^{-1}$). When these were plants grown at a PPFD of less than 1000 $\mu mol\ m^{-2}\ s^{-1}$, higher growth rates by elevated rhizosphere DIC were only observed in plants grown at high temperatures at 35°C or with salinity at 28°C. Our study with aeroponically-grown lettuce plants showed that elevated RZ [CO_2] stimulated photosynthetic CO_2 assimilation, with greater increase, at higher temperatures (He et al. 2004; 2010). The enhancement of plant growth under elevated DIC was due to the partial closure of stomata and lower WUE. In their study, Cramer and Richards (1999) also found that partial down-regulation of photosynthetic rate was observed under elevated rhizosphere DIC. Using aeroponically grown crisphead-type lettuce (*L. sativa* L. cv. 'Wintergreen'), we studied the diurnal changes of chlorophyll fluorescence F_v/F_m ratio of plants grown under fluctuating ambient temperatures and light

intensity, under different elevated RZ [CO_2] (He et al. 2007). As the chlorophyll fluorescence F_v/F_m ratio is used to estimate the efficiency of excitation energy captured by PSII reaction centers in dark-adapted leaves (He and Lee 2004), the lowest F_v/F_m ratios were obtained between 1200h and 1500h when solar irradiation and air temperatures reached the highest values of 2100 μmol m^{-2} s^{-1} and 36°C, respectively, on sunny days (Figure 7). F_v/F_m ratios were much lower in plants grown under ambient than elevated RZ [CO_2] during the period of 1300h and 1500h. Among the plants grown under elevated RZ [CO_2], midday depression of F_v/F_m ratios was greater at 360 ppm and 2000 ppm of RZ [CO_2] than at 10 000 and 50 000 ppm of RZ [CO_2]. These findings indicated that dynamic photoinhibition (reduced midday F_v/F_m ratio) caused by high PPFD and temperatures was alleviated by elevated RZ [CO_2]. In the same experiment discussed earlier (Figure 4), the higher A and lower g_s found in plants grown at elevated than ambient RZ [CO_2] suggested that g_s was not directly involved in the alleviation of dynamic photoinhibition (Figure 7) during midday. Though it was previously reported that high atmospheric [CO_2] decreased photoinhibition (Hymus et al. 1999; Roden et al. 1999), Teskey and McGuire (2005) later found that the alleviation of photoinhibition in elevated RZ [CO_2] during midday may have resulted from high internal levels of CO_2 transported from roots instead. At high light intensity, when abundant energy is being supplied to the plants, one of the limitations on plant growth would have been the availability of CO_2 for photosynthesis due to diffusion (Cramer and Richards 1999). As such, mitigation of photoinhibition may be important in determining growth under high light intensities, especially when temperatures are high and stomata are partially closed as a result. High solar radiation and air temperatures in the tropics have been known to cause reduction in PSII efficiency and CO_2 assimilation in sun-exposed leaves (He et al. 1996). We have previously reported that cooling the lettuce plant roots to 20°C-RZT resulted in higher A_{sat} from reduced dynamic photoinhibition and preventing chronic photoinhibition (higher pre-dawn F_v/F_m ratio), when compared to plants grown under A-RZT, despite identical ambient environments of all plants. However, both dynamic and chronic photoinhibition were observed to have occurred in A-RZT grown plants (He et al. 2001). Recently, using the same lettuce plants (*L. sativa* L. cv. 'Wintergreen') grown aeroponically in a tropical greenhouse in Singapore, it was found that higher elevated RZ [CO_2] protected plants from both chronic and dynamic photoinhibition at both 20°C-RZT and hot A-RZT (Figure 8, He et al. 2013). Figure 8A shows F_v/F_m ratios were greater than 0.8, throughout the day, for plants grown under elevated RZ [CO_2] of both 10 000 and 50 000 ppm at 20°C-RZT. These results, thus, suggested that neither chronic nor dynamic photoinhibition occurred at these higher elevated RZ [CO_2] plants. However, for plants grown under RZ [CO_2] of 360 and 2000 ppm, reduced F_v/F_m ratios were observed from 1100h and lowest F_v/F_m ratios were obtained at 1300h. Increases of F_v/F_m ratios were observed, for these same plants, from 1500h onwards and values of greater than 0.8 were recorded only at 1900h. Thus, these findings implied that plants grown at 20°C-RZT under RZ [CO_2] of 360 and 2000 ppm, exhibited dynamic photoinhibition. In the case of A-RZT plants grown under RZ [CO_2] of 10 000 and 50 000 ppm, the F_v/F_m ratios recorded were greater than 0.8, at 0700h. Thereafter, decreased F_v/F_m ratios below 0.8 were obtained from plants grown under RZ [CO_2] of 360 and 2000 ppm. Plants had significantly lower F_v/F_m ratios under RZ [CO_2] of 360 ppm than other concentrations of elevated RZ [CO_2] throughout the day (Figure 8B). These data suggested that though both chronic and dynamic photoinhibition occurred in A-RZT plants under lower RZ [CO_2], there was a smaller effect on plants under RZ [CO_2] of 2000 ppm than RZ [CO_2] of 360 ppm. As previously discussed, lettuce plants grown under fluctuating air temperatures of 20–36°C under a maximal PPFD of 2100 μmol m^{-2} s^{-1} also showed less depression of midday F_v/F_m ratios under elevated RZ [CO_2] than ambient RZ [CO_2] (He et al. 2007).

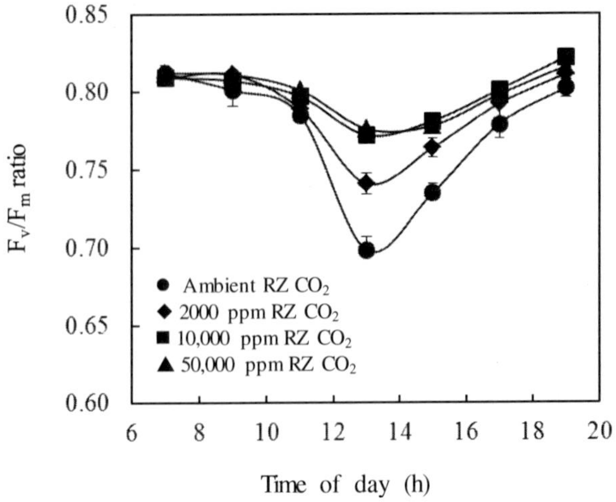

Figure 7. Diurnal changes of maximal efficiency of PSII electron transport (F_v/F_m ratio) of lettuce plants grown under different levels of elevated RZ [CO_2] on a sunny day (9 days after different elevated RZ [CO_2] treatments) in the greenhouse. Each point is the mean of 5 measurements from 5 different second leaves on a sunny day. Vertical bars represent the standard errors (redrawn from He et al. 2007).

On bright sunny days in the tropics, plants are usually exposed to lower irradiance in the morning but, at noon, they are exposed to both strong sunlight and high temperatures. Photoinhibition, ascribed to photoinactivation of PSII, occurred under the combination of high light and temperatures (He et al. 1996; 2001; He and Lee 2004). We have previously found that midday F_v/F_m ratios were much lower in plants grown at high A-RZT (ranging from 24-38°C), than 20°C-RZT, in a tropical greenhouse (He et al. 2001). At high A-RZT and growth irradiances, low internal [CO_2] reduced the photosynthetic utilization of radiant energy. Thus, cooling the roots to 20°C-RZT allowed higher A_{sat} by reducing dynamic photoinhibition, and preventing chronic photoinhibition, when compared to A-RZT plants (He et al. 2001). Higher elevated RZ [CO_2] could also protect plants from both chronic and dynamic photoinhibition at 20°C-RZT, as well as, alleviate both chronic and dynamic at A-RZT (He et al. 2013).

Reduced photosynthetic electron consumption is a direct consequence of decrease in CO_2 supply, causing a down-regulation of photosynthesis and increasing alternate requirements for additional electron and O_2 consumption such as photorespiration and the Mehler reaction (Cornic and Fresneau 2002). In our previous study, it was found that high A-RZT caused shoot water deficit by altering the balance between water uptake by the root system and water loss from the shoot of lettuce plants (He et al. 2001). Stomatal closure induced by water deficit depletes intercellular and chloroplastic CO_2, reducing photosynthetic CO_2 assimilation (Chaves and Oliveira 2004). When stomatal limitation of photosynthesis was more severe in A-RZT plants (He et al. 2001) more endogenous electron acceptors become reduced and O_2 may be the main available electron acceptor that could result in photodamage of the

photosynthetic apparatus. It has been reported that, under natural conditions during midday, 60% of the total photosynthetic electron flow was used for carbon assimilation while about 40% was devoted to photorespiration in oak trees (*Quercus cerris* L.) due to water deficit (Valentini et al. 1995). Cramer and Richards (1999) reported that carbon transported from the roots to shoot via xylem could alleviate not only photoinhibition but also photorespiration. According to Cornic and Fresneau (2002), when leaf RWC < 70%, it is likely that stomatal closure is the main cause in the decline of leaf photosynthesis with CO_2 concentration in the chloroplasts decreasing, when stomata close in drying leaves. As a consequence, ribulose-1,5-bisphosphate oxygenation of C_3 plants increase and become the main sink for photosynthetic electrons. Based on these results, we assume that in our study, higher midday leaf RWC (>70%) of lettuce plants grown under elevated RZ $[CO_2]$ could help alleviate both photorespiration and photoinhibition (He et al. 2013).

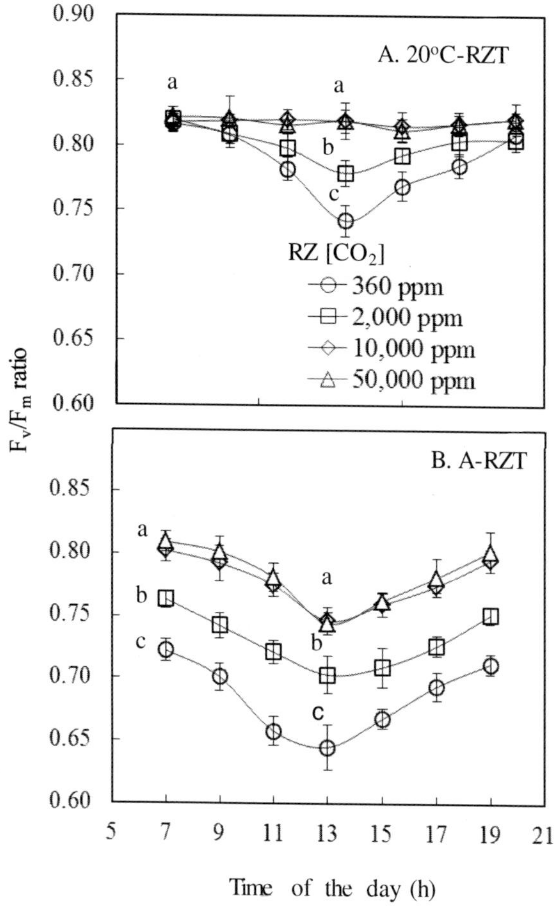

Figure 8. Diurnal changes of F_v/F_m ratio of lettuce plants grown under elevated RZ $[CO_2]$ at 20°C-RZT (A) and A-RZT (B) for 2 weeks. Each value is the mean of 5 measurements of 5 different leaves (n = 5). Vertical bars represent the standard errors. Means with different letters above the curves are statistically different ($p < 0.001$) as determined by Tukey's multiple comparison test (redrawn from He et al. 2013).

Elevated RZ [CO_2] Effects on Photosynthetic Nitrogen (N) Use Efficiency

The plant root system is the main organ involved in water and mineral uptake. They play an important role in root-shoot interaction to adjust nutrient, especially N, acquisition capacity to meet the shoot demand caused by environmental changes (BassiriRad et al. 2001). Nutrient could be more efficiently used in terms of, both, the rate of mineral uptake and assimilation under elevated [CO_2] (Khan et al. 2008). A sufficient supply of primary macronutrients, such as N, has to be sustained when plants grow faster and more vigorously (Khan et al. 2008). It was reported that acclimation of photosynthesis was connected with the N-status of the plant (Drake et al. 1997; Stitt and Krapp 1999), where leaf N can be found in components associated with photosynthesis and acclimation of photosynthesis was connected with the N-status of the plant. This, thus, explains chloroplast components representing a large proportion of the total leaf-N (Evans 1989; Lawlor et al. 2001). For different species, close relationships have been found between photosynthetic efficiency and leaf-N content (Reich et al. 1994). When plants exposed to ambient elevated [CO_2] they exhibited enhanced growth rates, creating an increased demand for nutrients, especially N (Rogers et al. 2006). When nitrate (NO_3^-) uptake and assimilation do not keep pace with photosynthesis and growth under elevated [CO_2], the decrease of N concentration is usually interpreted as the main reason (Pettersson and McDonald 1994). Leaf N concentration per unit area usually declined under elevated [CO_2] (Ellsworth et al. 2004), which may be due to larger leaf biomass requiring higher N investment into the important carbon fixation enzyme, Rubisco. Since elevated RZ [CO_2] affected stomatal function, mineral uptake such as that of N could have been affected. Since elevated RZ DIC has been shown to increase NO_3^- uptake (Cramer et al. 1993), NO_3^- uptake had also been reported to be increased under elevated RZ [CO_2] in the short-term (Van der Merwe and Cramer 2000). However, there was no change in the total tissue N with increased RZ [CO_2] (Viktor and Cramer 2003). Using crisphead-type lettuce (*L. sativa* L. cv. 'Wintergreen') grown aeroponically under different RZ [CO_2] and two different air temperatures, we studied the uptake of NO_3^- and overall N metabolism (He et al. 2010). NO_3^- and total reduced N concentrations of shoots were significantly higher in all plants under elevated RZ [CO_2] than under ambient RZ [CO_2] at both at 28°C/22°C and 36°C/30°C (Figure 9). No significant differences in NO_3^- and total reduced N concentrations of shoot were observed when plants were exposed to elevated [CO_2] of 10 000 and 50 000 ppm at both temperature regimes. At each RZ [CO_2], plants had significantly higher shoot NO_3^- and total reduced concentrations at 28°C/22°C than at 36°C/30°C (He et al. 2010). These results indicated that lettuce plants grown aeroponically with ample nutrient supply continuously sprayed on the larger root system under elevated RZ [CO_2], increased both leaf NO_3^- (Figure 9A) and total reduced N concentration (Figure 9B). As discussed earlier, the lower g_s but higher *A* of lettuce plants grown under elevated RZ [CO_2] could be partially due to the incorporation of DIC, allowing for the improved incorporation of N into amino acids in the roots, as a consequence of greater supplies of anaplerotic carbon for protein synthesis (Cramer and Lewis 1993; Cramer et al. 1993; Viktor and Cramer 2003). This was supported by the observed greater differences in *A* under higher PPFD (Figure 1) and the established positive correlation between maximal *A* and total reduced N of the shoots (Figure 10). Numerous studies have demonstrated a positive linear relationship between leaf N concentrations and Rubisco in mature leaves, where decreases in these two parameters result in reduced photosynthesis (Evans 1989; Li et al. 2009). On the other hand, non-linear correlation between maximal A and g_s under different elevated RZ [CO_2] (Figure 2) further confirmed that increase of A under elevated RZ [CO_2] could partially be due to the higher shoot total reduced N not g_s, as discussed earlier.

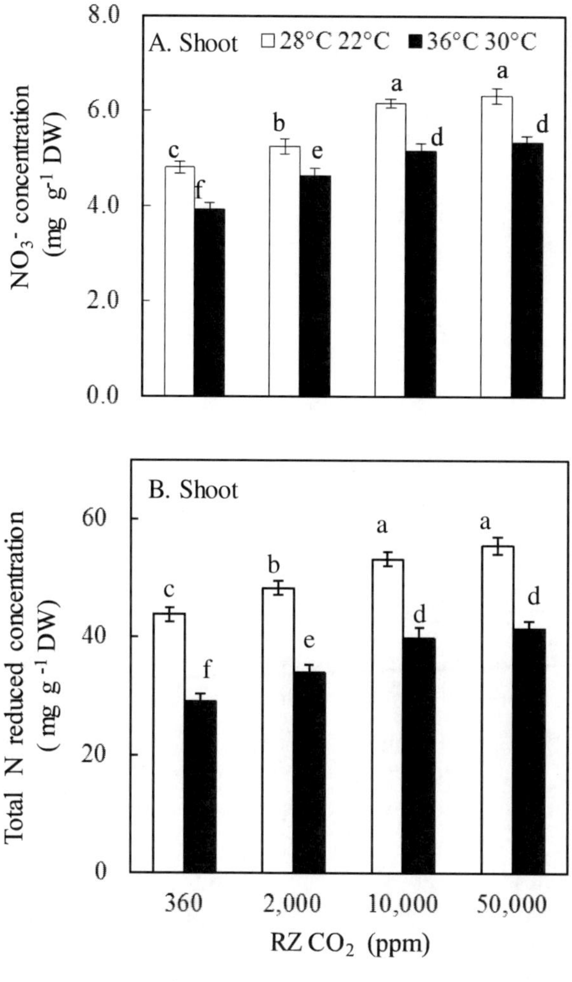

Figure 9. NO_3^- (A) and total reduced N (B) concentration of shoot of lettuce plants grown under different elevated RZ [CO_2] in two different environmentally controlled rooms at 28°C/22°C and 36°C/30°C (day/night ± 0.5°C) at a PAR of650 μmol m^{-2} s^{-1} (12h photoperiod) for three weeks. Each point is the mean of 5 measurements of 5 different plants from two different bins. Vertical bars represent the standard errors. Means with different letters above the columns are statistically different ($p < 0.001$) as determined by Tukey's multiple comparison test (redrawn from He et al. 2010).

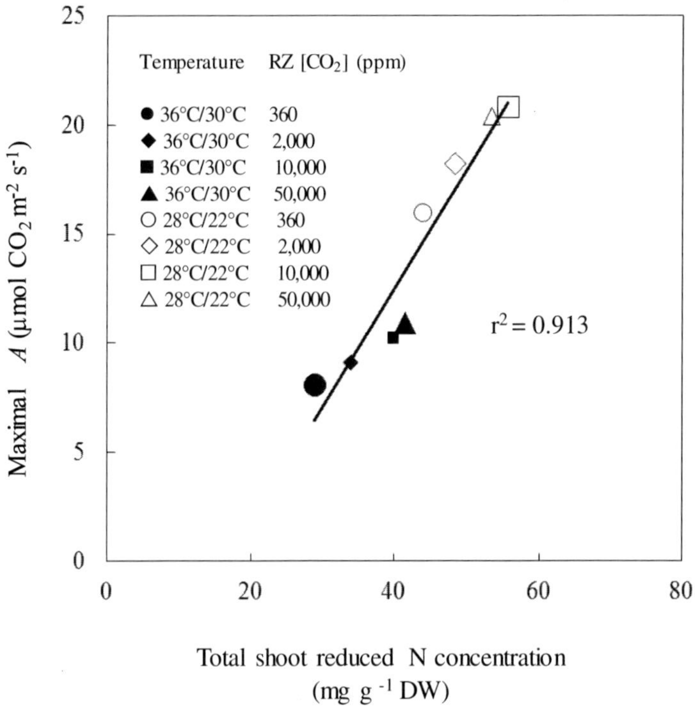

Figure 10. Correlations between maximal *A* and total reduced N concentration of shoot (data derived from Figure 1A and Figure 9B) of lettuce plants grown under different levels of elevated RZ [CO_2] at 28°C/22°C and 36°C/30°C (day / night, ± 0.5°C) at a PPFD of 650 μmol m^{-2} s^{-1} (12h photoperiod) for three weeks (redrawn from He et al. 2010).

In another experiment with lettuce (*L. sativa* L. cv. 'Wintergreen') grown in the tropical Singapore, we also found NO_3^- uptake and total reduced N concentration in plants grown under elevated RZ [CO_2] were higher than that under ambient RZ [CO_2] at both 20°C-RZT and hot A-RZT (He et al. 2013). Increased uptake of NO_3^- by roots under enriched RZ [CO_2] had also been reported in tomato plants grown hydroponically (van der Merwe and Cramer 2000).

A close linear correlation was established between A_{sat} and shoot total reduced N concentrations, total leaf soluble, and Rubisco protein concentrations (Figure 11) in plants grown under different RZ [CO_2] and RZTs.

Growing temperate lettuce crops in the tropics, our studies showed that cooling the RZ of plants alleviated not only stomatal limitations of photosynthesis but also non-stomatal limitations of photosynthesis even though their aerial parts were exposed to hot ambient temperatures (He et al. 2001; He and Lee 2001; 2004). Non-stomatal limitation of photosynthesis has been attributed to reduced carboxylation efficiency (He et al. 2001; Hikosaka et al. 2006) or decreased amount of Rubisco (Kanechi et al. 1996; He and Lee 2001). In our aeroponic system, NO_3^- was the only N source supplied into the nutrient solution. NO_3^- had to be converted to NH_4^+ before it was incorporated into amino acid and other organic N compounds, such as Rubisco protein. Non-stomatal limitations of

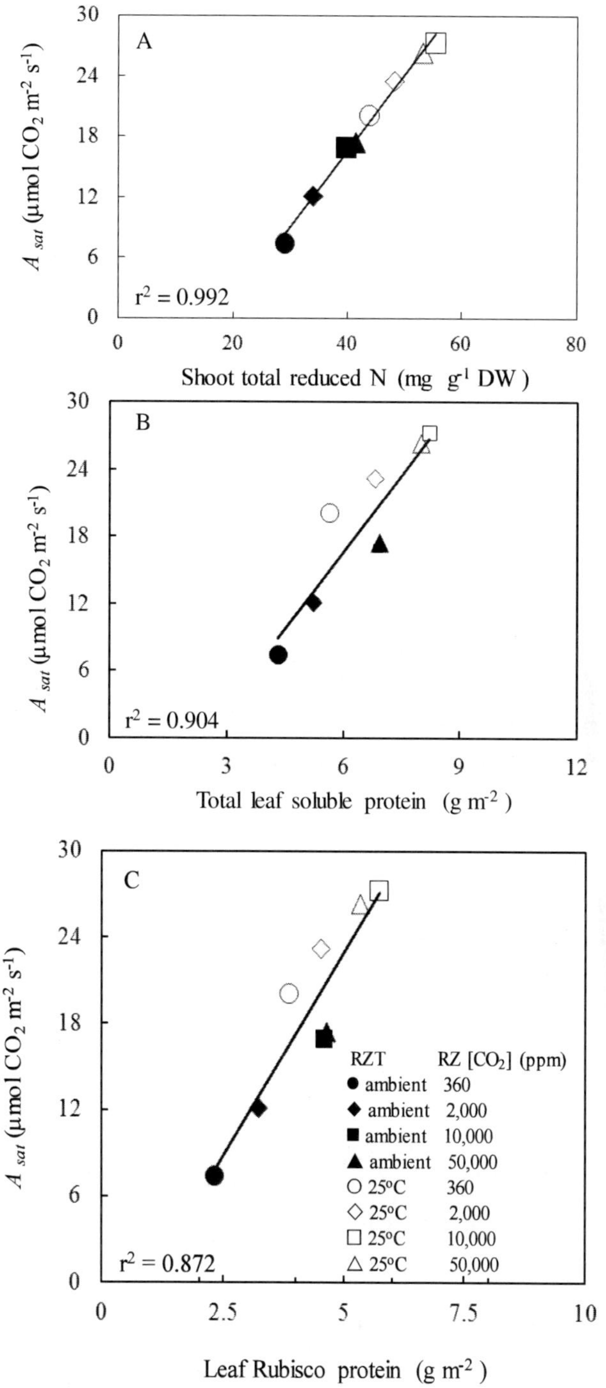

Figure 11. A_{sat} (data derived from Figure 5A) plotted against shoot total reduced N (data from He et al. 2013) (A); soluble protein (data from He et al. 2013) (B) and Rubisco protein (C) (data from He et al. 2013) (B) of lettuce plants grown under different levels of elevated RZ [CO_2] at 20°C- and A-RZT for three week (redrawn from He et al. 2013).

photosynthesis for lettuce plants grown at hot A-RZT could partly be due to poor root development which resulted in N deficiency (He et al. 2001). Hence, higher NO_3^- assimilation rate is required to support faster growth of temperate lettuce plants grown at cool-RZT. In our previous studies, it was found that NO_3^- uptake, transport, and NO_3^- reductase activity were increased by cooling the RZ of temperate and subtropical vegetable crops grown in the tropics (Tan et al. 2002; He et al. 2008; He 2010). Assimilation of NO_3^- into organic N containing compounds involved the reduction of NO_3^- to NO_2^- via the cytosolic enzyme nitrate reductase (NR), which was an NO_3^--inducible enzyme (Cookson et al. 2005). It had been demonstrated that total reduced N concentration decreased at high RZT (Du and Tachibana 1994; He et al. 2001). Reduction in NR activity was linked to the decrease of photosynthesis due to stomatal closure (Kaiser and Brendel-Benisch 1991). However, there was no linear relationship between A_{sat} and $g_{s\ sat}$ in plants grown under different RZ [CO_2] and RZTs (Figure 6). These results further supported our postulation that plants grown under elevated RZ [CO_2] had higher A_{sat}, at both RZTs, partially due to the improved incorporation of N into amino acids in the roots, resulting in greater supplies of anaplerotic carbon for protein synthesis (Cramer and Lewis 1993; Viktor and Cramer 2003).

Rubisco is widely thought to limit the light-saturated photosynthesis due to its relatively poor affinity for CO_2 and competing reactions with O_2 (Andrews and Lorimer 1987). The vast amount of Rubisco protein in the leaf partly offsets the low efficiency of Rubisco (Woodrow and Berry 1988) where Rubisco acting as a 'reserve ' protein could be advantageous in providing greater efficiency of photosynthesis in response to daily fluctuating conditions (Stitt and Schulze 1994). Stomatal closure in response to high A-RZT was observed immediately when lettuce plants were transferred from 20°C-RZT to A-RZT (He et al. 2001). However, decreases in A_{sat} cannot be easily recovered after increasing $g_{s\ sat}$ by transferring A-RZT lettuce plants back to 20°C-RZT as Rubisco activity and Rubisco protein were also affected by A-RZT due to the N deficiency observed in these plants (He and Lee 2001). Under elevated RZ [CO_2] at 20°C-RZT, lettuce plants increased not only shoot NO_3^- and total reduced N concentrations but also the leaf soluble and Rubisco protein (He et al. 2013). Long et al. (2004) suggested that almost all of the decline in leaf N could be accounted for by a decrease in Rubisco, which comprised 27% of total leaf N (Makino et al. 1997). Though reduced amounts of Rubisco resulted in lower photosynthetic CO_2 assimilation (Makino et al. 1997; Makino and Mae 2004), higher amounts of Rubisco was partially offset by lower Rubisco activation in high-N leaves (Ray et al. 2003). Nevertheless, light-saturated photosynthetic rate was generally higher in high-N than low-N leaves (Makino et al. 2003), where Li et al. (2009) reported that increases in shoot N, leaf soluble and Rubisco protein concentrations resulted in higher photosynthetic CO_2 assimilation. The positive linear relationships between A_{sat} and the shoot total reduced N concentration (Figure 11A), leaf soluble protein (Figure 11B), and Rubisco concentration (Figure 11C) further supported that increased A_{sat} under elevated RZ [CO_2] at both high A-RZT and 20°C-RZT is partially due to higher shoot total reduced N concentration, higher soluble and Rubisco proteins rather than high $g_{s\ sat}$ (Figure 12) (He et al. 2007; 2010).

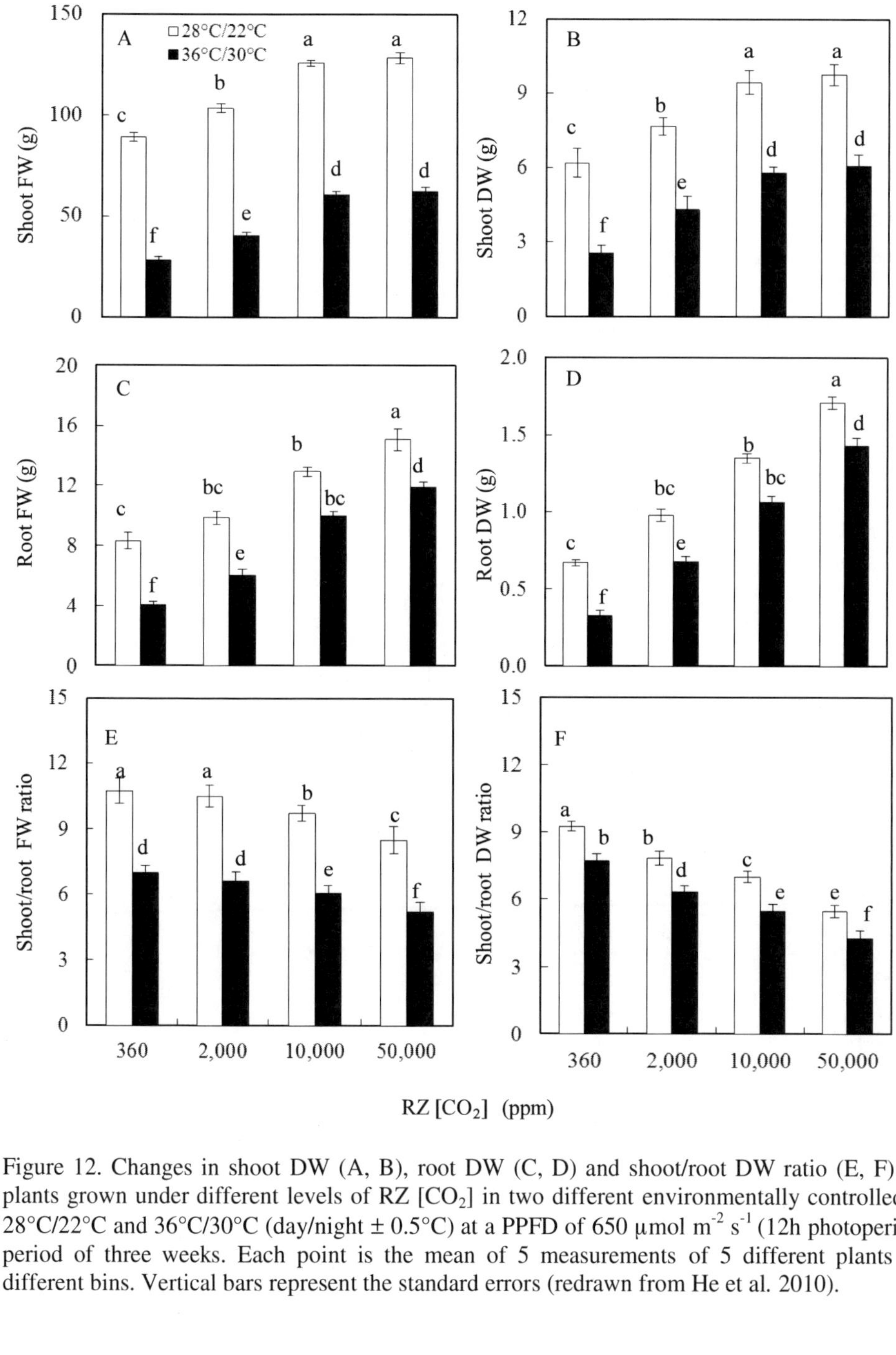

Figure 12. Changes in shoot DW (A, B), root DW (C, D) and shoot/root DW ratio (E, F) of lettuce plants grown under different levels of RZ [CO_2] in two different environmentally controlled rooms at 28°C/22°C and 36°C/30°C (day/night ± 0.5°C) at a PPFD of 650 μmol m^{-2} s^{-1} (12h photoperiod) over a period of three weeks. Each point is the mean of 5 measurements of 5 different plants from two different bins. Vertical bars represent the standard errors (redrawn from He et al. 2010).

ELEVATED RZ [CO_2] INFLUENCES ON SHOOT AND ROOT PRODUCTIVITY AND SINK CAPACITIES

Conflicting results of the effects of elevated RZ [CO_2] on plant shoot and root growth have been reported. Stolwijk and Thimann (1957) showed that root growth of pea (*Pisum sativum* L.) plants was inhibited by 80% by a 2% RZ [CO_2], whereas 65% RZ [CO_2] did not affect root growth of oat (*Avena sativa* L.) or barley (*Hordeum vulgare* L.). Bouma et al. (1997) reported that 2% RZ[CO_2], had no effect on both shoot and root growth of bean (*Phaseolusvulgaris* L.) and citrus (*Citrus volkameriana* Tan. & Pasq). However, Geisler (1967) reported that 10% [CO_2] stimulated root development of pea (*P. sativum* L.). These contradictory results could be due to the differences in plant species, RZ [CO_2] used or duration of RZ [CO_2] treatment.

It had been reported that elevated RZ [CO_2] enhanced plant growth, especially under stress conditions such as high salinity and temperature (Cramer and Richards 1999). Cramer and Richards (1999) found that both control and salinity-stressed tomato (*L. esculentum*) seedlings increased their biomasses in response to increased RZ [CO_2] at high temperature of 37°C under a PPFD of 1500 μmol m^{-2} s^{-1}. However, when these plants were grown under a PPFD of 1000 μmol m^{-2} s^{-1} or lower, the increases of plant growth by elevated RZ [CO_2] were only found in plants grown at 37°C without salinity or with salinity at lower shoot temperature of 28°C. In another study, Viktor and Cramer (2003) also found that there was significant increase in biomass accumulation in tomato (*L. esculentum*) seedlings, when they were grown under elevated RZ [CO_2].

Our study with crisphead-type lettuce (*L. sativa* L. cv. 'Wintergreen') grown under PPFD of 650 μmol m^{-2} s^{-1} (12 h photoperiod) showed that plants grown under elevated RZ [CO_2] exhibited faster leaf expansion, greater leaf area, higher shoot and root fresh weight (FW) than ambient RZ [CO_2] at both 28°C/22°C and 36°C/30°C. There were no significant differences in shoot and root FW when roots were exposed to 10 000 and 50 000 ppm RZ [CO_2] but they were greater than at 2000 ppm and ambient RZ [CO_2] (He et al. 2004). We also found that at 28°C/22°C, shoot/root FW ratios increased with RZ [CO_2] from ambient to 10 000 ppm RZ [CO_2]. Higher shoot/root FW ratio was due to higher relative increases in shoot FW. However, no further increase in shoot/root FW ratio was observed when plants were grown under 50 000 ppm of RZ [CO_2], due to relatively higher root production. Shoot/root FW ratios at 36°C/30°C were lower than those at 28°C/22°C at the same RZ [CO_2], indicating that high growth temperature resulted in relatively higher root production (He et al. 2004).

With the same crisphead-type lettuce (*L. sativa* L. cv. 'Wintergreen') under the same conditions (He et al. 2010), we obtained similar results as previously reported (He et al. 2004). Figure 12 summarizes FW and DW of shoot and root after three weeks of elevated RZ [CO_2] treatments. Shoot FW of lettuce plants which were exposed to elevated RZ [CO_2] had 15-31% and 30-55% increases, respectively at 28°C/22°C and 36°C/30°C compared to plants grown at ambient RZ [CO_2]. Though no significant differences were observed in shoot FW and DW between plants grown in 10 000 and 50 000 ppm RZ [CO_2] (Figures 12A, 12B), root FW and DW was significantly greater in plants grown at 50 000 ppm than 10 000 ppm (Figures 12C, 12D) elevated RZ [CO_2]. Though shoot and root DW had been significantly enhanced by elevated RZ [CO_2] after 3 weeks, a greater extent of increase was found in the roots. Thus, plants had lower shoot/root ratio at elevated RZ [CO_2] of 50 000 ppm than at other RZ [CO_2] (Figures 12E, 12F). In a further experiment, we focused on the growth kinetics of both shoot and roots. It was interesting to note that shoot/root FW and DW ratios of plants under elevated RZ [CO_2] were first decreased and then increased during the treatments. In this paper, only shoot/root DW ratio was presented. There was significant increase in shoot/root DW ratio over a period of three weeks in plants grown at 28°C/22°C under ambient RZ [CO_2] (Figure 13A). However, there was no significant change in this ratio for plants grown at 36°C /30°C under the same RZ [CO_2] (Figure 13B). The initial decreases in shoot/root DW ratio with increasing RZ [CO_2] was much greater at 36°C/30°C than at 28°C/22°C, suggesting that higher temperatures enhanced the effects of elevated RZ [CO_2] in the partitioning of more biomass to the roots. Shoot/root DW ratio gradually increased in all treatments only 8 days of the various treatments of elevated RZ [CO_2] and temperatures.

Photosynthetic acclimation to growth at elevated *ambient* [CO_2] depends on the source/sink ratio within the plant and 'sink capacity' (Davey et al. 2006; Ainsworth and Rogers 2007). As changes in environmental conditions influence sink demands, sink limitation would be reduced under warmer temperatures that stimulate growth rate and respiration. Thus, the effects of CO_2 enrichment is lesser at cooler than at warmer temperatures (Stitt 1991). In this study, shoot/root DW ratio decreased during the first week of elevated RZ [CO_2] treatments (Figure 13), indicating that roots as "sinks" immediately increased their strength under elevated RZ [CO_2]. Larger root systems increased water and mineral uptake and, thus, enhanced shoot growth.

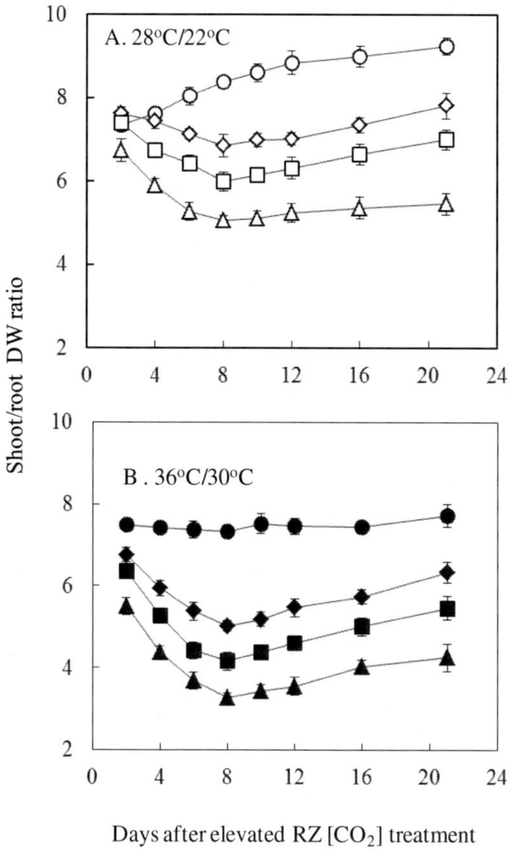

Figure 13. Shoot/root DW ratio of lettuce plants grown under different levels of elevated RZ [CO_2] in two different CE rooms at 28°C/22°C (A) and 36°C/30°C (B) (day/night ± 0.5°C) at a PPFD of 650 μmol m^{-2} s^{-1} (12h photoperiod) for three weeks. Each point is the mean of 5 measurements of 5 different leaves from two different bins. Vertical bars represent the standard errors. Means with different letters above the columns are statistically different ($p < 0.001$) as determined by Tukey's multiple comparison test (redrawn from He et al. 2010).

When the same lettuce plants (*L. sativa* L. cv. 'Wintergreen') grown under ambient temperatures ranging from 20-36°C and a maximal PPFD of 2100 μmol m^{-2} s^{-1} in a greenhouse in New Zealand, elevated RZ [CO_2] was found to enhance leaf area, shoot and root productivity (He et al. 2007). We postulated that increasing RZ [CO_2] alleviated midday depression of photosynthesis and hence increased the productivity of plants (He et al. 2007). It was also found the root FW and DW of plants grown at elevated RZ [CO_2] of 50 000 ppm were significantly greater than those plants grown at elevated RZ [CO_2] of 10 000 ppm. Higher root FW and DW resulted in lower shoot/root ratio at elevated RZ [CO_2] of 50 000 ppm compared to that of all the other plants. Under elevated atmospheric [CO_2], plants do not always show continued increase of photosynthesis compared to those grown under ambient [CO_2] (Clough et al. 1981; Hogan et al. 1991). In contrast, few studies have demonstrated continued stimulation of photosynthesis at elevated [CO_2], over an extended period of months or even years (Arp and Drake 1991; Eamus et al. 1993). Insensitivity of photosynthesis to elevated [CO_2] may be due to an imbalance between source capacity and sink development. Photosynthetic capacity could have been limited by the insufficiency of sinks for available carbohydrate storage presumably due to feedback inhibition and gene regulation of Rubisco (Stitt 1991; Webber et al. 1994). Since roots are a principle non-photosynthesizing sink for additional carbohydrate (Rogers et al. 1986; 1992), root growth and development play a key role in sustaining photosynthetic stimulation under elevated CO_2 environment (Rogers et al. 1992; Rogers and Runion 1994).

Rapid leaf extension and developing larger root systems may retain a larger proportion of their newly fixed carbon under elevated RZ [CO_2] (He et al. 2007). In our study, there were no restrictions in the root development of lettuce plants growing within the aeroponic system. Since the larger leaves and root systems were functioning as stronger sinks for additional carbohydrate synthesized (Eli 1996), plants were able to maintain the balance between source capacities and sink development under elevated RZ [CO_2]. Boru et al. (2003) reported that the root growth of soybean (*Glycine max* L), grown under elevated RZ [CO_2], was reduced due to O_2 deficiency. In our studies, O_2 supply to roots was adequate as the lettuce plants were aeroponically grown. Grown at A-RZT under ambient [CO_2], lettuce plants had a much smaller root system with fewer root tips and smaller leaf surface area and, thus, lower transpiration rate compared to those grown at 20°C-RZT (He et al. 2001; 2009). In our recent study with lettuce plants (*L. sativa* L. cv. 'Wintergreen') in a tropical greenhouse in Singapore under different elevated RZ [CO_2] and RZT, we found that A-RZT and 20°C-RZT plants had similar responses to elevated RZ [CO_2] where both resulted in significantly higher A_{sat} but much lower $g_{s\ sat}$ under elevated RZ [CO_2] (He et al. 2013). As discussed earlier, higher A_{sat} obtained from plants grown at elevated RZ [CO_2] may have partially resulted from using dissolved CO_2 that was transported via the transpiration stream from roots to the photosynthesizing leaves. Reduction of photoinhibition and photorespiration under elevated RZ [CO_2] may be important in determining growth and biomass accumulation, especially when stomata are partially closed at high midday temperatures. Our recent results showing greater accumulation of biomass in lettuce plants at elevated than ambient RZ [CO_2] at both RZTs support this postulation. Further, a greater increase in root than shoot biomass resulted in the observed lower shoot/root ratios at elevated than ambient RZ [CO_2]. The percentage increase in biomass under elevated RZ [CO_2] was greater at A-RZT than at 20°C-RZT although the total biomass was higher at 20°C-RZT (He et al. 2013), demonstrating that the temperature of a sink affects its metabolic rate and hence its capacity of carbohydrate utilization (Minchin et al.. 1996, Schurr et al. 2000, He et al. 2009). Our previous study showed that roots of 20°C-RZT lettuce plants had greater biomass, greater number of tips, longer total length and surface area, and a smaller average diameter compared to A-RZT plants (He et al. 2009). Though greater levels of ^{14}C-photoassimilates delivered to the lettuce root induced A-RZT, it was not accompanied by a larger root development (He et al. 2009). Higher levels of ^{14}C-photoassimilates delivered to the lettuce root at A-RZT could be attributed to the higher respiration rates in the roots which may require energy for the active uptake of nutrients and water in a poorly developed root system (Klock et al. 1997). As such, the high respiration rate may have taken place at the cost of root development at high A-RZT (Duand Tachibana 1994). Plants typically have more extensive and active root systems at elevated atmospheric [CO_2], than that at normal air concentrations (Curtis et al. 1994; Idso & Kimball 1992; Norby 1994), allowing them to explore larger volumes of soil in their search for nutrients (Norby et al. 1992; Rogers et al. 1992). We also observed that larger root systems with finer individual roots resulted in greater uptake of nutrient, such as NO_3^-, under elevated RZ [CO_2] (unpublished data). Thus, acclimation of photosynthesis to growth at elevated atmospheric [CO_2] was reported to depend on the source/sink ratio within the plant and 'sink capacity' (Makino & Mae 1999; Davey et al. 2006), where larger root systems resulted in increased capacities for utilizing photoassimilate (Davey et al. 2006; Ainsworth and Rogers 2007; He et al. 2009; He 2010).

ACKNOWLEDGMENT

All experimental work on temperate vegetables grown aeroponically was supported by the Academic Research Fund, Ministry of Education, Singapore.

REFERENCES

Ainsworth, E.A., Long, S.P., 2005. What have we learned from 15 years of free-air CO_2 enrichment (FACE)? A meta-analytic review of the responses of photosynthesis, canopy properties and plant production to rising CO_2. *New Phytol.*165, 351-372.

Ainsworth, E.A., Rogers, A., 2007. The response of photosynthesis and stomatal conductance to rising [CO_2]: mechanisms and environmental interactions. *Plant, Cell & Environ.* 30, 258-270.

Andrews, J.T., Lorimer, G.M., 1987. Rubisco: structure, mechanism and prospects for improvement. In: Hatch, M.D., Boardman, N.K., (ed.): The biochemistry of plants. Vol. 10, Pp. 132-219. Academic press, New York.

Arp, W.J., Drake, B.G., 1991. Increased photosynthetic capacity of *Scirpus olneyi* after 4 years of exposure to elevated CO_2. *Plant, Cell & Environ.*14, 1003–1006.

BassiriRad, H., Gutschick, V. P., Lussenhop, J., 2001. Root system adjustments: regulation of plant nutrient uptake and growth responses to elevated CO_2. *Oecologia,* 126, 305-320.

Bernacchi, C.J., Calfapietra, C., Davey, P.A., Wittig, V.E., Scarascia-Mugnozza, G.E., Raines C.A., Long, S.P., 2003. Photosynthesis and stomatal conductance responses of poplars to free air CO_2 enrichment (PopFACE) during the first growth cycle and immediately following coppice. *New Phytol.*159, 609–621

Berveiller, D., Damesin, C., 2008. Carbon assimilation by tree stems: potential involvement of phosphoenolpyruvate carboxylase. *Trees* 22, 149-157.

Bialczyk, J., Lechowski, Z., Libik, A., 1996. Fruiting of tomato cultivated on medium enriched with bicarbonate. *J. Plant Nutri.* 19, 305-321.

Björkman, O., Powles, S. B., 1984. Inhibition of photosynthetic reactions under water stress: interaction with light level. *Planta, 161,* 490-504.

Bloemen, J., McGuire, M.A., Aubrey, D.P., Teshey, R.O., Steppe, K., 2013. Assimilation of xylem-transported CO_2 is dependent on transpiration rate but is small relative to atmospheric fixation. *J. Exp. Bot.* 64, 2129-2138

Boru, G., Vantoai, T., Alves, J., Hua, D., Knee, M., 2003. Responses of soybean to oxygen deficiency and elevated root-zone carbon dioxide concentration. *Ann. Bot.* 91, 447-53.

Bouma, T.J., Nielson, K.L., Eissenstat, D.M., Lynch, J.P., 1997. Soil CO_2 concentration does not affect growth or root respiration in beans or citrus. *Plant, Cell & Environ.* 20, 1495–1505.

Chang, K., Roberts, J.K.M., 1992. Quantitation of Rates of Transport, Metabolic Fluxes, and Cytoplasmic Levels of Inorganic Carbon in Maize Root Tips during K^+ Ion Uptake. *Plant Physiol.* 99, 291-297.

Chaves, M.M., Oliveira, M.M., 2004. Mechanisms underlying plant resilience to water deficits: prospects for water-saving agriculture. *J. Exp. Bot.* 55, 2365-2384.

Clifford, S.C., Stronach, I.M., Mohamed, A.D., Azam-Ali, S.N., Crout, N.M., 1993. The effects of elevated atmospheric carbon dioxide and water stress on light interception, dry matter production and yield in stands of groundnut (*Arachis hypogaea* L.). *J. Exp. Bot.* 44, 1763-1770.

Clough, J.M., Peet, M.M. Kramer, P.J., 1981. Effects of High Atmospheric CO_2 and sink size on rates of photosynthesis of a soybean cultivar. *Plant Physiol.* 67, 1007-1010. Cookson, S.J., Williams, L.E., Miller, A.J., 2005. Light-dark changes in cytosolic nitrate pools depend on nitrate reductase activity in *Arabidopsis* leaf cells. *Plant Physiol.*138**,** 1097-1105.

Cornic, G., Fresneau, C., 2000. Photosynthetic carbon reduction and carbon oxidation cycles are the main electron sinks for photosystem II activity during a mild drought. *Ann. Bot.* 89, 887-894.

Cramer, M.D., 2002. Inorganic Carbon Utilization by Root Systems in Plant Roots. In: Waisel Y., Eshel, A., Kafkafi, U. (eds.). The Hidden Half 3rd Ed. Marcel Dekker, Inc.

Cramer, M.D., Lewis, O.A.M., Lips, S.H. 1993. Inorganic carbon dioxide fixation and metabolism in maize roots as affected by nitrate and ammonium nutrition. *Physiol. Plant.* 89, 632-639.

Cramer, M. D., Lips, S.H., 1995. Enriched rhizosphere CO_2 concentrations can ameliorate the influence of salinity on hydroponically grown tomato plants. *Physiol. Plant.* 94, 425-432.

Cramer, M. D., Richards, M.B., 1999. The effect of rhizosphere dissolved inorganic carbon on gas exchange characteristics and growth rates of tomato seedlings. *J. Exp. Bot.* 50, 79-87.

Cramer, M.D., Shane, M.W., Lambers, H., 2005. Physiological changes in white lupin associated with variation in root-zone CO_2 concentration and cluster-root P mobilization. *Plant, Cell & Environ.* 28, 1203-1217.

Cramer, M.D., Titus, C.H.A., 2001. Elevated root zone dissolved inorganic carbon can ameliorate aluminium toxicity in tomato seedlings. *New Phytol.*152, 29–39.

Crous, K.Y., Zaragoza-Castells, J.,. Ellsworth, D.S., Duursma, R.A., Löw, M., Tissue D.T., Atkin, O.K., 2012. Light inhibition of leaf respiration in field-grown *Eucalyptus saligna* in whole-tree chambers under elevated atmospheric CO_2 and summer drought. *Plant, Cell & Environ.* 35, 966–981.

Curtis, P.S., Wang, X., 1998. A meta-analysis of elevated CO_2 effects on woody plant mass, form, and physiology. *Oecologia* 113, 299–313.

Curtis, P.S., Zak, D.R., Pregitzer, K.S., Terri, J.A., 1994. Above- and below-ground response of *Populus grandidentata* to elevated atmospheric CO_2 and soil N availability. *Plant Soil* 165, 45-51.

Davey, P.A., Olcer, H., Zakhleniuk, O., Bernacchi, C.J., Calfapietra, C., Long, S.P., Raines, C.A., 2006. Can fast-growing plantation trees escape biochemical down-regulation of photosynthesis when grown throughout their complete production cycle in the open air under elevated carbon dioxide? *Plant, Cell & Environ* 29**,** 1235–1244.

De Jong, E., Schappert, H.J.V., 1972. Calculation of soil respiration and activity from CO_2 profile in the soil. *Soil Sci.* 113, 328 – 333.

DeLucia, E.H., Hamilton, J.G., Naidu, S.L., Thomas, R.B., Andrews, J.A., Finzi, A., Lavine, M., Matamala, R., Mohan, J.E., Hendrey, G.R., Schlesinger, W.H., 1999. Net primary productivity of a forest ecosystem with experimental CO_2 enrichment. *Science* 284, 1177–1179.

Demmig, B., Björkman, O., 1987. Comparison of the effects of excessive light on chlorophyll fluorescence (77K) and photon yield of oxygen evolution in leaves of higher plants. *Planta* 171, 171-184.

Dimou, M., Paunescu, A., Aivalakis, G., Flemetakis, E., Katinakis,P., 2009. Co-localization of Carbonic Anhydrase and Phosphoenolpyruvate Carboxylase and Localization of Pyruvate Kinase in Roots and Hypocotyls of Etiolated *Glycine max* Seedlings. *Inter. J. Mol .Sci.*10, 2896-2910.

Drake, B.G., Gonzàlez-Meler, M.A., Long, S.P., 1997. More Efficient Plants: A consequence of rising atmospheric CO_2? *Ann. Rev. Plant Physiol. Plant Mol. Biol.* 48, 609-39.

Du, Y.C., Tachibana, S., 1994. Effect of supraoptimal root temperature on the growth, root respiration and sugar content of cucumber plants. *Sci. Hort.* 58**,** 289-301.

Edwards, S., Nguyen, B., Do, B., Roberts, J.K.M., 1998. Contribution of Malic Enzyme, Pyruvate Kinase, Phosphoenolpyruvate Carboxylase, and the Krebs Cycle to Respiration and Biosynthesis and to Intracellular pH regulation during hypoxia in Maize Root Tips Observed by Nuclear Magnetic Resonance Imaging and Gas Chromatography-Mass Spectrometry. *Plant Physiol.*116, 1073-1081.

Eamus, D., Berryman, C.A., Duff, G.A., 1993. Assimilation, stomatal conductance, specific leaf area and chlorophyll responses to elevated CO_2 of *Maranthes corymbosa,* a tropical monsoon rain forest species. *Aust. J. Plant Physiol.* 20, 741-755.

Eli, Z., 1996. Photoassimilate distribution in plants and crops: source-sink relationships. Publisher: Marcel Dekker.

Ellsworth, D., Reich, P., Naumburg, E., Koch, G., Kubiske, M., Smith, S., 2004. Photosynthesis, carboxylation and leaf nitrogen responses of 16 species to elevated pCO_2 across four free-air CO_2 enrichment experiments in forest, grassland and desert. *Global*

Change Biol. 10, 2121-2138.Evans, J.R., 1989. Photosynthesis and nitrogen relationships in leaves of C_3 plants. *Oecologia* 78, 9–19

Evans, J.R., 2013. Improving photosynthesis. *Plant Physiol.* 162, 1780–1793.

Farquhar, G.D., Sharkey, T.D., 1982. Stomatal conductance and photosynthesis. *Annu. Rev. Plant Physiol.* 33, 317–345

Ford, C.R., Wurzburger, N., Hendrick, R.L., Teskey, R.O., 2007. Soil DIC uptake and fixation in *Pinus taeda* seedlings and its C contribution to plant tissues and ectomycorrhizal fungi. *Tree Physiol.* 27, 375–383.

Freudenberger, H., 1940. Die Reaktion der Schliesszellen auf Kohlensaüre and Sauerstoff Entzug. *Protoplasma* 35, 15–34.

Geisler, G., 1967. The interactive effect of CO_2 and O_2 in soil on root and top growth of barley and peas. *Plant Physiol.* 42, 305–307.

He, J., 2009. Impact of root-zone temperature on photosynthetic efficiency of aeroponically grown temperate and subtropical vegetable crops in the tropics. In: Buchner, Th. B., Ewingen, N.H. (eds.), *Theory and Applications in Energy, Biotechnology and Nanotechnology.* Chapter 4. Nova Science Publishers, Inc., New York.

He, J., 2010. Mineral nutrition of aeroponically grown subtropical and temperate crops in the tropics with manipulation of root-zone temperature at different growth irradiances. *Plant stress*, 4, 14-30.

He, J., Austin, P.T., Lee, S.K., 2010. Effects of elevated root zone CO_2 and air temperature on photosynthetic gas exchange, nitrate uptake, and total reduced nitrogen content in aeroponically grown lettuce plants. *J. Exp. Bot.* 61, 3959-3969.

He, J., Austin, P.T., Nichols, M.A., 2004. Effect of root-zone CO_2 on productivity and photosynthesis in aeroponically grown lettuce plants. *Acta Horticul.* 648,39-45.

He, J., Austin, P.T., Nichols, M.A., Lee, S.K., 2007. Elevated root-zone CO_2 protects lettuce plants from midday depression of photosynthesis. *Environ. Exp. Bot.* 61, 94-110.

He, J., Chee C.W., Goh C.J., 1996. "Photoinhibition" of *Heliconia* under natural tropical conditions - Importance of leaf orientation for light interception and leaf temperature. *Plant,Cell & Environ.* 19, 1238-1248.

He, J., Chua, N.Y.A., Qin, L. 2008. Interaction between iron stress and root-zone temperature on physiological aspects of aeroponically grown Chinese broccoli (*Brassica alboglabra*). *J. Plant Nutri. 31,* 1-20.

He, J., Lee, S.K., 1998a. Growth and photosynthetic characteristics of lettuce (*Lactuca sativa* L.) grown under fluctuating hot ambient temperatures with the manipulation of cool rootzone temperature. *J. Plant Physiol.* 152, 387-391.

He, J., Lee, S.K., 1998b. Growth and photosynthetic responses of three aeroponically grown lettuce cultivars (*Lactuca sativa* L.) to different rootzone temperatures and growth irradiances under tropical aerial condition. *J. Horti. Sci.Biot. 73*, 173-180.

He, J., Lee, S.K., 2001. Relationship among photosynthesis, Ribulose-1,5-Bisphosphate Carboxylase (Rubisco) and water relations of subtropical vegetable Chinese broccoli grown in the tropics by manipulation of root-zone temperature. *Environ. Exp. Bot. 46*, 119-128.

He, J., Lee S.K., 2004. Photosynthetic utilization of radiant energy by temperate lettuce grown under natural tropical condition with manipulation of root-zone temperature. *Photosynthetica*, *42*, 457-463.

He, J., Lee, S.K., Dodd, I.C., 2001. Limitations to photosynthesis of lettuce grown under tropical conditions: alleviation by root-zone cooling. *J. Exp. Bot. 52*, 1323-1330.

He, J., Qin, L., Lee, S.K., 2013. Root-zone CO_2 and root-zone temperature effects on photosynthesis and nitrogen metabolism of aeroponically grown lettuce (*Lactuca sativa* L.) plants in the tropics. *Photosynthetica*, 52, 330-340

He, J., Tan, L.P., Lee S.K., 2009. Root-zone temperature effects on photosynthesis, ^{14}C-photoassimilate partitioning and growth of temperate lettuce (*Lactuca sativa* cv. 'Panama') grown in the tropics. *Photosynthetica*, 47, 95-103.

Hibberd, J.M., Quick, W.P., 2002. Characteristics of C_4 photosynthesis in stems and petioles of C_3 flowering plants. *Nature* 415, 451–454.

Hogan, K.P., Smith, A.P., Ziska, L.H.,199*1*. Potential effects of elevated CO_2 and changes in temperature on tropical plants. *Plant, Cell & Environ.* **14**, 763-778.

Horton, P., Ruban, A.V., Walters, R.G., 1996. Regulation of light harvesting in green plants. *Annu Rev Plant Physiol Plant Mol Biol.* 47, 655-684, 1996.

Hymus, G.J., Baker, N.R., Long, S.P., 2001, Growth in elevated CO_2 can both increase and decrease photochemistry and photoinhibition of photosynthesis in a predictable manner. *Dactylis glomerata* grown in two levels of nitrogen nutrition. *Plant Physiol.*127, 1204–1211.

Hymus, G.J., Ellsworth, D.S., Baker, N.R., Long, S.P., 1999. Does free-air carbon dioxide enrichment affect photochemical energy use by evergreen trees in different seasons? A chlorophyll fluorescence study of mature loblolly pine. *Plant Physiol.*120, 1183–1191.

Idso, S.B., Kimball, B.A. 1992. Seasonal fine-root biomass development of sour orange trees grown in atmospheres of ambient and elevated CO_2 concentration. *Plant, Cell & Environ.* 15, 337-41.

Kaiser, W.M., Brendle-Behnisch, E., 1991. Rapid modulation of spinach leaf nitrate reductase activity by photosynthesis. I. Modulation in vivo by CO_2 availability. *Plant Physiol.* 96**,** 363–367.

Khan, F.N., Lukac, M., Turner, G., Godbold, D.L., 2008. Elevated atmospheric CO_2 changes phosphorus fractions in soils under a short rotation poplar plantation (EuroFACE). *Soil Biol. Biochem.* 40, 1716-1723.

Kitao, M., Lei, T.T., Koike, T., Kayama, M., Tobita, H., Maruyama, Y., 2007. Interaction of drought and elevated CO_2 concentration on photosynthetic downregulation and susceptibility to photoinhibition in Japanese white birch seedlings grown with limited N availability. *Tree Physiol.* 27, 727-35.

Klock, K.A., Taber, H.G., Graves, W.R., 1997. Root respiration and phosphorus nutrition of tomato plants grown at a 36°C root-zone temperature. *J. Am. Soc. Horti. Sci.*122, 175-178

Kontunen-Soppela, K. Riikonen, J., Ruhanen, H., Brosche, M., Somervuo, P., Peltonen, P., Kangasjarvi, J., Auvinen, P., Paulin, L., Keinänen, M., Oksanen, E., Vapaavuori, E., 2010. Differential gene expression in senescing leaves of two silver birch genotypes in response to elevated CO_2 and tropospheric ozone. *Plant, Cell &Environ.* 33, 1016–1028.

Lake, J.A., Quick, W.P., Beerling, D.J., 2001. Signals from mature to new leaves. *Nature* 411, 154-155.

Lawlor, D.W., Lemaire, G., Gastal, F., 2001. Nitrogen, plant growth and crop yield. In: Lea, P.J., Morot-Gaudry, J.F. (eds.) Plant Nitrogen, Berlin, Springer-Verlag, pp. 343-367.

Li, Y., Gao, Y., Xu, X., Shen, Q., Guo, S., 2009. Light-saturated photosynthetic rate in high-nitrogen rice (*Oryza sativa* L.) leaves is related to chloroplastic CO_2 concentration. *J. Exp. Bot.* 60, 2351–2360.

Lobell, D.B., Gourdji S.M., 2012. Focus issue on the plant physiology of global change : The Influence of Climate Change on Global Crop Productivity. *Plant Physiol.* 160, 1686-1697.

Long, S.P., Ainsworth, E.A., Rogers, A., Ort, D.R.,2004. Rising atmospheric carbon dioxide: plants FACE the future. *Annu. Rev. Plant Biol.* 55, 591–628.

Makino, A., Mae, T., 2004. Photosynthesis and Plant Growth at Elevated Levels of CO_2. *Plant Cel, Physiol.* 40**,** 999-1006.

Makino, A., Sakuma, H., Sudo, E., Mae, T., 2003. Differences between maize and rice in N-use efficiency for photosynthesis and protein allocation. *Plant Cell Physiol.* 44, 952-956.

Makino, A., Shimada, T., Takumi, S., Kaneko, K., Matsuoka, M., Shimamoto, K., Nakano, H., Miyao-Tokutomi, M., Mae, T., Yamamoto, N., 1997. Does decrease in ribulose-1,5-bisphosphate carboxylase by antisense rbcS lead to a higher N-use efficiency of photosynthesis under conditions of saturating CO_2 and light in rice plants? *Plant Physiol.* 114, 483-491.

McGuire, M.A., Marshall, J.D., Teskey, R.O., 2009. Assimilation of xylem-transported 13C-labelled CO_2 in leaves and branches of sycamore (*Platanus occidentalis* L.). *J. Exp. Bot.* 60, 3809-3817.

Minchin, P.E.H, Frarrar, J.F., Thorpe, M.R. 1996. What determines carbon partitioning between competing sinks? *J. Exp. Bot.* 47, 1293-1296.

Norby, R.J., 1994. Issues and perspectives for investigating root responses to elevated atmospheric carbon dioxide. *Plant Soil* 165, 9–20.

Norby, R.J., Gunderson, C.A., Wullschleger, S.D., O'Neill, E.G., McCracken, M.K., 1992. Productivity and compensatory responses of yellow-poplar trees in elevated CO_2. *Nature* 357, 322-24.

Norstadt, F.A., Porter L.K., 1984. Soil gases and temperatures: a beef cattle feedlot compared to alfalfa. *Soil Sci. Soc. Am. J.* 48, 783-789.

Nowak, R. S., Ellsworth, D. S., Smith, S. D., 2004. Functional responses of plants to elevated atmospheric CO_2: do photosynthetic and productivity data from FACE experiments support early predictions? *New Phytol.* 162, 253–280.

Osmond, C. B., 1994. What is photoinhibition? Some insights from comparisons of shade and sun plants. In: Baker, N.R., Bowyer J.R. (Eds.). Photoinhibition of photosynthesis: From Molecular Mechanisms to the field. Bios, Oxford. Pp. 1–24.

Pereira, J.S., Chaves, M.M., 1993. Plant water deficits in Mediterranean ecosystems. In: Smith, J.A.C., Griffiths, H. (eds.), Water deficits. Plant responses from cell to community. Oxford: ßios Scientific Publishers, pp. 237–251.

Peterhansel, C., Maurino, V.G., 2011. Photorespiration redesigned. *Plant Physiol.* 155, 49–55.

Peterson, R.B., 1983. Estimation of photorespiration based on the initial rate of postillumination CO_2 release: II. Effects of O_2, CO_2, and temperature. *Plant Physiol.* 73, 983-988.

Pettersson, R., McDonald, J.S., 1994. Effects of nitrogen supply on the acclimation of photosynthesis to elevated CO_2. *Photosynth Res.* 39, 389–400.

Pons, T.L., Welschen, R.A.M., 2003. Midday depression of net photosynthesis in the tropical rainforest tree *Eperua grandiflora*: contribution of stomatal and internal conductance, respiration and Rubisco functioning. *Tree physiol.* 23, 937-947.

Qin, L. He, J., Lee,'S. K., Dodd, I. C. 2007. An assessment of ethylene mediation of lettuce (*Lactuca sativa*) root growth at high temperatures. *J. Exp. Bot.* 58, 3017-3024.

Raich, J.W., Potter, C.S., Bhagawati, D., 2002. Interannual variability in global soil respiration, 1980-94. *Global Change Biol.* 8, 800–812.

Ray, D., Sheshshayee, M.S., Mukhopadhyay, K., Bindumadhava, H., Prasad, T.G., Kumar, M.U., 2003. High nitrogen use efficiency in rice genotypes is associated with higher net photosynthetic rate at lower Rubisco content. *Biol.Plant.* 46, 251-256.

Reich P.B., Kloeppel, B.D., Ellsworth, D.S., 1995. Different photosynthesis–nitrogen relations in deciduous hardwood and evergreen coniferous tree species. *Oceologia* 104, 24-30.

Retallack, G.L., 2001. A 300-million-year record of atmospheric carbon dioxide from fossil plant cuticles. *Nature* 411, 287-290.

Roden, J.S., Egerton, J.J.G., Ball, M.C., 1999. Effect of elevated [CO_2] on photosynthesis and growth of snow gum (*Eucalyptus pauciflora*) seedlings during winter and spring. *Aust. J. Plant Physiol.* 26, 37–46.

Rogers, H.H., Cure, J.D., Smith, J.M., 1986. Soybean growth and yield response to elevated carbon dioxide. *Agric. Ecosys. Environ.*16, 112-128.

Rogers, A., Gibon, Y., Stitt, M., Morga, P.B., Bernacchi, C.J. , Ort, D.R. , Long, S.P. 2006. Increased C availability at elevated carbon dioxide concentration improves N assimilation in a legume. *Plant, Cell & Environ.* 29, 1651–1658.

Rogers, H.H., Peterson, C.M., McCrimmon, J.N., Cure, J.D., 1992. Response of plant roots to elevated atmospheric carbon dioxide. *Plant Cell & Environ.*15,749-752.

Rogers, H.H., Runion, G.B., 1994. Plant response to atmospheric CO_2 enrichment with emphasis on root and the rhizosphere. *Environ. Pollu.* 83, 155-189.

Ruiz-Vera, U.M., Siebers, M., Gray, S.B., Drag, D., Rosenthal, D.M., Kimball, B.A., Ort, D.R., Bernacchi, C.J., 2013. Global Warming Can Negate the Expected CO_2 Stimulation in Photosynthesis and Productivity for Soybean Grown in the Midwestern United States. *Plant Physiol.* 162, 410-423.

Schlesinger, W.H., 1997. Biogeochemistry: An Analysis of Global Change. Academic Press, New York.

Schurr, U., Heckenberger, U., Herdel,K., Walter, K., Feil, R., 2000. Leaf development in *Ricinus communis* during drought stress: dynamics of growth processes, of cellular structure and of sink-source transition. *J. Exp. Bot.* 51, 1515-1529.

Stitt, M. 1991. Rising CO_2 levels and their potential significance for carbon flow in photosynthetic cells. *Plant, Cell & Environ.***14**, 741–62.

Stitt, M., Krapp, A., 1999. The interaction between elevated carbon dioxide and nitrogen nutrition: the physiological and molecular background. *Plant Cell & Environ* 22, 583-621.

Stitt, M., Schulze, D., 1994. Does Rubisco control the rate of photosynthesis and plant growth? An exercise in molecular ecophysiology. *Plant Cell & Environ.* 17, 465-487.

Stolwijk, J.A.J., Thimann, K.V., 1957. On the uptake of carbon dioxide and bicarbonate by roots, and its influence on growth. *Plant Physiol.* 32, 13–20.

Tan, L.P., He,J., Lee, S.K., 2002. Effects of Root-Zone Temperature on the Root Development and Nutrient Uptake of *Lactuca sativa* L. cv 'Panama' Grown in an Aeroponic System in the Tropics. *J. Plant Nutri.* 25, *297-314.*

Taub, D., Miller, B., Allen, H., 2008. Effects of elevated CO_2 on the protein concentration of food crops: a meta-analysis. *Global Change Biol.*14, 565-575.

Teskey, R.O., McGuire, M.A., 2002. Carbon dioxide transport in xylem causes errors in estimation of rates of respiration in stems and branches of trees. *Plant Cell & Environ.* 25, 1571–1577.

Teskey, R.O., McGuire, M.A., 2005. CO_2 transported in xylem sap affects CO_2 efflux from *Liquidambar styraciflua* and *Platanus occidentalis* stems, and contributes to observed wound respiration phenomena. *Trees Struct. Funct.* 19, 357–362.

Teskey, R.O., McGuire, M.A., 2007. Measurement of stem respiration of sycamore (*Platanus occidentalis* L.) trees involves internal and external fluxes of CO_2 and possible transport of CO_2 from roots. *Plant, Cell & Environ.* 30, 570–579.

Tetu, S.G., 2007. The *Flaveria bidentis* β-Carbonic Anhydrase Gene Family encodes Cytosolic and Chloroplastic Isoforms demonstrating Distinct Organ-Specific Expression Patterns. *Plant Physiol.* 144, 1316-1327.

Tezara, W., Mitchel, V., Driscoll, S.P., Lawlor, D.W., 2002. Effects of water deficit and its interaction with CO_2 supply on the biochemistry and physiology of photosynthesis in sunflower. *J. Exp. Bot.* 53, 1781 - 1791.

Valentini, R., Epron, D., Angelis, P.D., Matteucci, G., Dreyer, E., 1995. *In situ* estimation of net CO_2 assimilation, photosynthetic electron flow and photorespiration of Turkey oak (*Q. cerris* L.) leaves: diurnal cycles under different water supply. *Plant, Cell & Environ.* 18, 631–640.

van der Merwe, C.A., Cramer, M.D. 2000. The influence of dissolved inorganic carbon in the root-zone on nitrogen uptake and the interaction between carbon and nitrogen metabolism. In: Louçao M.A. , Lips, S.H. (eds.) Nitrogen in a sustainable ecosystem – from the cell to the plant. Leiden, The Netherlands: Backhuys Publishers. pp 145–151.

Vapaavuori, E.M., Pelkonen, P., 1985. HCO_3^- uptake through the roots and its effect on productivity of willow cuttings. *Plant, Cell & Environ.* 8, 531-534.

Viktor, A., Cramer, M.D., 2003. Variation in root-zone CO_2 concentration modifies isotopic fractionation of carnon and nitrogen in tomato seedlings. *New Phytol.* 157, 45–54.

von Caemmerer, S., Evans, J.R., 2010. Enhancing C_3 photosynthesis. *Plant Physiol.* 154, 589–592.

Vuorinen, A. H., Vapaavuori, E. M., Raatikainen, O., Lapinjoki, S. P., 1992. Metabolism of Inorganic Carbon Taken Up by Roots in *Salix* Plants. *J. Exp. Bot.*43, 789-795.

Webber, A.N., Nie, G.Y., Long, S.P., 1994. Acclimation of photosynthetic proteins to rising atmospheric CO_2. *Photosynth. Res.* 39, 413-425.

Woodrow, I.E., Berry, J.A., 1998. Enzymatic regulation of photosynthetic CO_2 fixation in C_3 plants. *Annu. Rev. Plant Physiol. Plant Mol. Biol.* 39, 533-594.

Yin, X., 2002. Responses of leaf nitrogen concentration and specific leaf area to atmospheric CO_2 enrichment: a retrospective synthesis across 62 species. *Global Change Biol.* 8, 631-642.

In: Photosynthesis
Editor: Nafees Khan
ISBN: 978-1-63463-304-8

Chapter 3

CHLOROPHYLL *A* FLUORESCENCE, PHOTOSYNTHESIS ACTIVITY AND CHLOROPHYLL CONTENT IN SENESCING LEAVES

Marina Cavaiuolo, Roberta Bulgari, Giacomo Cocetta and Antonio Ferrante*

Dept. Agricultural and Environmental Sciences – Production, Landscape, Agroenergy, Università degli Studi di Milano, Italy

ABSTRACT

Leaf senescence is the last step of leaf development and can be induced by many genetic and environmental factors. The chloroplast and many leaf pigments undergo a progressive degeneration. The leaf functionality is slowed down and several physiological processes are altered. The chlorophyll *a* fluorescence and its derivate indexes can be used to monitor leaf senescence. The loss of leaf functionality is tightly correlated with the photosynthetic activity. The chlorophyll *a* fluorescence and its derived parameters can be suggested as tool to monitor the leaf senescence. Natural and induced senescence such as dark incubation, detachment, salinity stress, ethylene exposure, heavy metal stress etc. will be taken in consideration. The physiological, biochemical and molecular basis connected with the photosynthesis and chlorophyll fluorescence will be described and discussed.

INTRODUCTION

Leaf senescence is the last step of the leaf development and it is characterized by several physiological and biochemical changes that lead to cell death of leaves and the whole plant. The leaf senescence is a genetically regulated and coordinated event, which can be triggered by natural development or induced by external factors (harvest, pathogen attacks, abiotic

* Corresponding Author address; antonio.ferrante@unimi.it

stresses, etc.) (Woo et al., 2013). In trees, leaf senescence can occur on annual basis and it is regulated by the seasons, which guarantee the turnover of the photosynthetic apparatus such as occurs in the deciduous plants. In evergreens the leaf senescence is instead regulated by internal mechanisms and the turnover can usually occur after two-three years. In the annual plants the leaf senescence often coincides with plant death (Lim et al., 2007).

The natural senescence in deciduous plants can be also considered as a sort of "hardening" to face the winter, since the plants slow down all metabolic processes and activate the re-allocation of nutrients from senescing leaves to storage tissues to overcome the bad season (Buchanan-Wollaston et al., 2003; 2007). At cellular level, the senescence is characterized by a reduction of soluble proteins, lipids, sugars and nucleic acids. The visible symptoms of leaf senescence are usually associated to chlorophyll degradation and yellowing. In detached organs or whole plant, the symptoms of leaf senescence can be different from natural senescence and they can be induced by storage conditions (dark), lack of nutrients, internal hormone unbalance (Trivellini et al., 2012).

In the most part of the ornamental species harvest consists in the cut of whole plants (carnation, lilies, etc.) or shoots with flower (roses). Detached organs do not receive nutrients and hormones such as the cytokinins, which are preferentially synthetized in the roots. The lack of cytokinins is usually associated to leaf yellowing and exogenous application of this plant hormone or substances with cytokinins-like activity usually delay leaf senescence and yellowing (Ferrante et al., 2002; 2012).

The initiation of leaf senescence can be detected and assessed by several physiological, biochemical and molecular markers. At a molecular level the activation of leaf senescence can be assessed by monitoring the induction of specific genes associated with this phenomenon. The most part of these senescence associated genes (SAGs) are cysteine proteases (Weaver et al., 1998).

However, the leaf senescence is tightly correlated to chlorophyll turnover and associated structures or molecules such as chloroplasts, carotenoids and other leaf pigments (Christ and Hörtensteiner, 2014). Therefore, the chlorophyll *a* fluorescence and photosynthesis measurements could represent non-destructive methods for assessing the progress of leaf senescence *in vivo*.

Chlorophyll *A* Fluorescence and Photosynthesis Changes in Senescing Leaves

The measurement of the chlorophyll *a* fluorescence is a common parameter which can be used in plant physiology to assess the stress conditions and the senescence progress in leaves through the performance of the photosynthetic system. The spectrum and kinetic of the fluorescence emitted from a leaf are biological parameters that can be rapidly measured. The fluorescence emitted from the chlorophyll *a* of leaves falls in the range from 680 to 740 nm of the electromagnetic radiations. This fluorescence is completely emitted from the photosystem II (PSII). This energy dissipated from the photosynthetic apparatus has been associated with the health status of the leaves and their functionality (Strasser et al., 1995). The energy of light absorbed from the chlorophyll molecules can be used for the photosynthetic activity of the leaf, can be dissipated as heat or re-emitted as fluorescence. These three possible fates of the energy are in competition among them, the increase of one of them induces the reduction of the others.

Therefore, through the measurement of the chlorophyll *a* is a possible to obtain information on the energetic dissipation that occurs in a leaf (Lazár, 1999). A minimum part of the absorbed light, about 1-2%, can be emitted as fluorescence.

The first correlation studies between photosynthesis and fluorescence of chlorophyll *a* were conducted in the 1931, when Kautsky and Hirsh found out that leaves exposed to high light intensity after a dark period show a strong increase of fluorescence, which if reported on a graph in time function describes a polyphasic curve reaching a maximum value (Figure 1). After this peak the fluorescence declines and reaches a stationary state. At this stage the declining of the fluorescence is correlated with the assimilation of carbon dioxide.

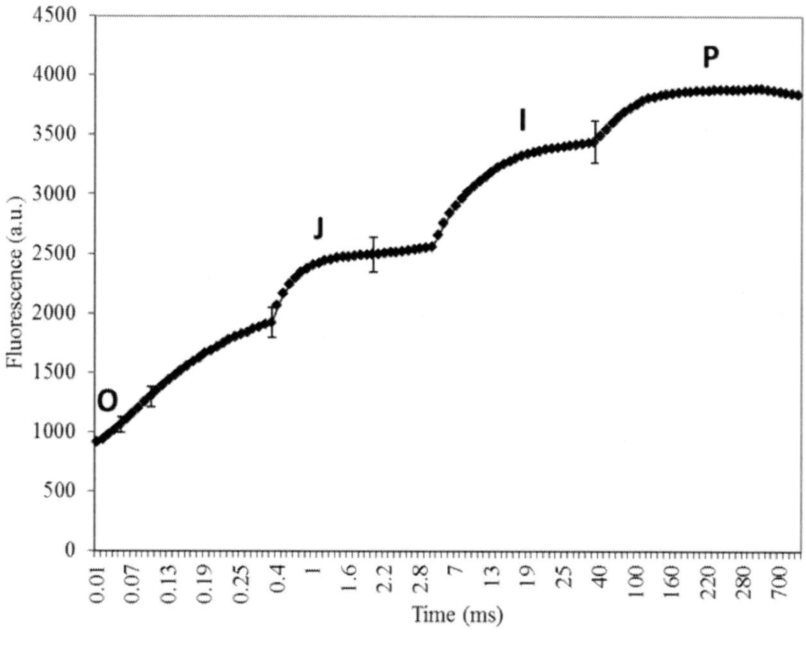

Figure 1. Chlorophyll *a* fluorescence induction curve measured in dark adapted leaf and exposed to saturation light, using Handy PEA (Hansatech, UK) portable fluorimeter.

This curve is called induction curve of the chlorophyll *a* fluorescence which is obtained when leaves after an adequate dark period are suddenly illuminated. The use of the chlorophyll *a* fluorescence in the eco-physiological studies and the number of publications regarding the application of this technique have been increasing since portable fluorimeters have become available.

In this chapter, the chlorophyll *a* fluorescence of dark adapted leaves are reported and fluorescence parameters are referred to Handy Plant Efficiency Analyser (PEA) data. The Handy PEA instrument provides raw data and basic calculated parameters (Table 1). The raw data can be used for further calculations and identify specific indexes for the stress of interest.

Table 1. The main JIP-test parameters with description and changes during leaf senescence

Measured fluorescence values		
Parameters	**Description**	**Changes during senescence**
Fo = O point of the induction curve (I.C.)	Fluorescence intensity at 20 μs (Handy PEA) or 50 μs (PEA)	Increase
F_M =P point of the I.C.	F_M, maximal fluorescence intensity	Decrease
J	Fluorescence intensity at 2 ms	Increase
I	Fluorescence intensity at 30 ms	Increase
$F_{300\mu s}$	Fluorescence intensity at 300 μs	Decrease
Calculated parameters		
F_V	Maximal variable fluorescence, (F_M – Fo)	Decrease
Specific fluxes, reaction centre parameters		
ABS/RC	Average antenna size, absorbed energy per reaction centre (RC)	Increase
TR_0/RC	Trapping energy per active RC. Maximal trapping rate of PS II, (M_0 / VJ)	Increase
ET_0/RC	Electron transport in an active RC	Increase
DI_0/RC	Effective dissipation of an active RC	Increase
PI	Performance Index: Multi-parametric expression of these three independent steps contributing to photosynthesis.	Decrease
Phenomenological fluxes, cross-section parameters		
ABS/CS	Absorbed energy flux per cross section (CS	Increase
TRo/CS	Trapped energy flux per CS	Increase
ETo/CS	Electron transport flux per CS	Decrease
DIo/CS	Dissipation of energy per CS	Increase
RC/CSo	Density of reaction centres at Fo	Decrease
RC/CSm	Density of reaction centres at F_M	Decrease

Leaves are the site where all of the most important photochemistry reactions take place and the senescence affects the vitality and functionality of the leaf, hence also the chlorophyll *a* fluorescence and the photosynthesis activities. Therefore, the chlorophyll *a* fluorescence can be used to assess the senescence stage through the estimation of the vitality and functionality of the leaves. The senescence can be induced by aging (natural senescence) or by stresses such as harvest or detachment in the vegetables and ornamentals. In leafy vegetables, senescence cannot be esteemed by simply measuring the chlorophyll losses while it can be easily assessed by chlorophyll *a* fluorescence measurement (Ferrante and Maggiore, 2007).

In dark adapted leaves, the leaf senescence can be appreciated by the JIP test parameters (Table 1) derived from the elaboration of the intermediate points of the fluorescence induction curve (Strasser et al., 1995). In leafy vegetables, the leaf senescence is very well described by the performance index (PI), dissipation energy (DIo/RC or DIo/CS) and density of active reaction center per cross section (RC/CSm) at the maximal fluorescence (F_M). The intensity of the signals depends on the tissue chlorophyll concentration. It has been demonstrated by monitoring the leaf senescence of lettuce (*Lactuca sativa* L.) and spinach (*Spinacia oleracea* L.) stored at 4 or 8 °C. Most of the parameters related to chlorophyll fluorescence were higher or even doubled in the spinach compared to the lettuce as well as the chlorophyll content (Baldassarre et al., 2011).

The JIP test parameters were used to probe the senescence in cotyledons of pea by studying the alterations of the PSII (Prakash et al., 2003).

Lamb's lettuce placed at 4 and 10 °C had different senescence kinetics; the chlorophyll *a* fluorescence parameters and JIP indexes were used to assess the quality and senescence. Different indexes such as PI, DIo/CS, ABS/DIo, ETo/DIo and RC/CSm were able to highlight differences and described the quality losses. Among the different indexes, the PI had the higher correlation coefficient with the chlorophyll and carotenoids content. The induction curve during senescence is lowered and the I point of the curve is not identifiable (Ferrante and Maggiore, 2007).

All parameters which can be measured or calculated from the chlorophyll *a* fluorescence data are affected by senescence. The differences can be appreciated in terms of the magnitude of the responses. The maximum quantum efficiency of PSII (F_V/F_M) slowly declines and can be hardly observed at the onsets of senescence, even if there are differences among plant species. The Fo, minimal fluorescence, represents the fluorescence intensity at 20 μs, when plastoquinone electron acceptor pool (Qa) is fully oxidized. The increase of Fo can be related to the damage of PSII which is not able to discharge all electrons. The JIP test is particularly used to monitor the leaf senescence and quality. Therefore, they might also have practical applications in control the quality during the distribution chain of the horticultural produce.

The net photosynthesis activity is the most important physiological process in the green organisms for providing carbohydrates and energy. In plants, the net photosynthesis follows a diurnal pattern and changes during leaf development. The maximum value of the net photosynthesis is reached when the leaf achieves the fully expanded stage and then it progressively declines during senescence. In horticultural crops such as tomato, the older leaves are removed from plants since the net photosynthesis declines while the respiration continues reducing the energy sources for plant growth and productivity. The photosynthesis activity represents the leaf functionality by directly measuring the CO_2 assimilation in leaves. During senescence of cut stock flowers, the chlorophyll content, F_V/F_M and net photosynthesis were tightly correlated (Ferrante et al., 2009). The net photosynthesis declined throughout the vase life and higher values were found in treatments that inhibited leaf yellowing such as thidiazuron. In the leaf, the ribulose-1,5-bisphosphate carboxylase (RuBisCo) represents more than 50% of the total soluble proteins and during senescence the reduction of protein content, hence also RuBisCo induces a decline of the photosynthesis activity (Matile et al., 1996).

The chlorophyll and protein content of leaves can be used as indicators of leaf senescence and photosynthetic activity potential of plants. High nitrogen fertilization acts by delaying the net photosynthesis reduction during senescence, since it extends the plant life and enhances the protein content in leaves (Peng et al., 2014).

CHANGES IN EXPRESSION PROFILES OF GENES RELATED TO CHLOROPLAST AND PSII FUNCTIONALITY

The first hallmark of leaf senescence is the loss of chlorophyll and the breakdown of the chloroplast. This organelle contains up to the 70% of the total leaf protein and constitutes the major source of carbon and nitrogen for the redistribution and recycling of nutrients in senescent leaves (Morita,1980; Abdallah et al., 2000).

During senescence thylakoid membranes are first disassembled and grana are unstacked leading to the accumulation of differently sized plastoglobuli. Following dismantling, chloroplasts undergo degradation of macromolecules and loss of carbon assimilation and primary photochemical reactions (Grover, 1993; Biswal, 1995; Wingler, 1998). These degradation events are exerted by enzymes which are specific either to the chloroplast or to the vacuole. It was shown that through the generation of spherical autophagicRuBisCo-containing bodies (RCBs), the chloroplasts devoid of thylakoid membranes are mobilized to the vacuole for the final degradation of the remaining chloroplast constituents (Chiba et al., 2003; Ishida et al., 2008).

Chloroplast dismantling is accompanied by the down-regulation of photosynthesis-associated genes (PAGs) and by the up-regulation of genes which are necessary for the degradation of chloroplast DNA (cpDNA), lipids, pigments, proteins and enzymes and for the transport of nutrients out of the senescing tissue.

Enucleation studies in the protoplast leaf cells of *Elodea densa* showed that in chloroplast senescence is under the direct control of nuclear genes. Chloroplasts in nucleated cells showed a remarkable senescence, while chloroplasts in enucleated cells preserved the green color and the structure (Yoshida et al., 1961; Gan et al., 1997). In general, chloroplast development is under the control of both chloroplast and nuclear genes. However, the regulation of gene expression occurs at the post-transcriptional and translational level through the action of nuclear encoded factors, mostly belonging to the PPR family in plants that control the chloroplast mRNA stability, maturation, translation and degradation. Their action is also dependent on the presence of environmental or stressing factors. In addition, many stress responsive chloroplast small RNAs (csRNA) were identified in *Arabidopsis* and Chinese cabbage with roles in the maintenance of the photosynthetic activity and structures of the chloroplast (Wang et al., 2011).

Throughout senescence the expression of the genes involved in all chloroplast activities is subjected to a high-resolution timed regulative process in which the genes for chlorophyll biosynthesis are first down-regulated followed by carbon fixation and other genes involved in photosynthesis (Breeze et al., 2011). However, while the mRNA levels of nuclear genes are decreased, the expression of the chloroplast genes remains constant (Bate et al., 1991). In chloroplasts, the 16S rrn operon, which encode the D1 protein of photosystem II (PSII) and the ribosomal RNAs (rRNA), the psbA that encodes the large subunit (SLU) of rRNA and the rbcL that encodes the large subunit of ribulose-1,5-bisphosphate carboxylase/oxygenase (RuBisCo), did not show significant variations in the expression levels. By contrast, nuclear PAG genes such as the cab, which encodes chlorophyll a/b-binding protein, and the rbcS, which encodes the small subunit of RuBisCo, showed a progressive down-regulation along the progressing of senescence (Bate et al., 1991).

Nuclear senescence associated down-regulated genes include genes linked to plastid and thylakoid structure as well as to the activity of the PSI and PSII, carbon fixation and amino acid metabolism, chlorophyll biosynthesis and tetrapyrrole binding (Breeze, 2011; Kong et al., 2013). The high-chlorophyll-fluorescence (HCF) genes with functions in maintaining the stability of chloroplast-encoded mRNAs (Meurer et al., 1996; Meierhoff et al., 2003) were also shown to be down-regulated and allowed an enhanced degradation of photosynthetically related transcripts in the chloroplast.

Work carried out on maize showed that the expression patterns of phosphoenolpyruvate carboxylase (PEPC), superoxide dismutase (SOD), senescence enhanced gene 1 (See1) and senescence enhanced gene 2 (See2) were up-regulated (He et al., 2004). The See1 and See2 genes encoding cysteine proteases were specifically expressed during leaf senescence and seed germination respectively with putative roles in the activation of proteins required for senescence such as other proteases (Kingston-Smith et al., 1999).

Recently, the rapid leaf senescence 1 (RLS1) gene encoding a NB (nucleotide-binding site)-containing protein with an ARM (armadillo) domain at the carboxyl terminus was identified in rice. The expression of RLS1 was enhanced during leaf senescence and was associated with an autophagy-like process of chloroplast degradation (Jiao et al., 2012). The gene was found also in sorghum, grape, and black cottonwood genomes although the function has to be validated in these species.

cpDNA fragmentation was reported in many plant species and variation of its content was associated not only with chloroplast development and leaf senescence (Rowan and Bendich, 2009) but also with environmental stresses (Sodmergen et al., 1991). For example, oxidative stress induces the partial degradation of the large subunit of ribulosebisphosphate carboxylase/oxygenase in the chloroplasts of barley (Desimone, 1996). The destruction of chloroplast DNA is probably dependent on the activation of some metallo-nucleases, in particular five Ca^{2+}, four Zn^{2+}, and four Mn^{2+} dependent nucleases whose activity was detected in the leaf blades of rice (Sodmergen et al., 1991). Moreover, a DPD1 exonuclease that is dual-targeted to chloroplasts and mitochondria was identified and showed a higher expression during leaf senescence (Sakamoto et al., 2014). In the woody species *Acer pseudoplatanus* L. senescence-associated protein factors strongly interacting with arbcL-containing cpDNA fragment of 10kbp were identified and such interaction would control the rbcL expression and the RuBisCo degradation (Fulgosi et al., 2012).

Several chloroplast localized endopeptidases (EP) were discovered and purified from pea chloroplast including four stroma-localized EP and one thylakoid-localized EP (Musgrove et al., 1989; Liu and Jagendorf, 1986). In particular, EP1 was shown to degrade the large subunit of ribulose-1,5-bisphosphate carboxylase/oxygenase (LSU) into a 36-kDa fragment (Kokubun et al., 2002). A DegP2 gene encoding a chloroplast homologue of the prokaryotic trypsin-type Deg/Htr serine proteases was also found in Arabidopsis and targets the damaged D1 protein of PSII complex for proteolysis. DegP2 is associated with the thylakoid membranes and is highly expressed in response to salinity, drought and high intensity to light exposure and it is likely induced during leaf senescence (Haußühl et al., 2001).

Both nuclear and chloroplast encoded caseinolyticprotease (Clp) subunits were detected in the chloroplasts of senescing Arabidopsis leaves (Shanklin et al., 1995). In plants, the Clp family comprise a plastid-encoded pClpP gene, six nuclear-encoded nClpP (1-6) genes and nuclear encoded chaperones ClpB, -C, -X, and ClpD, all with functions in the ATP-dependent degradation of misfolded or unassembled proteins (Halperin, 1996). pClpP, nClpP1, nClpP2, nClpP4, nClpP6 and nClpC were reported to be constitutively expressed and their transcripts and protein levels stayed unchanged during various stress treatments (Shanklin et al., 1996). By contrast, the two nClp3 and nClp5 were up-regulated during stresses and senescence (Miller et al., 1999). The nuclear early responsive dehydration gene 1 (ERD1) encoding a chloroplast-targeted putative ATPase subunit (ClpA) of Clp protease was reported to be induced by water stress and also up-regulated during leaf senescence and to interacts with ClpP in the degradation of chloroplast proteins in Arabidopsis thaliana. Moreover, CND41, a DNA-binding aspartic protease, acts as a protease and degrades RuBisCo (Kato et al., 2004).

A massive and extensive transcriptome profiling by RNA-seq of the chloroplast gene expression in leaves and/or other senescing tissues are missing. The rapid advancement and availability of different sequencing technologies and platforms combined with bioinformatics analyses will make easier the study of chloroplast gene expression in function of the changes associated with nuclear expression variations that are specific to plant developmental stage, senescence process and presence of environmental stresses.

BIOCHEMISTRY OF PHOTOSYNTHESIS IN SENESCING LEAVES

Senescence is a coordinated, deteriorative growth process where biochemical processes are well orchestrated at molecular level.

To characterize a leaf that is undergoing senescence, it is necessary to identify the sequential order in which various parameters change. Biochemical changes that occur during leaf senescence follow a subsection of events: loss of photosynthetic pigments, degradation of protein, and reabsorption of mineral nutrients (Zhang et al., 2006).

Changes in the biochemistry of photosynthesis were reported to be associated with changes in the ultra-structure of the chloroplasts in *Ginko biloba* leaves during development (Yang et al., 2010).

The clearest sign of leaf physiological decline during senescence is the fading of the photosynthetic activity. Since photosynthesis is the primary source of dry matter and energy, its reduction leads first to senescence and then, likely to death of the plant. The decrement in the photosynthetic activity is mainly due to the degradation of chlorophyll rather than a decrease in the efficiency of the pigment itself (Buchanan et al., 2000). Other phenomena like the loss of activity of Photosystem II (PSII), of the CF and Cyt f proteins, cytochrome b6f as well as of the electron transport complex between photosystem I (PSI) and PSII, and of the enzymes like RuBisCo are also features of leaf senescence (Ford and Shibles, 1988; Hidema et al., 1991).

The synthesis of proteins (and enzymes) decreases fairly early during senescence. The degeneration of the photosynthetic machinery happens through chlorophyll degradation and the progressive loss of proteins in the chloroplast, such as RuBisCo and chlorophyll a/b binding protein (CAB). Experimental evidences indicate that the RuBP carboxylase level can be a limiting factor in photosynthesis in leaf (Makino et al., 1983). Changes in activities of photosynthetic enzymes and photochemical processes were followed during aging of leaves of wheat (*Triticum aestivum* L. cv Roy).

Activities of stromal enzymes began to decline prior to photochemical activities. In general, total soluble proteins and the activities of RuBisCo and NADPtriose- phosphate dehydrogenase declined in parallel and earlier than chlorophyll, photosynthesis rate, and photosynthetic electron transport activity. Leaves appeared to lose whole chloroplasts as opposed to a general degradation of all chloroplasts. Chloroplasts undergo some modifications before being lost, but the reduction in leaf photosynthesis is apparently caused by a loss of the whole chloroplast (Camp et al., 1982).

The RuBisCo-containing bodies (RCBs), which derive from the degradation of the enzyme were found in the cytoplasm and in the vacuole as a result of the proteolitic activity against the enzyme (Miller and Huffaker, 1981; Wittenbach et al., 1982; Thayer and Huffaker, 1984; Yoshida and Minamikawa, 1996). The nitrogen resulting from the enzyme degradation is translocated into growing organs and affects nitrogen economy (Mae et al., 1983). Many reports have shown that the decrease of RuBisCo content is faster than the decrease in the number of chloroplasts in leaves during natural or dark-induced senescence (Martinoia et al., 1983; Mae et al., 1984;Wardley et al., 1984; Ford and Shibles, 1988; Ono et al., 1995). Thus it can be supposed that RuBisCo starts to be degraded into small nitrogenous molecules such as oligo peptides and/or amino acids within the chloroplasts at the beginning of senescence and that is then degraded by vacuolar proteases in the process of autolysis during the last stage of senescence.

Several factors such as developmental and environmental conditions, as well as the genetic diversity can differentially affect the way by which photosynthesis changes in senescent leaves. During senescence leaves of *Lolium temulentum* grown under different light radiations showed a steady decline in amount of RuBisCo, light-harvesting chlorophyll a/b, protein of photosystem II (LHCII), 65 kDa protein of photosystem I (PSI), cytochrome f (Cytf) and coupling factor 1 (CF1) in unshaded leave (Mae et al., 1993).The content of ATP, photophosphorylation, Ca^{2+}-ATPase, the electron transport activities and the activity of photosynthetic O_2 evolution, are among the biochemical indices of photosynthetic functionality and so their progressive decline is typically observed along the senescence progression (Yang et a., 2013). The decrease in electron transport along photosystem II may be due to the loss of activity of the of the PSII reaction center complex, but also to the inhibition of energy transfer from carotenoids to chlorophyll (Lu et al., 2002).

CHLOROPHYLL DEGRADATION AND PSII FUNCTIONALITY

The decrease of chlorophyll is the most frequently used parameter to describe the yellowing of senescing leaves (Matile et al., 1999). During leaf senescence some changes in the structure and function of chloroplasts occur. Chloroplasts lose volume and their shape changes. The first step of chlorophyll degradation is the conversion of chlorophyll *b* to chlorophyll *a* (Tanaka and Tanaka, 2006). During plant senescence, chlorophyll (Chl) is degraded to non-fluorescent Chlcatabolites (NCCs) (Hörtensteiner, 2004; Christ and Hörtensteiner, 2014).

The chlorophyll catabolism begins with the conversion of Chl *a* to chlorophyllide in a reaction catalyzed by chlorophyllase. This enzyme produces also phytol, which acts by removing Mg from chlorophyllide is required the enzyme Mg-dechelatase, yielding pheophorbide *a*. Both chlorophyllide and pheophorbide retain their intact porphyrin ring structures and have green color. Pheophorbide *a* oxygenase (PaO) and red chlorophyll catabolite reductase (RCCR) are the two enzymes responsible for the conversion of pheophorbide *a* to red chlorophyll catabolites (RCC); this conversion results in opening the ring and this step is associated with the loss of green color (Matile et al., 1996). PaO is located at the inner envelope membrane of senescing chloroplasts (gerontoplast) (Matile and Schellenberg, 1996); RCCR is a soluble chloroplast protein (Wüthrich et al., 2000). PaO reaction requires O_2 and involves Fe, which operates in a redox cycle driven by reduced ferredoxin. PaO uses pheophorbide *a* but not pheophorbide *b* as a substrate, so chlorophyll *b* must be converted to chlorophyllide *a* or pheophorbide *a* before it can be catabolized. RCC does not accumulate in plants but is metabolized by RCC reductase, which catalyzes the ferredoxin-dependent reduction of a double bond in the pyrrole system of RCC to produce an almost colorless tetrapyrrole with a strong blue fluorescence (FCC – fluorescent chlorophyll catabolite) (Buchanan et al., 2000). The final destination of catabolites is the vacuole. The end products of these modifications are various non-fluorescent catabolites (NCCs) which vary greatly in number and type, and moreover some structural variants in the family of NCCs remain to be discovered (Matile et al., 1999).

The PSII is the pigment protein complex situated in the thylakoid membrane of higher plants, algae and cyanobacteria. The PSII uses light energy to catalyze a series of electron transfer reactions resulting in the splitting of water into molecular oxygen, protons, and electrons (Hankamer et al., 1997). PSII absorbs light at an optimal wavelength of 680 nm and contains two pheophythin molecules, an iron atom and two plastoquinone molecules, QA and QB, which act as the terminal electron acceptors.

Senescence results in a decrease in photochemical activities of both PSI and PSII. However, PSII is more susceptible to senescence than PSI and a greater decrease in PSII activity is observed during senescence of leaves (Lu and Zangh, 1998). The decrease in PSII electron transport induced by senescence should lead to some alterations in the primary photochemical processes of PSII; information about the performance of PSII was derived from measurements of *in vivo* variable PSII-Chl fluorescence (Havaux, 1992).

PSII is affected to various abiotic stresses and is the most vulnerable part of the photosynthetic apparatus to light-induced damage. Damage to PSII will often be the first manifestation of stress in a leaf (Maxwell and Johnson, 2000). All things considered, calculations of the F_V/F_M and F_V/F_O ratios (where F_M is the maximum total fluorescence and F_V the maximum variable fluorescence emission, $F_V = F_M - F_O$) can be used to measure the sensitivity of plants to cold, heat or saline stress (Panda et al., 2008) and to have a direct assessments of plant health (Force et al., 2003). In fact, stresses that affect PSII efficiency lead to a characteristic decrease in F_V/F_M value (Krause and Weis, 1991).

Conclusion

Leaf senescence is a fascinating biological event that is highly regulated from both internal and external factors. The initiation and the progress of leaf senescence are not very well clear yet and the recent molecular tools can help in better understanding what lay underneath this last step of leaf development. However, the availability of instruments able to monitor leaf functionality can immediately provide information on plant physiological status. The high time resolution allows collecting a large amount of data that can be correlated to plant responses under various stress conditions. The elaboration of data combined with the physiological and biochemical changes during senescence can be used for many practical applications such as the evaluation of quality in vegetables and ornamentals during the postharvest stage.

REFERENCES

Abdallah, F., Salamini, F., Leister, D., (2000). A prediction of the size and evolutionary origin of the proteome of chloroplasts of Arabidopsis. *Trends Plant Sci.* 5, 141-142.

Baldassarre, V., Cabassi, G., Ferrante, A., (2011). Use of chlorophyll *a* fluorescence for evaluating senescence in leafy vegetables. *Aus. J. Crop Sci.* 5, 735-741.

Bate, N.J., Rothstein, S.J., Thompson, J.E., (1991). Expression of nuclear and chloroplast photosynthesis-specific genes during leaf senescence. *J. Exp. Bot.* 42, 801-811.

Biswal, B., (1995). Carotenoid catabolism during leaf senescence and its control by light. *J. Photochem. Photobiol. B: Biol.* 30, 3-13.

Breeze, M., Harrison, E., McHattie, S., Hughes, L., Hickman, R., (2011). High-resolution temporal profiling of transcripts during arabidopsis leaf senescence reveals a distinct chronology of processes and regulation. *Plant Cell.* 23, 873-894.

Buchanan, B. B., Gruissem, W., Jones, R. L., (2000). Biochemistry and molecular biology of plants. 40. Rockville: Am. Soc. Plant Physiol.

Buchanan-Wollaston, V. Earl, S., Harrison, E., Mathas, E., Navabpour, S., Page T., Pink, D., (2003). The molecular analysis of leaf senescence-a genomics approach. *Plant Biotech. J.* 1, 3-22.

Buchanan-Wollaston, V., Page, T., Harrison, E., Breeze, E., Lim, P. O., Nam, H. G., Lin, J. F., Wu, S. H., Swidzinski, J., Ishizaki, K., Leaver, C.J., (2005). Comparative transcriptome analysis reveals significant differences in gene expression and signalling pathways between developmental and dark/starvation-induced senescence in Arabidopsis. *Plant J.* 42, 567-585.

Camp, P. J., Huber, S. C., Burke, J. J., Moreland, D. E., (1982). Biochemical changes that occur during senescence of wheat leaves I. Basis for the reduction of photosynthesis. *Plant Physiol.* 70, 1641-1646.

Chiba, A., Ishida, H., Nishizawa, N. K., Makino, A., Mae, T., (2003). Exclusion of ribulose-1, 5-bisphosphate carboxylase/oxygenase from chloroplasts by specific bodies in naturally senescing leaves of wheat. *Plant Cell Physiol.* 44, 914-921.

Christ, B., Hörtensteiner, S., (2014). Mechanism and Significance of Chlorophyll Breakdown. *J. Plant Growth Regul.* 33, 4 -20.

Desimone, M., Henke, A., Wagner, E., (1996). Oxidative stress induces partial degradation of the large subunit of ribulosebisphosphate carboxylase/oxygenase in isolated chloroplasts of barley. *Plant Physiol.* 111, 789-796.

Ferrante A., Mensuali-Sodi A., Serra, G., (2009). Effect of thidiazuron and gibberellic acid on leaf yellowing of cut stock flowers. *Cent. European J. Biol.* 4, 461-468.

Ferrante, A., Trivellini, A., Mensuali-Sodi, A., (2012). Interaction of 1-methylcyclopropene and thidiazuron on cut stock flowers vase life. *Open Hort. J.* 5, 1-5.

Ferrante,, A., Hunter D.A., Wesley, P.H., Reid, M.S., (2002). Thidiazuron – a potent inhibitor of leaf senescence in Alstroemeria. *Post. Biol. Tech.* 25, 333-338.

Force, L., Critchley, C., van Rensen, J. J., (2003). New fluorescence parameters for monitoring photosynthesis in plants.*Photosyn. Res.* 78, 17-33.

Ford, D. M., Shibles, R., (1988). Photosynthesis and other traits in relation to chloroplast number during soybean leaf senescence. *Plant Physiol.* 86, 108-111.

Fulgosi, H., Ježić, M., Lepeduš, H., Štefanić, P.P., Ćurković-Perica, M., Cesar V., (2012). Degradation of chloroplast DNA during natural senescence of maple leaves. *Tree Physiol.* 32, 346-354.

Gan, S. and Amasino, R.M., (1997). Making Sense of Senescence. Molecular Genetic Regulation and Manipulation of Leaf Senescence. *Plant Physiol.* 113, 313-319.

Grover, A., (1993). How do senescing leaves lose photosynthesis activity? *Curr. Sci.* 64, 226-234.

Halperin, T., Adam, Z., (1996). Degradation of mistargeted OEE33 in the chloroplast stroma. *Plant Mol. Biol.* 30, 925–933.

Hankamer, B., Barber, J., Boekema, E. J., (1997). Structure and membrane organization of photosystem II in green plants. *Ann. Rev. Plant Biol.* 48, 641-671.

Haußühl, K., Andersson, B., Adamska I., (2001). A chloroplast DegP2 protease performs the primary cleavage of the photodamaged D1 protein in plant photosystem II. *EMBO J.* 20, 713-722.

Havaux, M., (1992). Stress tolerance of photosystem II in vivo antagonistic effects of water, heat, and photoinhibition stresses. *Plant Physiol.* 100, 424-432.

He, P., Osaki, M., Takebe, M., Shinano, T., Wasaki, J., (2005). Endogenous hormones and expression of senescence-related genes in different senescent types of maize. *J. Exp. Bot.*56, 1117-1128.

Hidema, J., Makino, A., Mae, T., Ojima, K., (1991). Photosynthetic characteristics of rice leaves aged under different irradiances from full expansion through senescence. *Plant Physiol.* 97, 1287-1293.

Hikosaka, K., (1996). Effects of leaf age, nitrogen nutrition and photon flux density on the organization of the photosynthetic apparatus in leaves of a vine (*Ipomoea tricolor* Cav.) grown horizontally to avoid mutual shading of leaves. *Planta.* 198, 144-150.

Hörtensteiner, S., (2004). The loss of green color during chlorophyll degradation—a prerequisite to prevent cell death? *Planta.* 219, 191-194.

Ishida, H., K. Yoshimoto, M. Izumi, D. Reisen, Y. Yano, A. Makino, Y. Ohsumi, M.R. Hanson, T. Mae., (2008). Mobilization of rubisco and stroma-localized fluorescent proteins of chloroplasts to the vacuole by an ATG gene-dependent autophagic process. *Plant Physiol.* 148, 142-155.

Jiao, B.B., Wang, J.J., Zhuc, X.D., Zenga, L.G., Lia, Q., Hea, Z.H., (2012). A novel protein RLS1 with NB–ARM domains is involved in chloroplast degradation during leaf senescence in rice.*Mol. Plant.* 5, 205-217.

Kato, Y., Murakami, S., Yamamoto, Y., Chatani, H., Kondo, Y., Nakano, T., Yokota, A., Sato, F., (2004). The DNA-binding protease, CND41, and the degradation of ribulose-1,5-bisphosphate carboxylase/oxygenase in senescent leaves of tobacco. *Planta.* 220, 97–104.

Kingston-Smith, A., Donnison, I., Robson, P., Theodorou, M., (1999). Plant proteases: the link between plant death and animal life? In: Bridges I, ed. IGER innovations. The Netherlands: Aberystwyth, 42–45.

Kokubun, N., Ishida, H., Makino, A., Mae T., (2002). The Degradation of the Large Subunit of Ribulose-1,5-bisphosphate Carboxylase/oxygenase into the 44-kDa Fragment in the Lysates of Chloroplasts Incubated in Darkness. *Plant Cell Physiol.* 43, 1390-1395.

Kong, X., Luo, Z., Dong, H., Eneji, A.E., Li, W., Lu, H., (2013). Gene expression profiles deciphering leaf senescence variation between early- and late-senescence cotton lines. *PLoS One.* 29, e69847.

Krause, G. H., Weis, E., (1991). Chlorophyll fluorescence and photosynthesis: the basics. *Ann. Rev. Plant Biol.* 42, 313-349.

Lim, P. O., Kim, H. J., Nam, H. G., (2007). Leaf senescence. *Ann. Rev. Plant Biol.* 58,115-136.

Lu, C. and Zhang, J. (1998). Changes in photosystem II function during senescence of wheat leaves. *Physiol. Plant.* 104, 239-247.

Lu, Q., Lu, C., Zhang, J., Kuang, T., (2002). Photosynthesis and chlorophyll *a* fluorescence during flag leaf senescence of field-grown wheat plants. *J. Plant Physiol.* 159, 1173-1178.

Mae, T., Kai, N., Makino, A., Ohira, K., (1984). Relation between ribulosebisphosphate carboxylase content and chloroplast number in naturally senescing primary leaves of wheat. *Plant Cell Physiol.* 25, 333-336.

Mae, T., Makino, A., Ohira, K., (1983) Changes in the amounts of ribulosebisphosphate carboxylase synthesized and degraded during the life span of rice leaf (*Oryza sativa* L.). *Plant Cell Physiol.* 24, 1079-1086.

Mae, T., Thomas, H., Gay, A.P., Makino, A. and Hidema, J., (1993) Leaf development in *Loliumtemulentum* – photosynthesis and photosynthetic proteins in leaves senescing under different irradiances. *Plant Cell Physiol.* 34, 391-399.

Makino, A., Mae, T.,Ohira, K., (1983). Photosynthesis and ribulose 1, 5-bisphosphate carboxylase in rice leaves changes in photosynthesis and enzymes involved in carbon assimilation from leaf development through senescence. *Plant Physiol.* 73, 1002-1007.

Martinoia, E., Heck, U., J Dalling, M., Matile, P. H., (1983). Changes in chloroplast number and chloroplast constituents in senescing barley leaves. *Biochem. Physiol. Pfla.*, 178, 147-155.

Matile, P. and Schellenberg, M. (1996). The cleavage of phaeophorbidea is located in the envelope of barley gerontoplasts. *Plant Physiol. Biochem.* 34, 55-59.

Matile, P., Hörtensteiner, S., Thomas, H., (1999). Chlorophyll degradation. *Ann. Rev. Plant Biol.* 50, 67-95.

Matile, P., Hortensteiner, S., Thomas, H., Krautler, B., (1996). Chlorophyll breakdown in senescent leaves. *Plant Physiol.* 112, 1403.

Maxwell, K. and Johnson, G.N., (2000). Chlorophyll fluorescence-a practical guide. *J. Exp. Bot.* 51, 659-668.

Meierhoff, K., Felder, S., Nakamura, T., Bechtold, N., Schuster, G. (2003). HCF152, an Arabidopsis RNA binding pentatricopeptide repeat protein involved in the processing of chloroplast psbB-psbT-psbH-petB-petD RNAs. *Plant Cell.* 15, 1480-1495.

Meurer, J., Meierhoff, K., Westhoff, P., (1996). Isolation of high-chlorophyll-fluorescence mutants of *Arabidopsis thaliana* and their characterization by spectroscopy, immunoblotting and northern hybridisation. *Planta* 198, 385-396.

Miller, B.L. and Huffaker, R.C., (1981) Partial purification and characterization of endoproteinases from senescing barley leaves. *Plant Physiol.* 68, 930-936.

Miller, J.D., Arteca, R.N., Pell, E.J., (1999) Senescence-associated gene expression during ozone-induced leaf senescence in Arabidopsis. *Plant Physiol.* 120, 1015-1024.

Morita, (1980). Release of nitrogen from chloroplasts during leaf senescence in rice (*Oryza sativa* L.). *Ann. Bot.* 46, 297-302.

Musgrove, J.E., Elderfield, P.D., Robinson, C., (1989). Endopeptidases in the stroma and thylakoids of pea chloroplasts. *Plant Physiol.* 90, 1616-1621.

Ono, K., Hashimoto, H., Katoh, S., (1995) Changes in the number and size of chloroplasts during senescence of primary leaves of wheat grown under different conditions. *Plant Cell Physiol.* 36, 9-17.

Panda, D., Sharma, S. G., Sarkar, R. K., (2008). Chlorophyll fluorescence parameters, CO_2 photosynthetic rate and regeneration capacity as a result of complete submergence and subsequent re-emergence in rice (*Oryza sativa* L.). *Aqu. Bot.* 88, 127-133.

Peng, Y., Li, C., Fritschi, F.B., (2014). Diurnal dynamics of maize leaf photosynthesis and carbohydrate concentrations in response to differential N availability. *Environ. Exp. Bot.* 99, 18-27.

Prakash, J.S.S., Srivastava A., Strasser R.J., Mohanty P., (2003). Senescence-induced alterations in the photosystem II functions of *Cucumissativus* cotyledons: probing of senescence driven alterations of photosystem II by chlorophyll a fluorescence induction OJIP transients. *Indian J. Biochem. Biophy.* 40, 160-168.

Rowan, B.A, Bendich, A.J., (2009). The loss of DNA from chloroplasts as leaves mature: fact or artefact? *J. Exp. Bot.* 60, 3005-3010.

Sakamoto, W. and Takami, T., (2014). Nucleases in higher plants and their possible involvement in DNA degradation during leaf senescence. *J. Exp. Bot.*doi: 10.1093/jxb/eru091.

Shanklin, J., Dewitt, N.D., Flanagan, J.M., (1995) Thestroma of higher plant plastids contain ClpP and ClpC, functional homologs of *Escherichia coli*ClpP and ClpA: an archetypal two-component ATP-dependent protease. *Plant Cell* 7, 1713-1722.

Sodmergen, S., Kawano, S.T., Kuroiwa, T., (1991). Degradation of chloroplast DNA in second leaves of rice (*Oryza sativa*) before leaf yellowing. *Protoplasma.* 160, 89-98.

Strasser, R.J., Srivastava, A., Govindjee, (1995). Polyphasic chlorophyll a fluorescence transient in plants and cyanobacteria. *Photochem. Photobiol.* 61, 32-42.

Tanaka, A., Tanaka, R., (2006). Chlorophyll metabolism. *Curr. Opi. Plant Biol.* 9, 248-255.

Thayer, S.S., Huffaker, R.C., (1984) Vacuolar localization of endoproteinases EP1 and EP2 in barley mesophyll cells. *Plant Physiol.* 75, 70-73.

Trivellini,, A., Jibran R., Watson, L.M., O'Donoghue, E., Ferrante, A., Sullivan, K., Dijkwel, P., Hunter, D.A., (2012). Carbon-deprivation-driven transcriptome reprogramming in detached developmentally-arresting Arabidopsis inflorescences. *Plant Physiol.* 160, 1357-1372.

vanWijk, K.J., (2001) Regulatory aspects of photosynthesis: Advances in photosynthesis. edsAndersson B., Aro E.M.

Wang, L., Yu, X., Wang, H., Lu, Y.Z., de Ruiter, M., Prins, M., He, Y.K., (2011). A novel class of heat-responsive small RNAs derived from the chloroplast genome of Chinese cabbage (*Brassica rapa*). *BMC Genom.* 12, 289.

Wardley, T. M., Bhalla, P. L., Dalling, M. J., (1984). Changes in the number and composition of chloroplasts during senescence of mesophyll cells of attached and detached primary leaves of wheat (*Triticumaestivum* L.). *Plant Physiol.* 75, 421-424.

Weaver, L.M., Gan, S., Quirino, B., Amasino, R.M., (1998). A comparison of the expression patterns of several senescence-associated genes in response to stress and hormone treatment. *Plant Mol. Biol.* 37, 455-469.

Wingler, A., Schaewen, A.V., Leegood, R.C., Lea, P.J., Quick W.P., (1998). Regulation of leaf senescence by cytokinin, sugars, and light. *Plant Physiol.* 116, 329-335.

Wittenbach, V.A., Lin, W., Herbert, R.R., (1982). Vacuolar localization of proteases and degradation of chloroplasts in mesophyll protoplasts from senescing primary wheat leaves. *Plant Physiol.* 69, 98-102.

Woo, H.R., Kim, H.J., Nam, H.G., Lim, P.O., (2013). Plant leaf senescence and death–regulation by multiple layers of control and implications for aging in general. *J. Cell Sci.* 126, 4823-4833.

Wüthrich, K.L., Bovet, L., Hunziker, P.E., Donnison, I.S., Hörtensteiner, S., (2000). Molecular cloning, functional expression and characterisation of RCC reductase involved in chlorophyll catabolism. *Plant J.* 21, 189-198.

Yang, X.S., Chen, G.X., Xie, K.B., Wei, X.D., Zhang, M.P., Gao, Z.P., Han, B., (2010). Photosynthetic, biochemical, and ultra-structural changes in the chloroplasts of two ginkgo (*Ginkgo biloba* L.) cultivars during leaf development. *J Horti. Sci. Biotech.* 85, 323.

Yang, X.S., Chen, G.X., Yuan, Z.Y., (2013). Photosynthetic decline in Ginkgo leaves during natural senescence. *Pak. J. Bot.* 45, 1537-1540.

Yoshida Y., (1961). Nuclear control of chloroplast activity in Elodea leaf cells. *Protoplasma* 54, 476-492.

Yoshida, T., Minamikawa, T., (1996). Successive amino-terminal proteolysis of the large subunit of ribulose 1,5-bisphosphate carboxylase/oxygenase by vacuolar enzymes from French bean leaves. *European J. Biochem.* 238, 317–324.

Zhang, C.J., Chen, G.X., Gao, X.X., Chu, C.J., (2006). Photosynthetic decline in flag leaves of two field-grown spring wheat cultivars with different senescence properties. *South Afr. J. Bot.* 72, 15-23.

In: Photosynthesis
Editor: Nafees Khan

ISBN: 978-1-63463-304-8

Chapter 4

CARBON ISOTOPE HETEROGENEITY IN PHOTOSYNTHESIZING BIOMASS AND PERSPECTIVES OF ITS APPLICATION IN BIOLOGICAL STUDIES

Alexander A. Ivlev*

Russian State Agrarian University – "MSKHA of K. A.Timirjazev,"
Moscow, Russian Federation

ABSTRACT

The phenomenon of carbon isotope heterogeneity in a photosynthetic biomass became known at the beginning of 1960s, after the pioneering works of Abelson and Hoering (1960) and Park and Epstein (1960, 1961), but the real mechanism of carbon isotope fractionation leading to uneven carbon isotope distribution in a biomass and the reasons for the isotope patterns of the biomolecules remain vague. In this chapter, it is shown that ^{13}C distribution is tightly bound to the oscillatory character of metabolic processes and to the real biosynthetic pathways of metabolites. It was found that due to oscillations, photosynthesizing cells function in regime of filling/depleting metabolic pools. In the phase of depletion the isotope effect-generating reactions are followed by a Raleigh effect, resulting in the dependence of the isotopic patterns of the molecules on the sequence of their syntheses, i.e., in the strict temporal organization of cell processes. It is shown that oscillations in photosynthesis consist of a phase of CO_2 assimilation and a phase of photorespiration. Each phase is followed by carbon isotope fractionation having the effects of opposite signs. As a result, carbon isotope heterogeneity emerges. Post-photosynthetic metabolism, mainly presented as the metabolism of the glycolytic chain, is also presented through oscillations that have two phases: glycolysis and gluconeogenesis. In glycolysis, carbon isotope fractionation occurs. This is one more reason for the isotope heterogeneity of a biomass and works in combination with biosynthetic pathways to determine the isotopic patterns of post-photosynthetic metabolites. Knowledge of the nature of carbon isotope heterogeneity provides a new

* Corresponding author address: aa.ivlev@list.ru.

delicate tool for studies on living organisms. Some possible fields for carbon-isotope data applications with the presented model are considered.

Keywords: Carbon isotope fractionation, isotopic patterns of metabolites, metabolic oscillations, CO_2 assimilation, photorespiration, RuBisCo, Raleigh effect, kinetic isotope effect, temporal organization of cell processes, post-photosynthetic metabolism, glycolytic chain, glycolysis, gluconeogenesis

INTRODUCTION

Using the carbon isotopic characteristics of living organisms to get information about an organism and its metabolism has been attracting the growing attention of researchers. Isotope heterogeneity is a huge reservoir of valuable information about living organisms. Due to the pioneering works of Abelson and Hoering (1960), Park and Epstein (1960, 1961), Hayes and Monson (1982), and other researchers, it has become clear that isotope heterogeneity is an essential feature of any living organism and that ^{13}C distribution of the biocomponents is tightly bound to their biosynthetic pathways. The rapid development of instrumental techniques has provided an accumulation of considerable amounts of factual material, but understanding the mechanism of isotopic fractionation and its connection with the mechanisms of the cell functioning still remains vague. The phenomenon of isotope heterogeneity has turned out to be much more complex than just following some simple links.

Hereinafter I shall try to give a new explanation of the phenomenon and describe the links of ^{13}C distribution in terms of cell processes.

1. The Main Experimental Results of Carbon Isotope Fractionation Studies in Photosynthesizing Organisms

By the end of the seventies of the last century, due to the pioneering works of Park and Epstein (1960, 1961) and Abelson and Hoering (1960), it became evident that plants and other photosynthesizing organisms are capable not only of predominantly selecting $^{12}CO_2$ molecules from the surrounding air, but also of fractionating carbon isotopes directly in the cell, providing the isotope heterogeneity of a biomass. Their findings were confirmed in the works of a number of researchers (Degens et al., 1968; Hayes and Monson, 1982; Blair et al., 1985; and others). Isotope heterogeneity is manifested at different levels of organism studies—on a cellular, fractional, and intramolecular scale.

Tables 1 and 2 illustrate the differences of isotopic characteristics in land and water plants, alga, and photosynthesizing microorganisms on a fractional scale regardless of the organism type.

Tables 3 and 4 show that isotopic heterogeneity is displayed as differences in the carbon isotope composition of close-cut fractions and individual metabolites. In addition, the more detailed the level of biomass examination is, the stronger the isotopic discrepancies are. They are displayed in each fraction.

Table 5 illustrates the ^{13}C distribution of amino acids from the protein fraction of a biomass for several photosynthesizing organisms (Abelson and Hoering, 1961). One can see great differences between the radical and carboxyl fragments of individual amino acids. Logic prompts that the isotopic differences on higher levels of biomass examination reflect isotope fractionation occurring on a molecular level.

All the material presented in Tables 1–5 support the conclusion that isotope heterogeneity is an essential feature of living matter. The main questions are: What is the nature of carbon isotope effects arising in enzymatic reactions? Where does carbon isotope fractionation occur? And how are isotope characteristics linked to metabolic processes?

Table 1. Carbon isotope composition of biochemical fractions isolated from the biomass of several land plants (Park and Epstein, 1961)

Plant	$\delta^{13}C$			Reference
	Total biomass	Lipids	Carbohydrates + lipids	
Asplenium bulbiferum	- 29.2	-37.9	-30.7	Park and Epstein, 1961
Triticum vulgare	-28.5	-31.1	-30.7	Park and Epstein, 1961
Sequoia gigantean	-26.4	-28.3	-25.2	Park and Epstein, 1961
Solanum tuberosum	-25.6	-33.6	-25.6	Park and Epstein, 1961

Table 2. Carbon isotope composition of biochemical fractions isolated from the biomass of some water plants, alga, and photosynthesizing microorganisms

Plant, alga, and bacterium	$\delta^{13}C$				Reference
	Total biomass	Lipids	Proteins	Carbohydrate + lipids	
Ulva sp.	-15.7	-23.5	–	-15.4	Park and Epstein, 1961
Gelidium sp.	-20.5	-24.3	–	-19.9	Park and Epstein, 1961
Macrocystis pyrifera	-12.6	-21.0	–	-12.7	Park and Epstein, 1961
Laminaria saccharina	-18.4	-25.4	–	-17.0	Galimov, 1975
Anabaena variabilis	-27.5	-32.2	-26.4	-26.8	Galimov, 1975
Chlorella pyrenoidosa	-18.9	-22.8	-16.0	–	Abelson and Hoering, 1961
Anacystis nidulens	-14.8	-18.9	-13.0	–	Abelson and Hoering, 1961
Gracilaria sp.	-11.9	-17.3	-10.1	–	Abelson and Hoering, 1961
Euglena gracilis	-16.4	-18.5	-15.4	–	Abelson and Hoering, 1961

Table 3. Carbon isotope composition of biochemical fractions of lipids, proteins, and carbohydrates isolated from a marine alga biomass in samples collected from Atlantic Ocean water near the seashore of Ecuador and Peru (Degens et al., 1968)

		Proteins				Lipids			
Total carbon of biomass		Hydrolysis with HCl 6N		Hydrolysis with H_2SO_4 2N		Extraction with C_2H_5OH		Extraction with CH_3Cl	
from	to	from	to	from	to	from	to	from	to
-19.9	-20.4	-16.4	-10.1	-15.9	-18.8	-19.8	-22.5	-23.0	-37.4
		Carbohydrates							
Lignins		Hemicellulose		Cellulose		Pectins		Sugars	
from	to	from	to	from	to	from	to	from	to
-20.2	-25.3	-20.2	-26.7	-16.7	-21.1	-15.5	-17.5	-18.1	-21.5

Table 4. ^{13}C distribution in the protein fraction of some photosynthesizing microorganisms (Abelson and Hoering, 1961)

Microorganism / Amino acid	*Chlorella pyrenoidosa*	*Anacystis nidulens*	*Gracilaria* sp.	*Euglena gracilis*
Serine	-5.7	–	-14.3	-8.3
Glycine	-14.3	-10.0	-10.2	-12.8
Alanin	-10.3	-9.8	-15.2	-14.3
Aspartic acid	-6.6	-9.7	-14.4	-9.7
Glutamic acid	-18.7	-11.1	-17.2	-17.3
Leucine	-22.7	-17.3	-22.5	-23.5
Lysine	-17.0	–	–	-22.8

Table 5. ^{13}C distribution of individual amino acids from a biomass of some photosynthesizing microorganisms (Abelson and Hoering, 1961)

Microorganism / Amino acid	Chlorella pyrenoidosa	Anacystis nidulens	Gracilaria sp.	Euglena gracilis
Glycine				
Total C	-14.3	-10.0	-10.2	-12.8
Carboxyl	-5.3	-8.8	-12.8	-11.6
Radical	-23.3	-11.1	-7.7	-14.1
Aspartic acid				
Total C	-6.6	-9.7	-14.4	-9.6
Carboxyl	+2.2	+2.8	-5.3	+1.0
Radical	-15.6	-22.1	-18.4	-20.3
Glutamic acid				
Total C	-18.7	-11.1	-17.2	-17.3
Carboxyl	-8.8	+11.7	-8.4	-0.2
Radical	-21.1	-16.7	-19.3	-21.5
Leucine				
Total C	-22.7	-17.3	-22.5	-23.5
Carboxyl	-16.3	-15.0	-24.1	-13.6
Radical	-26.0	-17.8	-22.1	-25.6

2. Attempts to Explain the Causes of the Carbon Isotope Heterogeneity of a Photosynthesizing Biomass

Many attempts have been undertaken to explain the reasons for isotope heterogeneity. Degens (1969) assumed that isotopic distinctions within protein, lipid, and carbohydrate fractions are associated with isotope fractionation following the interconversion of biocomponents. He ascribed the isotope differences of the fractions to fractionation at respiration, protein—carbohydrate and lipid-carbohydrate turnover.

Abelson and Hoering (1961) assumed that some intracellular isotopic differences were linked with different pathways of CO_2 assimilation. This idea looked reasonable since it was found that the carbon fixed in ribulose bisphosphate carboxylase (RuBisCo), the first step in CO_2 assimilation, is always enriched in ^{12}C as compared to the environmental CO_2 (Park and Epstein, 1961). Sometime later, it was revealed that different pathways of CO_2 assimilation are followed by different carbon isotope effects (Smith and Epstein, 1971; Deleens and Garnier-Dardart, 1977). De Niro and Epstein (1977) explained the isotopic discrepancies of lipids and proteins with the carbon isotope effect of pyruvate decarboxylation since most of the lipids are synthesized via C_2-fragments (acetyl-CoA) produced in the reaction.

Taking into account the observed link of isotope heterogeneity with metabolic pathways, all the above researchers believed that the isotope effects in a cell have a kinetic nature.

Galimov (1985) and after him Schmidt (2003) have suggested quite another explanation for isotope heterogeneity based on a thermodynamic consideration. They have asserted that ^{13}C distribution is mainly determined by the isotope exchange of carbon occurring in metabolic reactions and they do not link them with a biosynthesis pathway or the mechanism of metabolic processes. The discussion on the nature of isotope effects in cell processes that has arisen (Ivlev and Shnoll, 1978; O'leary and Yapp, 1978; Monson and Hayes, 1982; Varshavsky, 1988; Buchachenko, 2003) has shown that the hypothesis on the thermodynamic nature of the effects and the authors' assertion about the independence of ^{13}C distribution in biomolecules on cell mechanisms and metabolic pathways have no serious theoretical foundation and are incompatible with the known irreversibility of cell processes. Thus far, the experimental data completely denies thermodynamic ^{13}C distribution and conversely are in good correspondence with the kinetic nature of the effects (see later).

3. Carbon Isotope Fractionation in Photosynthesis. Oscillatory Character of Photosynthesis and Carbon Isotope Heterogeneity

Since Park and Epstein had revealed that in the reaction of ribulosebisphosphate (RuBP) carboxylase the fixed carbon is always enriched with ^{12}C relative to the environmental CO_2, most researchers believed that the carbon isotope composition of primary assimilates are mainly determined by the isotope effect in CO_2 assimilation. Besides, it was known that different CO_2 assimilation pathways (C_3 and C_4 pathways) are followed by different isotope fractionation. Nevertheless step by step new facts have appeared evidencing that this simple explanation is not enough to describe many facts.

In gas exchange experiments using CO_2 enriched in ^{13}C, it was revealed that the residual CO_2 in a closed chamber got enriched in a "light" isotope. This means that the leaves of plants (kidney bean, poplar, maize) under study advantageously assimilate "heavy" molecules of CO_2 (Sanadze et al., 1978). Similar results were obtained by Voznesensky et al. (1982), who examined the photosynthesizing alga *Chlorella vulgaris* Beirger exposed to $^{14}CO_2$. The alga biomass displayed much more specific radioactivity than the assimilated CO_2. Ivanov et al. (1978) in their experiments with photosynthesizing purple sulfuric bacteria *Ectothiorhodospira shaposhnikovii* established that the extent and even the sign of biomass enrichment relative to the carbon source (bicarbonate) depended on the $^{13}C/^{12}C$ ratio in the initial substrate. These facts indicated that photorespiration was a putative reason for the observed enrichment of biomass in a "heavy" carbon. In fact, there are only two metabolic processes that could change the isotopic composition of biomass relative to the CO_2 assimilated. They are CO_2 assimilation and (photo) respiration. Since assimilation is known to be responsible for the ^{12}C enrichment of a biomass, logic prompts that the enrichment of a biomass in ^{13}C should be bound only with (photo) respiration.

The same conclusion stems from the analysis of the carbon isotope composition of metabolites linked with photorespiration. It was observed that photo-respiratory products are always enriched in ^{13}C relative to the total biomass and to other metabolites and fractions. Table 6 illustrates that typical products of photorespiration oxalates are significantly enriched in ^{13}C relative to the total biomass of different oxalate accumulating plants. The enrichment was observed under different conditions and for any type of photosynthesizing organisms.

Table 7 illustrates the ^{13}C enrichment of photo-respiratory products (labile sugars and proline) relative to lipid and protein fractions and to the total carbon of a biomass for different levels of water salinity.

Table 6. Carbon isotope ratio of a leaf biomass and oxalates of some oxalate-accumulating plants

Plant species	Plant type	Whole leaf	Oxalates	Reference
Spinaceae oleracea	C_3	- 27.5	- 11.9	Rivera and Smith, 1979
Pelagronium	C_3	- 31.0	- 12.4	Rivera and Smith, 1979
Mereurialis perennis	C_3	- 27.9	- 13.7	Rivera and Smith, 1979
Spinaceae oleracea	C_3	- 25.7	- 19.9	Raven et al., 1982
Echinomastus intertextus	CAM	- 13.4	- 7.3	Raven et al., 1982
Escobaria ruberoulosa	CAM	- 12.3	- 8.3	Raven et al., 1982
Opuntia euglemannii	CAM	- 13.3	- 8.5	Raven et al., 1982

Table 7 illustrates the ^{13}C enrichment of photo-respiratory products (labile sugars and proline) relative to lipid and protein fractions and to the total carbon of a biomass for different levels of water salinity.

The same picture was observed for tropical land plants of the C_3-CAM type *Clusia minor* in natural conditions (Table 8). Photo-respiratory fractions (soluble sugars and organic acids) of the plant display a pronounced ^{13}C enrichment compared to lipids and proteins at different water availability and at different levels of illumination.

Table 7. Distribution of ^{13}C in a biomass and the biochemical fractions of marine alga *Chorella stigmatophora*, grown under the effect of different environmental factors. $\delta^{13}C$ of ambient CO_2 is -21‰ (Ivlev and Kalinkina, 2001; Kalinkina and Udel'nova, 1990). $\delta^{13}C$ values are given in per mille relative to the PDB standard

Index	Concentration of NaCl in medium, mM		
	0	425	595
$\delta^{13}C$ of dry matter, ‰	- 61.6	- 59.0	- 64.5
$\delta^{13}C$ of lipids, ‰	- 66.0	- 65.0	- 63.8
$\delta^{13}C$ of proteins, ‰	- 42.1	- 40.9	- 47.3
$\delta^{3}C$ of labile sugars, ‰	- 30.0	–	- 30.5
$\delta^{13}C$ of proline, ‰	- 29.0	–	- 31.5

Table 8. Carbon isotope composition of biochemical fractions isolated from the leaves of *Clusia minor* under different environmental conditions (Borland et al., 1994). Samples were taken at dawn and dusk. $\delta^{13}C$ values are given in per mille relative to the PDB standard

N	FRACTION	WET SEASON		DRY SEASON	
		Exposed leaf	Shaded leaf	Exposed leaf	Shaded leaf
		Dawn	Dusk	Dawn	Dusk
1	Total carbon	-25.7	-30.3	-24.6	-29.1
2	Lipids and pigments	-28.7	-32.2	-27.7	-30.8
3	Amino acids	-31.7	-32.6	-31.3	-32.7
4	Soluble sugars	-21.2	-29.2	-17.9	-21.9
5	Organic acids	-22.3	-27.7	-21.1	-24.5

All the above facts allow the reasoning that photorespiration, like CO_2 assimilation, is followed by carbon isotope fractionation, but the sign of the isotope effect is opposite to that of the effect in CO_2 assimilation. The analysis shows that the most probable effect-generating reaction is glycine decarboxylation playing the key role in photorespiration (Ivlev, 1993; Ivlev et al., 1996, 1999).

The strongest direct arguments in favor of the idea about carbon isotope fractionation in photorespiration were obtained by Igamberdiev et al. (2001) in the examination of barley plants. The authors have examined wild type and mutants of those same plants. Mutants were the plants with an artificially inserted glycine decarboxylase (GDC) enzymatic complex with low activity; that is why they had low photorespiration. In accordance with theoretical expectations, the mutants' biomass was noticeably enriched in ^{12}C when compared to wild-type plants (Table 9).

Table 9. Carbon isotope composition of a leaf biomass taken from a barley plant of wild type and a GDC-mutant. $\delta^{13}C$ values are given in per mille relative to the PDB standard. n is the number of experiments

Plant	Age	$\delta^{13}C$, ‰
Barley, wild type	1 week	-39.0 (n = 4)
	6 weeks	-31.3 (n = 2)
Barley, GDC-mutant	1 week	-45.5 (n = 5)
	6 weeks	-37.5 (n = 3)

The same results were obtained (Igamberdiev et al., 2004) for the other plants (Table 10). Thus the obtained data allow the assertion that 1) photorespiration, like CO_2 assimilation, is accompanied by carbon isotope fractionation; 2) the isotope effect of photorespiration is of a reverse sign when compared to CO_2 assimilation; and 3) the glycine dehydrogenase reaction is, in effect, a generating one.

Table 10. Carbon isotope composition of a leaf biomass taken from different plants of wild type and GDC-mutants. $\delta^{13}C$ values are given in per mille relative to the PDB standard

Plant	$\delta^{13}C$, ‰	Reference
Barley, wild type	-39.0 ± 0.7	Igamberdiev et al., 2001
Barley, (GDC-mutant)	-45.5 ± 0.6	Igamberdiev et al., 2001
H. vulgare, wild type	-27.1 ± 0.3	Igamberdiev et al., 2001
H. vulgare, LaPr87/30 (GDC-mutant)	-32.8 ± 0.1	Igamberdiev et al., 2004
Arabidopsis thaliana, wild type	-29.1 ± 0.4	Igamberdiev et al., 2004
A. thaliana, CS 8012 (GDC-mutant)	-31.7 ± 0.3	Igamberdiev et al., 2004
Solanum tuberosum L., wild type	-26.6 ± 0.1	Igamberdiev et al., 2004
S. tuberosum L., P1 (GDC-mutant)	-29.1 ± 0.3	Igamberdiev et al., 2004
S. tuberosum L., P15 (GDC-mutant)	-29.2 ± 0.1	Igamberdiev et al., 2004

One of the consequences of isotope fractionation in photorespiration is the occurrence of intramolecular isotope heterogeneity in photo-respiratory metabolites. It is determined by the kinetic nature of the effect in glycine decarboxylation. One can see the mechanism of isotope heterogeneity emergence from the scheme of the glycine decarboxylation shown in Figure 1. Decarboxylation occurs through the C–C bond cleavage. If the process is followed by the carbon isotope fractionation (when the C–C bond cleavage is at a rate-limiting stage), according to the kinetic isotope effect theory (Melander and Saunders, 1983), carbon atoms of the products disposed at the ends of the broken bond get enriched in ^{12}C relative to the corresponding carbon atoms in the initial substrate, whereas the corresponding atoms in the residual substrate are enriched in ^{13}C. So the carbon atoms of the CO_2 and methylene group connected with tetrahydropholic acid (THFA), which transports C_1-fragment, are enriched in ^{12}C.

On the contrary, the corresponding atoms of unreacted glycine are enriched in ^{13}C. During the next step, the methylene group is transferred to the molecule of an unreacted substrate and serine, with an uneven carbon isotope distribution, is formed.

The following steps of transformation passing through trioses results in G6P formation, where "heavy" carbon atoms occupy positions C-3 and C-4 in the glucose skeleton whereas "light" carbon atoms occupy positions C-1 and C-6. If in the course of the reaction carbon substrate pool depletion takes place, then the so-called Raleigh effect arises and isotopic differences within the molecules become more pronounced.

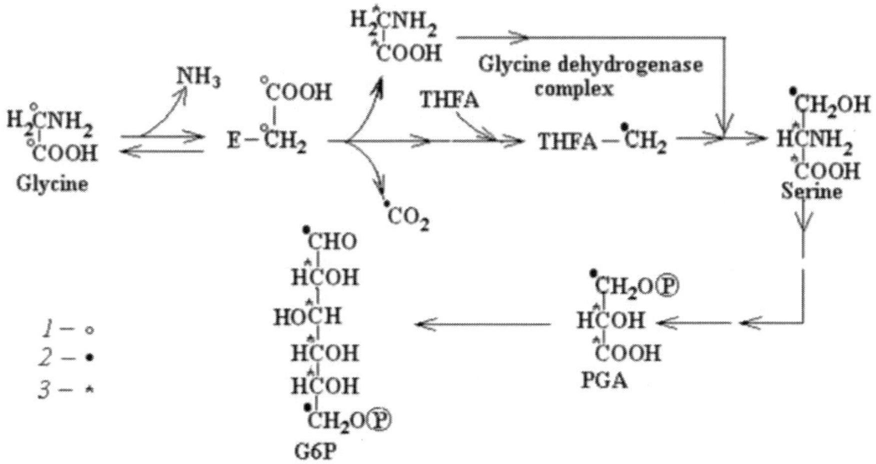

Figure 1. The emergence of the isotope inhomogeneity in glucose-6-phosphate (G6P) as a result of the carbon kinetic isotope effect in the GDC reaction. Empty circles (1) denote the carbon isotope composition of atoms disposed at the ends of the broken C–C bond, filled circles (2) denote carbon atoms enriched in ^{12}C in reaction products, and asterisks (3) denote carbon atoms corresponding to atoms disposed at the ends of the broken C–C bond.

Table 11 shows the results of a model calculation of ^{13}C distribution in glucose with an increase of turns of carbon flow in a photo-respiratory loop, i.e., with the increase of the extent of photo-respiratory substrate pool depletion.

Table 11. Isotopic compositions of the carbon atoms in the G6P, formed at the n-th turn in the photo-respiratory loop. Isotope composition of ambient G6P feeding a glycolate chain was taken to be equal to -20‰ with the uniform isotope distribution along the carbon skeleton (the explanation see later). The carbon isotope fractionation coefficient in the glycine decarboxylation reaction was taken to be equal to $\alpha = {}^{12}k/{}^{13}k = 1.020$. After each 10 turns of cycle, the extent of pool depletion is 10%

Atom position	Ambient G6P	Number of turns n 10	20	30	40	50	60	70	80	90
1	-20.0	-20.3	-20.7	-21.1	-21.4	-21.9	-22.4	-23.0	-23.7	-24.6
2	-20.0	-19.7	-19.4	-19.1	-18.7	-18.3	-17.8	-17.2	-16.4	-15.2
3	-20.0	-19.5	-18.8	-18.1	-17.3	-16.4	-15.2	-13.8	-11.9	-8.8
4	-20.0	-19.5	-18.8	-18.1	-17.3	-16.4	-15.2	-13.8	-11.9	-8.8
5	-20.0	-19.7	-19.4	-19.1	-18.7	-18.3	-17.8	-17.2	-16.4	-15.2
6	-20.0	-20.0	-20.7	-21.1	-21.4	-21.9	-22.4	-23.0	-23.0	-24.6

This isotopic pattern of photo-respiratory glucose is supported by the experimental data shown in Table 12, which illustrates carbon isotope distribution in the starch glucose of the storage organs of various plants. Through an independent method, it was proved (Ivlev, 2011) that the heterotrophic organs of plants, including the storage organs mainly, are made of labile carbohydrates that are synthesized in photorespiration. Similar results were obtained in experiments on glucose fermentation (Rossman et al., 1996; Weber et al., 1997).

$C_6H_{12}O_6 \rightarrow 2\ C_2H_5OH + 2\ CO_2$

Carbon dioxide, evolving in the process, inherits "heavy" carbon atoms from the C-3 and C-4 positions. The isotopic difference between a carbon of glucose and of CO_2 ranged from 4 to 8‰.

Table 12. Intramolecular carbon isotope distribution in the starch glucose of storage organs of various plants

Object	$\delta^{13}C$ of glucose	$\Delta^{13}C = \delta^{13}C - \delta^{13}C_{glucose}$, i - atom number					
		$OCH_{(1)}$-$HC_{(2)}OH$-$OHC_{(3)}H$-$HC_{(4)}OH$-$HC_{(5)}OH$-$C_{(6)}H_2OH$					
		C_1	C_2	C_3	C_4	C_5	C_6
Beta vulgaris, tuber (Rossman et al., 1991)	-25.2	-1.6	-0.4	+2.1	+6.3	-1.7	-5.1
Zea mays, seeds (Rossman et al., 1991)	-10.8	-1.7	-0.1	+ 1.1	+3.6	-0.2	-3.6
Zea mays, seeds (Ivlev et al., 1987)	-12.5	-3.1	+ 1.9				-1.9
Triticum aestivum, seeds (Galimov, 1985)	-23.1	-7.1	+3.5*				-7.1
Solanum tuberosum, tuber (Galimov, 1985)	-24.9	-9.1	+4.5*				-9.1
Oryza sativum, seeds (Galimov , 1985)	-26.1	-6.9	+3.5*				-6.9
Pisum sativum, seeds (Galimov, 1985)	-24.9	-4.1	+2.1*				-4.1

Note: The isotopic shifts of the carbon atoms $\Delta^{13}C$ are given relative to total glucose carbon. $\delta^{13}C$ values of glucose are given in PDB units. * The $\delta^{13}C$ values of C-3 and C-4 atoms were calculated according to Galimov (1985), assuming that the isotopic composition of the other carbon atoms equals to that of C-1 and C-6.

As one can see from Table 12, glucose from the starch of storage organs has a carbon isotope distribution characteristic of G6P in photorespiration. Atoms in positions C-3 and C-4 of glucose are enriched in ^{13}C, whereas atoms in positions C-1 and C-6 are enriched in ^{12}C.

The next question that needs to be answered is: how can two isotope effects with opposite signs coexist in the same photosynthesizing cell and result in differing carbon isotope compositions of metabolites? Carbon flow in CO_2 assimilation and that in photorespiration uses the same pieces of the Calvin cycle. If both flows entered the cycle simultaneously, no isotopic differences could arise because of mixing. The only way this paradox can be explained is to assume that the Calvin cycle works in two regimes separated in time. One regime corresponds to CO_2 assimilation; the other corresponds to photorespiration. This assumption seems reasonable if one takes into account that the key photosynthetic enzyme RuBisCo is capable of working in a cell as carboxylase or oxygenase depending on the CO_2/O_2 concentration ratio (McFadden, 1974; Laing et al., 1974), which periodically changes. The carboxylase phase of RuBisCo functioning coincides with CO_2 assimilation. Accordingly, the oxygenase phase coincides with photorespiration. In other words, this situation indicates the presence of the photosynthetic oscillations of carbon metabolism.

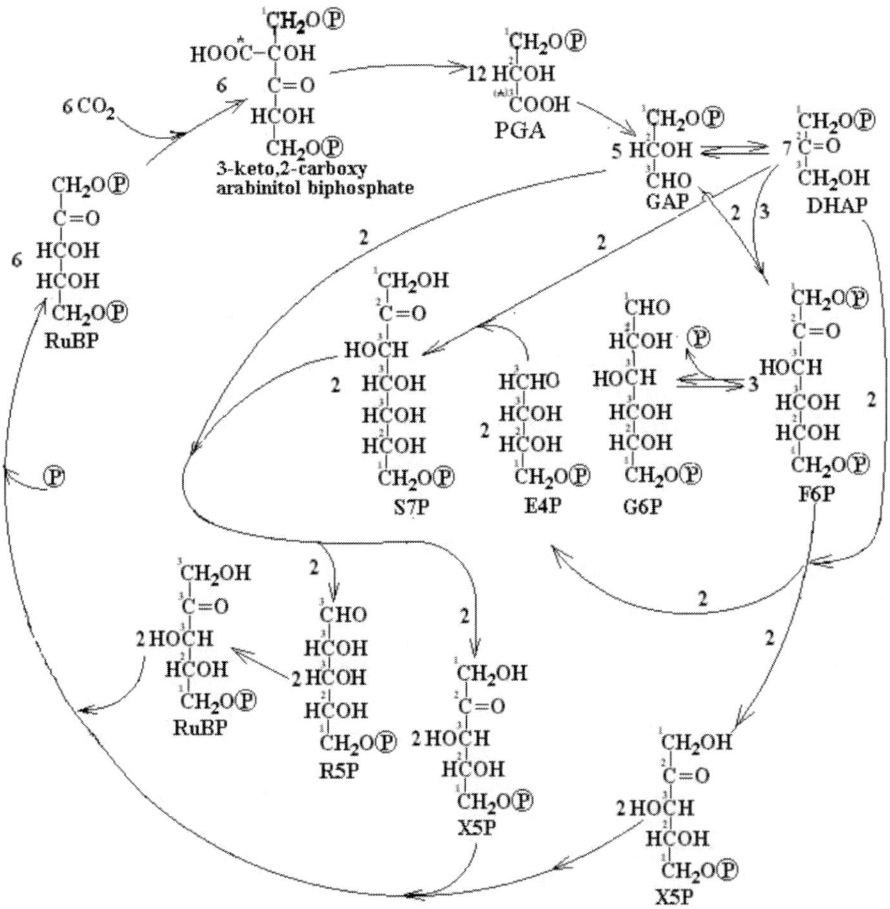

Figure 2. Calvin cycle work in carboxylase phase of RuBisCo functioning (Calvin cycle in regime of CO_2 assimilation).

Figure 2 shows the work of the Calvin cycle in the carboxylase phase of RuBisCo functioning (regime of CO_2 assimilation). In this regime, the Calvin cycle converts CO_2 into G6P and other products. G6P is mainly stored as an insoluble starch, which is used to supply carbon substrates with a glycolytic chain during the dark period (see later). Carbon isotope fractionation in the carboxylase phase is associated with RuBP carboxylation, which proceeds at the Calvin cycle entry. Thus each carbon atom, fixed in the reaction, becomes enriched in ^{12}C.

Unlike labile carbohydrates synthesized from photo-respiratory G6P (as shown above), G6P as well as the other products, synthesized in the carboxylase phase, have even carbon isotope distribution along the carbon skeleton. Transketolase and transaldolase reactions are the reason for that due to the randomization of the atoms in the cycle metabolites. So the carboxylase phase cycle works as a mixer (Hayes, 2001). The results of computational modeling of the ^{13}C distribution of G6P in the carboxylase phase are given Table 13. One can see from Table 13 that with every turn of the cycle, the initial differences between fixed CO_2 and acceptor RuBP are erased and ^{13}C distribution tends to be uniform.

Table 13. The results of computational modeling of ^{13}C distribution in G6P synthesized in the carboxylase phase of cycle functioning. The carbon isotope ratio of the fixed carbon was accepted equal to -20‰; the isotope ratio of acceptor RuBP is equal to -25‰

Number of turns n		1		5		10		15		20	
Atom position	Initial RuBP	G6P	RuBP	G6P	RuBP	G6P	RuBP	G6P	RuBP	G6P	RuBP
1	-25.0	-25.0	-25.0	-24.0	-22.9	-22.9	-21.2	-22.1	-20.5	-21.7	-20.2
2	-25.0	-25.0	-25.0	-24.0	-22.9	-22.9	-21.2	-22.1	-20.5	-21.7	-20.2
3	-25.0	-22.5	-22.5	-21.0	-20.1	-20.3	-20.0	-20.3	-20.0	-20.3	-20.0
4	-25.0	-22.5	-24.2	-21.0	-22.0	-20.3	-20.8	-20.3	-20.3	-20.3	-20.1
5	-25.0	-25.0	-24.2	-24.0	-22.0	-22.9	-20.8	-22.1	-20.3	-21.7	-20.1
6	–	-25.0	–	-24.0	–	-22.9	–	-22.1	–	-21.7	–

Figure 3 shows the work of the Calvin cycle in the oxygenase phase of RuBisCo functioning. After the carboxylase phase, residual G6P enters the cycle and via several transformations converts into glycolate, and the leaving chloroplast enters the peroxisome (Ivlev, 1993). In the peroxisome, it turns into glyoxylate, then into glycine. Further, glycine enters the mitochondria, where it undergoes decarboxylation. Glycine decarboxylation is followed by carbon isotope fractionation. As a result, serine with uneven carbon isotope distribution is produced, as was stated before. Finally, G6P and a carbohydrate pool of glucose with an irregular isotope distribution are formed.

Unlike the glucose produced in the carboxylase phase, uneven isotope distribution in photo-respiratory glucose does not disappear with the number of turns of carbon flow in the photo-respiratory loop, but intramolecular differences are reproduced and become more pronounced because of the Raleigh effect. Indirect evidence for that is the anomalous "heavy" carbon isotope composition of CO_2 that has evolved in the light-enhanced dark respiration (LEDR) and is dependent on the light intensity in the light period (Ivlev and Dubinsky, 2011; Ivlev, 2011).

Hence, as a consequence of oscillations in photosynthesis, two carbohydrate pools differing in their carbon isotope ratio and having specific carbon isotope distribution are formed. It is important to stress that these pools are used further for different metabolic needs, resulting in the emergence of two isotopically different carbon flows. In particular, the carbohydrate pool formed in the carboxylase phase is used to provide substrates with a glycolytic chain in the dark. The pool of labile carbohydrates is used to provide substrates with the syntheses of heterotrophic organs of plants, organic acids, etc.

To substantiate the idea of photosynthetic oscillations, we have performed model calculations using the simple photosynthetic scheme presented in Figure 4.

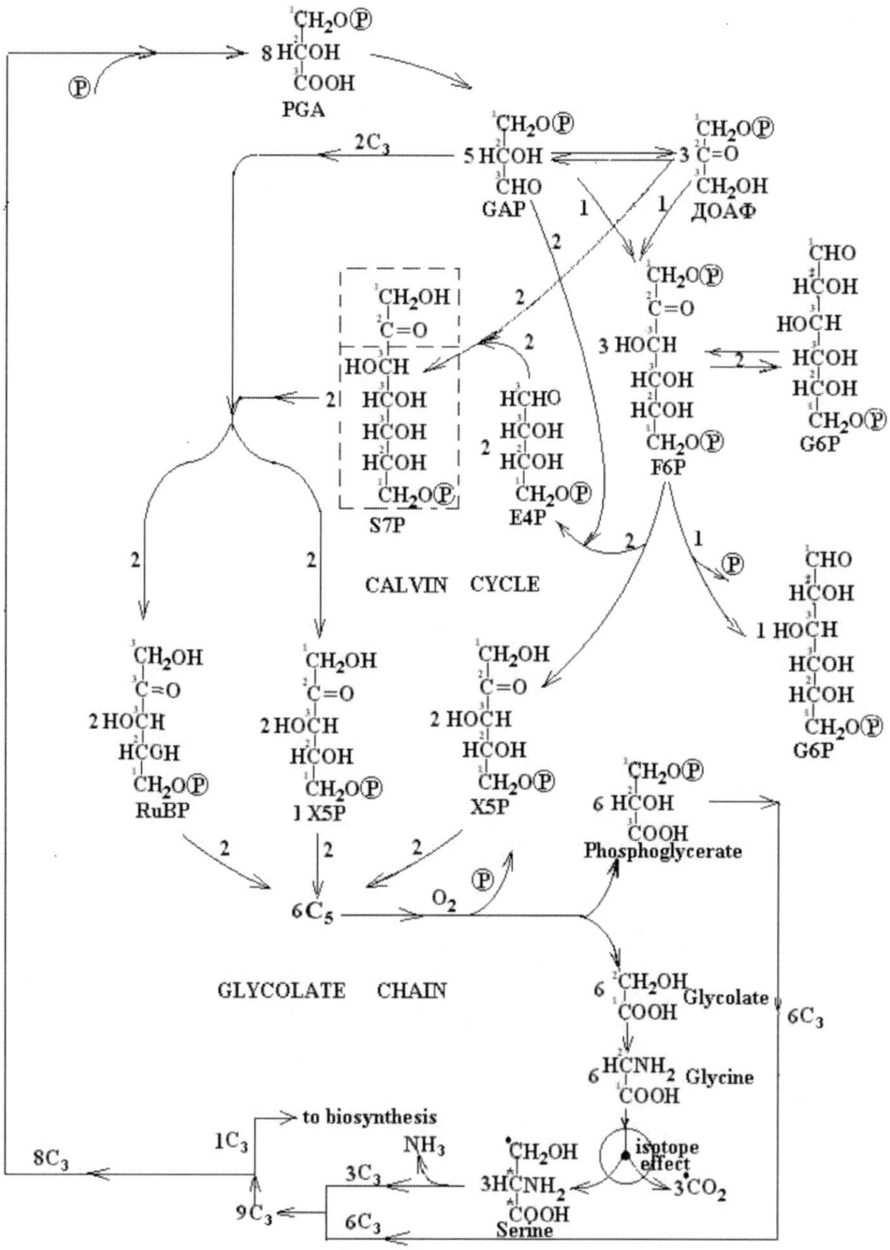

Figure 3. Calvin cycle work in the oxygenase phase of RuBisCo functioning (Calvin cycle in regime of photorespiration).

The scheme takes into account all the processes, substrates, and products of photosynthesis that provide the full carboxylase and oxygenase activity of RuBisCo. Furthermore, the calculations were performed with real photosynthetic cell parameters taken from the literature (Dybinsky and Ivlev, 2012). It was proved that photosynthetic oscillations may exist in the limits of the values of the cell parameters. In addition, it was revealed that photosynthetic oscillations are short-term events, with the period of oscillations lasting from 1 to 3 s depending on the set of parameters.

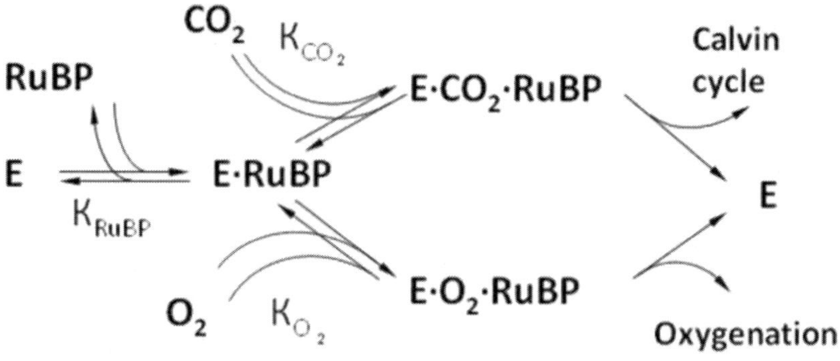

Figure 4. The simplified scheme of photosynthesis used for model calculations.

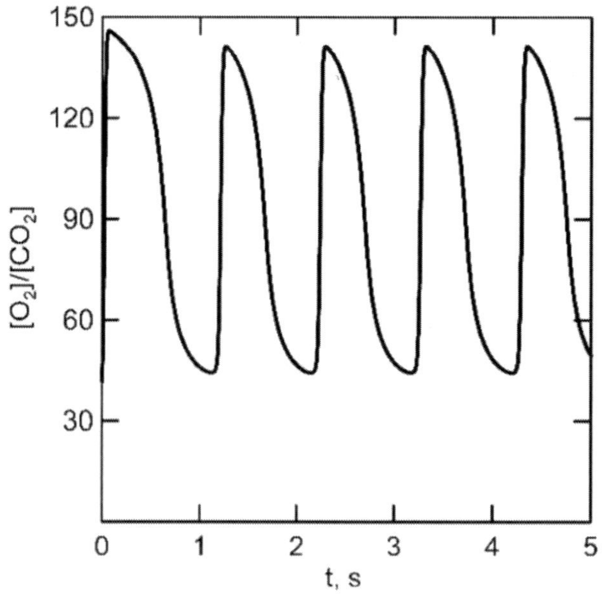

Figure 5. Oscillations of CO_2/O_2 concentration ratio obtained from model calculations according to the scheme presented in Figure 4.

Figure 5 illustrates the oscillations of the O_2/CO_2 concentration ratio. When the ratio reaches the upper meanings, RuBisCo switches from the Calvin cycle into the photorespiration regime. When it falls to the minimal meanings, the Calvin cycle is switched over to the CO_2 assimilation regime. The theoretical basis for the existence of the oscillatory regime in photosynthesis was created in the work of Igamberdiev and Roussel (2011). They classified different oscillatory mechanisms from the standpoint of nonlinear dynamics and showed that they might be bound to the delayed feedback between CO_2 assimilation and CO_2 release in photorespiration.

Roussel et al. (2007) provided strong support for the oscillation idea through their experimental data. The authors have entertained an attempt to reveal photosynthetic oscillations in the substomatal space of tobacco leaves (Figure 6).

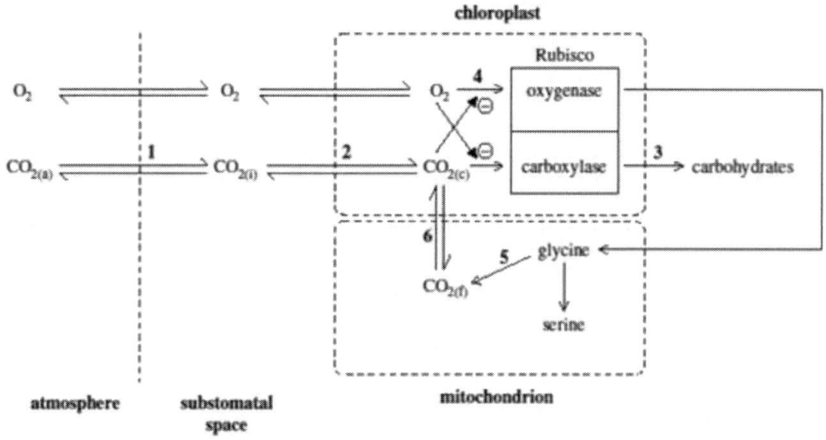

Figure 6. The simplified scheme for carbon assimilation and photorespiration in a photosynthesizing cell of tobacco leaf. Dashed lines indicate compartmental boundaries. Carbon dioxide in the atmosphere passes through stomata and enters the substomatal space. Eventually, it reaches chloroplasts, where the carbon is fixed. Oxygen may participate in photorespiration leading to the appearance of glycine in mitochondria and thus to photorespiration.

The measurements have been carried out in the range of CO_2 concentrations close to the compensation point where oscillations may be found the easiest way. They really revealed CO_2 fluctuations with an amplitude of 2–4 μLL^{-1}, which were much more than the experimental errors. The period was of the order of a few seconds. The authors explained them as a feedback interaction between CO_2 fixation and photorespiration.

Thus the oscillatory regime of photosynthesis results in the isotope heterogeneity of a biomass due to the existence of carbon isotope effects, with opposite signs arising in the carboxylase and oxygenase phases, correspondingly, and also due to the formation of two isotopically different carbohydrate pools utilized for different metabolic needs.

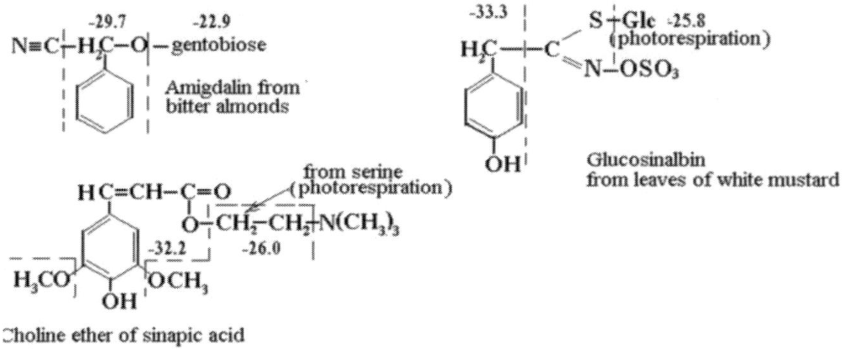

Figure 7. Carbon isotope distribution in some plant glucosilonates. Figures nearby the molecules indicate $\delta^{13}C$ of the examined parts. Data on intramolecular carbon isotope ratio are taken from the work of Schmidt et al. (1995).

The use of carbohydrate pools lead to the emergence of two isotopically different carbon flows in a cell. It has been shown that uneven carbon isotope distribution may be a result of the kinetic nature of the carbon isotope effect in glycine decarboxylation, but it also may be associated with the use of isotopically different carbohydrate pools in the synthesis of separate pieces of complex molecules. Figure 7 demonstrates the intramolecular carbon isotope distribution caused by the existence of two carbohydrate pools in photosynthesizing cells (Ivlev, 2008, 2010).

One can see that the two designated parts of the molecules have distinct differences in carbon isotope composition. It was shown that one part of the molecules, which includes the aromatic nucleus and an adjacent methylene carbon atom, has a carbon isotope composition enriched in ^{12}C and is close to that of carbohydrates synthesized in the carboxylase phase (Ivlev, 2008, 2010). The other part of these molecules is enriched in ^{13}C. Two upper molecules have carbohydrates fragments of photo-respiratory origin. The bottom molecule includes the serine fragment, which is also linked with photorespiration. Thus isotope heterogeneity in photosynthesis may be of different origin.

4. Carbon Isotope Fractionation in Post-Photosynthetic Metabolism. Metabolic Oscillations in the Glycolytic Chain. Isotope Heterogeneity in Metabolites of the Glycolytic Chain

Post-photosynthetic metabolism is mainly associated with the metabolism of the glycolytic chain. It has been studied by Sel'kov et al. (Pye, 1969; Sel'kov, 1973, 1975; Sel'kov and Betz, 1973; Dynnik and Sel'kov, 1975; Sel'kov and Shevelev, 1989). They proved the oscillatory character of metabolism in the glycolytic chain. Unlike the photosynthetic oscillations, these are long-term events, with a period lasting from several hours to 12 hours. The fructose bisphosphate futile cycle was found to be the regulator of glycolytic oscillations. It can work either towards phosphorilation or towards hydrolysis, directing carbon flow "up" or "down" the glycolytic chain. The glycolytic oscillations consist of two phases—glycolysis and gluconeogenesis.

In glycolysis, which corresponds to the dark period of photosynthesis, carbohydrates stored in the carboxylase phase of photosynthetic oscillations feed the glycolytic chain and the carbon flow goes "down" (Figure 8). After that, the carbohydrates are destroyed and turn into lipids and proteins. Glycolysis is coupled with dark respiration, lipids, and proteins syntheses.

In gluconeogenesis, which corresponds to the light period, carbon flow goes "up." Lipids and proteins are destroyed, whereas carbohydrates are resynthesized. Gluconeogenesis is coupled with lipid—and protein—carbohydrate turnover.

Thus, glycolytic oscillations are a kind of metabolic swing. Carbon isotope fractionation in post-photosynthetic metabolism is mainly linked with pyruvate decarboxylation, which is an effect-generating reaction. Pyruvate decarboxylation is the key reaction of post-photosynthetic metabolism.

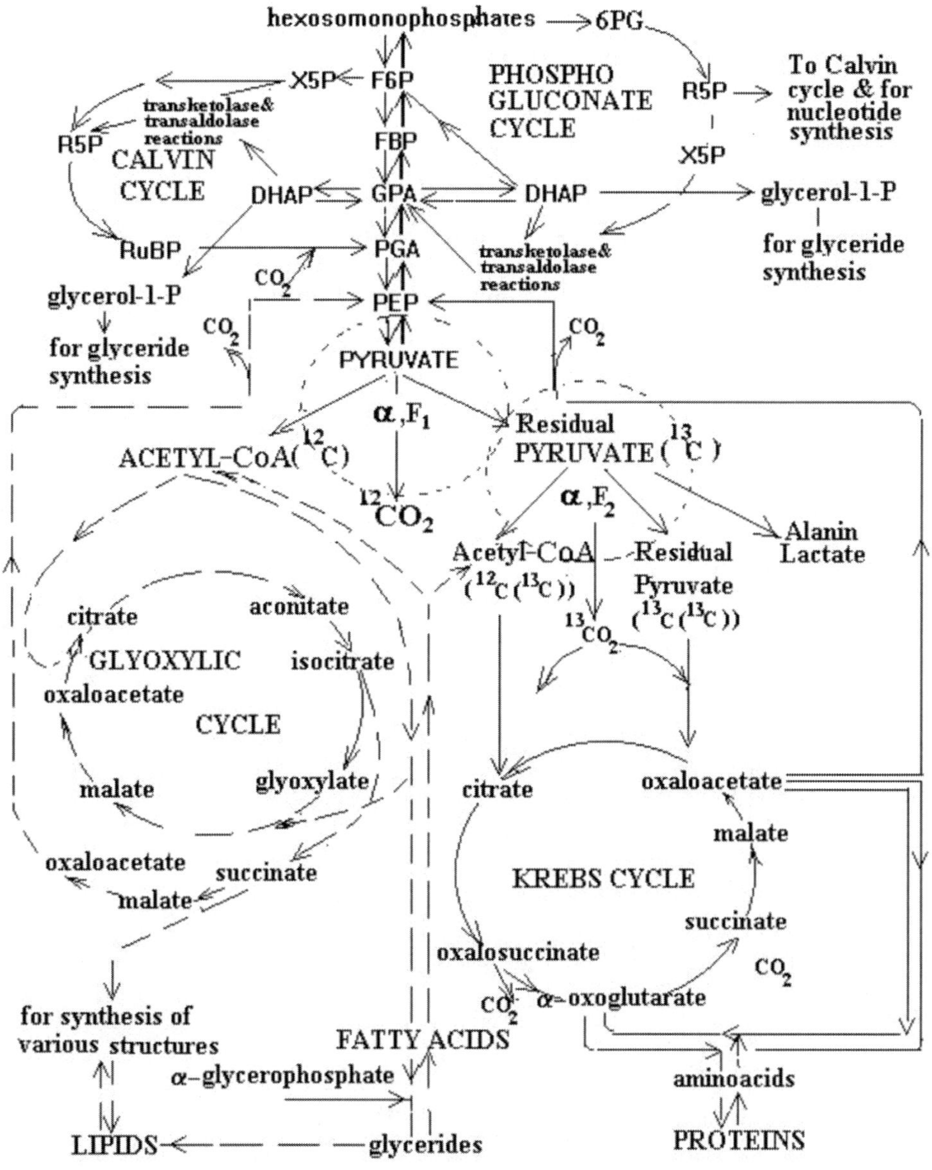

Figure 8. The scheme of carbon isotope fractionation in post-photosynthetic metabolism. Dashed lines show carbon substrate flows enriched in ^{12}C relative to carbohydrates feeding the glycolytic chain. Solid lines denote carbon substrate flows enriched in ^{13}C. .

Abbreviations: F6P – fructose-6-phosphate, FBP – fructose 1,6-bisphosphate, GPA – phosphoglyceric acid, DHAP – dihydroxyacetone phosphate, PEP – phosphoenolpyruvate, X5P – xylose-5-phosphate, R5P – ribose-5-phosphate.

A special place of the reaction in a metabolic system is explained by the fact that it is the cross-point of the central metabolic pathways, where all catabolic and anabolic processes commence. Another reason stems from the fact that all the products of the reaction (CO_2 and acetyl-CoA) and the unreacted substrate (pyruvate) are the structural units used for the synthesis for practically all the components of post-photosynthetic metabolism. Hence the carbon isotope composition of the structural units, combined with metabolic pathways, determines the isotope composition of all the components and fractions. For convenience here and further, we denote the structural units as C_1-fragments for CO_2, C_2-fragments for acetyl-CoA, and C_3-fragments for unreacted pyruvate.

The special role of pyruvate decarboxylation in carbon isotope fractionation was initially demonstrated by De Niro and Epstein (1977) in their experiments on growing *E. coli* on different substrates (Table 14).

Table 14. Carbon isotope composition of lipids formed by fermentation of *E. coli* grown on different substrates (De Niro and Epstein, 1977). Carbon isotope composition is given relative to the PDB standard

Substrate	Carbon source $\delta^{13}C$, ‰	Lipids $\delta^{13}C$, ‰	Isotopic shift $\Delta^{13}C$, ‰
Glucose	9.5	-15.7	-6.2
		-16.3	-7.8
CH_3COONa	-20.5	-28.9	-8.4
CH_3COONa	-20.1	-21.5	-1.4
		-20.7	-0.6

$\Delta^{13}C = \delta^{13}C_{Lip} - \delta^{13}C_{Sub}$, where $\delta^{13}C_{Lip}$ and $\delta^{13}C_{Sub}$ are the carbon isotope composition of lipids and substrates, respectively.

They showed that in blocking the entry of C_2-fragments from pyruvate decarboxylation, by introducing them into the environment using exogenous acetate, the isotopic shifts dramatically fall.

The same occurrence was observed in the experiments with *Corynebacteria*, when molasses (carbohydrates) were exchanged for acetate. The intramolecular shifts caused by the carbon isotope effect of pyruvate decarboxylation were sharply reduced (Ivlev et al., 1990, 1991).

The results of Tables 14 and 15 prove that pyruvate decarboxylation is the main effect-generating reaction of post-photosynthetic metabolism.

Table 15. Changes in intramolecular carbon isotope distribution in amino acids produced by *Corynebacteria* grown on different substrates (Ivlev et al., 1990, 1991). Carbon isotope composition is given relative to the PDB standard

Bacteria	Substrate	$\delta^{13}C$, ‰	Amino acid	$\delta^{13}C$, ‰	$\Delta^{13}C$, ‰
B. flavum E-531	Molasses	-23.6	Lysine	-24.8	-6.3
	Molasses +	-23.6	Lysine	-29.7	-2.7
	Acetate*	-31.5			
	Molasses	-22.2	Glutamic acid	-28.0	+7.4
	Acetate	-32.2	Glutamic acid	-26.7	-2.7

Note: $\Delta^{13}C = \delta^{13}C_{carboxyl} - \delta^{13}C_{total}$, where $\delta^{13}C_{carboxyl}$ and $\delta^{13}C_{total}$ are the carbon isotope composition of carboxyl and the total carbon of amino acid. *Nutrition mixture: 80% molasses + 20% acetate.

The next important finding in the post-photosynthetic metabolism was the discovery of a Raleigh effect accompanying the effect-generating reaction. This means that oscillations are coupled with the filling/depleting of the substrate pool. As in photorespiration, the Raleigh effect occurs in glycolysis in the course of pyruvate pool depletion. To substantiate the Raleigh effect it is necessary to consider the isotope distribution in structural units and to estimate the range of isotopic variations for each atom of the structural units. The pyruvate decarboxylation and structural units produced in the reaction are shown on Figure 9.

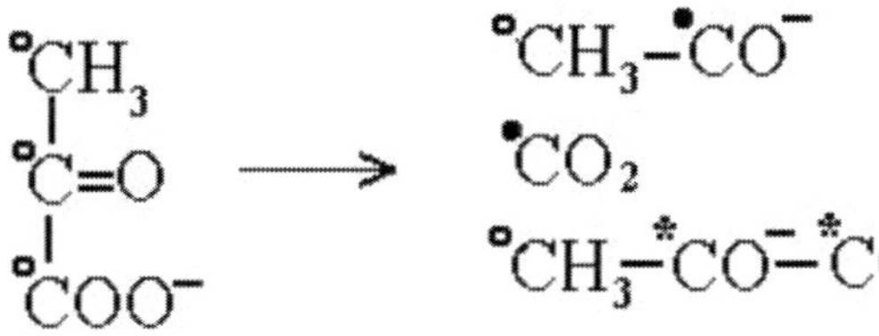

Figure 9. The scheme of pyruvate decarboxylation reaction and carbon isotope distribution in the products and the unreacted substrate resulting in the course of decarboxylation due to kinetic isotope effect following C – C cleavage. Symbol of atoms are the same as in the reaction of glycine decarboxylation (see Figure 1).

As in glycine decarboxylation, and in accordance with the kinetic isotope effect theory, the atoms disposed at the ends of the broken C–C bond in the products are enriched in ^{12}C (denoted with filled circles) relative to the atoms in the initial substrate or to the atoms that do not participate in bond destruction (denoted by the empty circles). On the contrary, the atoms at the ends of the broken bond in the unreacted substrate are enriched in ^{13}C.

Considering the Raleigh effect, it is convenient to subdivide the atoms in the structural units into three types.

Type 1 includes methyl atoms of C_2-fragments and C_3-fragments that do not participate in the C–C bond destruction and do not change their isotope composition in biosynthetic reactions. Taking into account that a glycolytic chain is supplied with carbohydrates synthesized in the carboxylase phase of photosynthetic oscillations and built of atoms with an isotopic composition determined by the isotope effect of CO_2 assimilation (RuBP carboxylation), it should be underlined that atoms of this type have the same carbon isotope composition, which may be taken as an internal standard.

Type 2 includes the carbonyl atoms of CO_2 and of C_2-fragments disposed at the ends of the cleaved C–C bond. In accordance with the kinetic isotope effect theory, depending on the extent of the pyruvate pool depletion F, their carbon isotope composition can be either enriched in ^{12}C relative to the atoms of the initial substrate, if F is less than 0.5, or enriched in ^{13}C, if F is more than 0.5. So they are characterized with a wide range of changeability.

Type 3 includes carboxyl and the neighboring carbonyl atoms of the C_3-fragments. Their carbon isotope composition at any F is enriched in ^{13}C relative to atoms of other fragments of types 1 and 2.

The first indication of the existence of a Raleigh effect was obtained by Hayes and Monson and their colleagues (Voglar and Hayes, 1979; Hayes and Monson, 1982; Monson and Hayes 1982) in the examination of the carbon isotope composition of total carbon and the carboxyl carbon of individual fatty acids isolated from different organisms (Table 16).

It was well known that the total carbon of fatty acids is always enriched in ^{12}C relative to that of protein fraction and of the biomass of an organism. But the examination of individual fatty acids isolated from the lipid fraction of the organisms showed that they were different in terms of carbon isotope composition. Furthermore, their carboxyl carbon is of different carbon isotope composition (Table 16). This isotope heterogeneity is explained by considering the fact that fatty acids are synthesized from C_2-fragments according to the "head to tail" principle and by assuming that the uneven atoms belong to type 2. This also explains why a carboxyl atom may be either enriched relative to the total carbon in ^{12}C or in ^{13}C. The enrichment, as said before, depends on the extent of the substrate pool depletion. This assumption was confirmed by the results of the experiments that were carried out by Monson and Hayes (1982) and Blair et al. (1985) on *E. coli* grown on the glucose of a known carbon isotope composition (Tables 17 and 18).

Table 16. ^{13}C distribution in fatty acids isolated from the lipid fraction of different organisms

Organism	Fatty acids	$\delta^{13}C$ of fatty acid	
		Total carbon	Carboxyl carbon
E. coli Monson and Hayes, 1982	Myristic 14:0	-13.7	-27.1
	Palmitic 16:0	-12.2	-15.2
	Palmitoleic 16:1	-13.0	-19.2
	9,10-Methylenepalmitic 17:cycle	-13.7	-20.3
	Vaccenic 18:1	-12.5	-13.9
S. cereviseae Hayes and Monson, 1982	Myristic 14:0		-5.1
	Myristic 14:1		-0.7
	Palmitic 16:0		-3.1
	Palmitoleic 16:1		+1.1
	Stearic 18:0		-17.4
	Vaccenic 18:1		-5.1
Wheat seeds Vogler and Hayes, 1979	Palmitic 16:0	-18.1	-20.2
	Vaccenic 18:1	-16.0	-12.3
	Linoleic 18:2	-15.4	-6.5
Soybeans Vogler and Hayes, 1979	Palmitic 16:0	-30.3	-40.1
	Vaccenic 18:1	-29.4	-31.5
	Linoleic 18:2	-29.7	-28.5
Nutmeg Vogler and Hayes, 1979	Myristic 14:0	-29.9	-38.8

$\delta^{13}C$ values are given in per mille relative to the PDB standard.

As one can see from Table 17, odd atoms belonging to type 2 are of different isotope composition, which provides evidence that they are synthesized at different extents of the substrate pool depletion. This is the most convincing argument that even atoms belonging to type 1 have a carbon isotope composition close to the carbon isotope composition of the glucose feeding *E. coli*.

Table 18 illustrates the isotopic differences in fractions and compounds synthesized from C_2-fragments. Fatty acids have close isotope composition to lipid fraction. It is evident that fatty acids mainly contribute to this fraction.

Table 17. Distribution of carbon isotopes in the fatty acids of the lipid fraction of *E. coli* grown on the glucose of the known isotope composition ($\delta^{13}C_{gl}$ = -9.96‰, N is the number of carbon atoms) (Monson and Hayes, 1982)

Fatty acids	Total carbon $\delta^{13}C$‰	Odd atoms N	Odd atoms $\delta^{13}C$‰	Even atoms N	Even atoms $\delta^{13}C$‰
Myristic 14 : 0	- 13.75	1	- 27.1		
Palmitic 16 : 0	- 12.23	1	- 15.2		
Palmitoleic 16 : 1	- 13.01	1	- 19.2		
		9	- 16.0	10	- 9.5
9,10-Methylene - palmitic 17 : cycle	- 13.71	1	- 20.3		
Vaccenic 18 : 1	- 12.58	1	- 13.9		
		11	- 15.8	12	- 9.5

*Total number of carbon atoms: number of double bonds.

Table 18. Carbon isotope fractionation in *E. coli* grown on glucose with a known carbon isotopic composition ($\delta^{13}C_{glucose}$ = - 9.0‰) Blair et al. (1985). $\delta^{13}C$ values are given in per mille relative to the PDB standard

Lipids	Fatty acids	^{13}C distribution in acetate Total	Carboxyl	Methyl
-11.7	-12.1	+3.3	+15.0	-8.8

The most interesting fact is that the carbon isotope distribution of acetate is also a C_2-fragment. One can observe the same situation as that found in the previous case. The carboxyl atom belonging to type 2 is very "heavy," whereas the methyl atom belonging to type 1 has the same carbon isotope composition as the feeding glucose. The above results completely correspond to what would be expected from a theoretical Raleigh effect. The Raleigh effect means that work in post-photosynthetic metabolism cells occurs in the regime of filling/depleting the substrate pools. Hence the Raleigh effect corresponds well with the oscillatory hypothesis. The other arguments in favor of the oscillatory dynamics of carbon metabolism were presented in our previous works (Ivlev, 2008, 2010, 2012).

5. Perspectives on the Application of the Metabolic Carbon Isotope Effects Method to Living Organism Studies

As the above presentations have appeared quite recently, they have not yet found a wide application. But it is already evident that isotopic heterogeneity linked with metabolic pathways, associated with the functional state of the organism, and responding to the environmental factors, should include a lot of useful information that may be used in living organism studies. Here we try to designate some promising directions based on the model presentations described above.

Metabolic pathway studies. The tight relationship between the isotopic patterns of metabolites and metabolic pathways is the basis of such an examination. To illustrate these relationships, let us review some examples.

As said before, to find out the metabolic pathway one needs to know the isotope distributions of structural units, their dependence on the extent of pool depletion, and the assumed pathway. The coincidence of the assumed carbon isotope distribution with the measured one gives a positive reply.

To show how it works, we will compare the reconstructed model isotopic pattern with the real isotopic distribution in a case where the metabolic pathway is known.

Example 1

As shown on Figure 10, acetoin is produced from glucose via the condensation of two C_2-fragments. The measured ^{13}C distribution is shown in the bottom of Figure 10. It includes two atoms of type 1 and two atoms of type 2. The comparison of the real ^{13}C distribution with the distribution expected from the model shows full coincidence. In fact, atoms belonging to type 1 demonstrate close proximity in terms of carbon isotope composition, while atoms belonging to type 2 have different carbon isotope composition ranging across a wide interval. The difference in carbon isotope composition evidences that two C_2-fragments, composing the molecule, are formed at different moments of the pyruvate pool depletion.

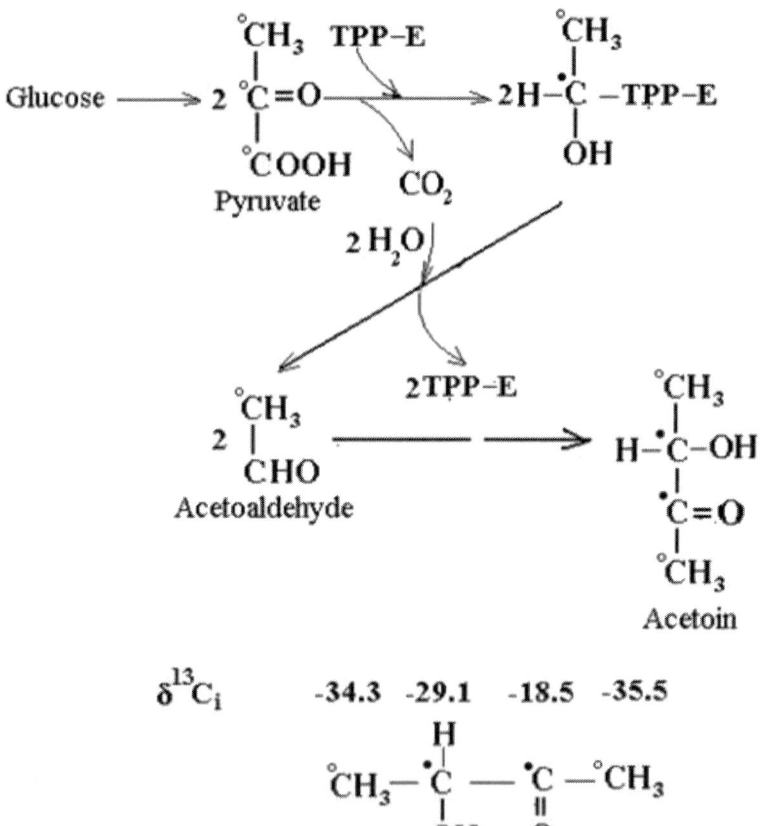

Figure 10. The scheme of pyruvate decarboxylation reaction and carbon isotope distribution in the products and in the unreacted substrate resulting in the course of decarboxylation due to the kinetic isotope effect following the C–C bond cleavage. Symbols for the atoms are the same as in the reaction of glycine decarboxylation (Figure 1).

Example 2

There are three possible metabolic pathways for the branched amino acids syntheses (Figure 11). They are the isoleucine pathway, valine pathway, and leucine pathway. From the work of Abelson and Hoering (1961) it is known that the real ^{13}C distribution of branched amino acids in photosynthesizing alga is an inverse one. In some cases, their carboxyl groups are enriched in ^{12}C relative to the radical; in other cases, they are enriched in ^{13}C. Comparing the ^{13}C distribution corresponding to each pathway with the result expected from the model (Figure 11), we can explain their discrepancies and choose the pathway corresponding to the measured ^{13}C distribution.

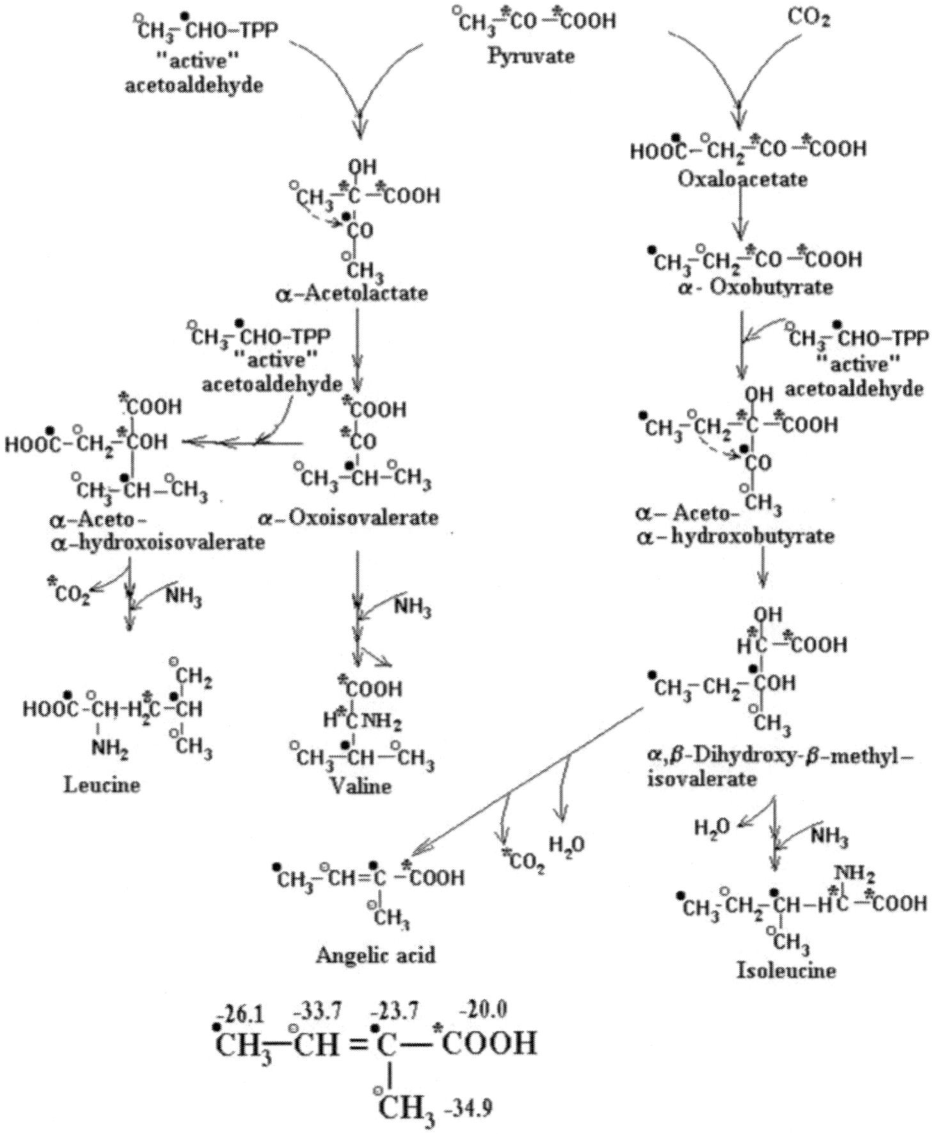

Figure 11. *Acetoin synthesis*. Metabolic pathway and ^{13}C pattern. The $\delta^{13}C$ values of the carbon atoms were taken from work of Rinaldi et al. (1974).

According to the known biosynthesis pathway of branched amino acids, in the case of leucine, the carboxyl group inherits an atom of type 2. The model allows leucine two possibilities. The isotopic composition of the carboxyl group can be either “light” or “heavy” depending on the extent of the pool depletion. Carboxyl atoms of valine and isoleucine belong to type 3 and are always enriched in ^{13}C relative to atoms of types 1 and 2. A detailed examination of the ^{13}C distribution of angelic acid (Schmidt et al., 1985) confirms the validity of such an explanation.

In fact, the synthesis of angelic acid occurs according to an isoleucine pathway. This molecule includes two atoms of type 1, two atoms of type 2, and one atom of type 3. In accordance with the model, atoms of type 1 have a very similar carbon isotope composition. Atoms of type 2 have different carbon isotope compositions, reflecting the difference in the extent of pool depletion. At the same time, atoms of type 2 are “lighter” than atoms of type 3. Thus direct isotope measurements fully correspond to the theoretical description.

Example 3

Figure 12 shows the synthesis pathway of sinigrin, a plant glucosilonate isolated from black mustard seeds. The complex molecule sinigrin consists of two parts: aglycon and carbohydrate. The latter joins the aglycon piece during the last steps of synthesis.

The molecular structure of sinigrin includes two atoms of type 1 and one atom of types 2 and 3 each. As in the previous case, the atoms of type 1 have a close carbon isotope composition. The atom of type 2 is of intermediate carbon isotope composition while the atom of type 3 is mostly enriched in ^{13}C.

There are some differences between the atoms of type 1 in comparison to the previous cases. The reason for the indicated differences is not the real isotopic shifts but represent a difference in accuracy for the measurement techniques used. The carbon atom of type 1 in the skeleton, disposed at the end of the double C=C bond was measured by means of the NMR technique, while the methyl atom of type 1 was measured by the mass-spectrometric technique.

It is worth mentioning that the carbon isotope composition of the atoms of type 1 is inherited from carbohydrates synthesized in the carboxylase phase of photosynthetic oscillations. If compared to the carbon isotope composition of the carbohydrate piece of the molecule, it will be seen that the latter is more enriched in ^{13}C than the atoms of type 1. This allows the conclusion that the carbohydrate part was synthesized in photorespiration.

Temporal organization of cell processes studies. One of the most interesting fields of possible application for carbon isotope heterogeneity is in the study of the temporal organization of cell processes. The basis for such studies is the Raleigh effect, which arises due to the filling/depleting regime of cells functioning in a phase of pool depletion and is associated with the isotope effect-generating reaction. As an example, let us consider the Raleigh effect following pyruvate decarboxylation proceeding into glycolysis. The C_2-fragments appearing in the reaction have carbonyl atoms that are progressively enriched in ^{13}C with pyruvate pool depletion. Therefore the metabolites that are built from C_2-fragments at greater extents of pool depletion should be “heavier” relative to those produced at earlier stages.

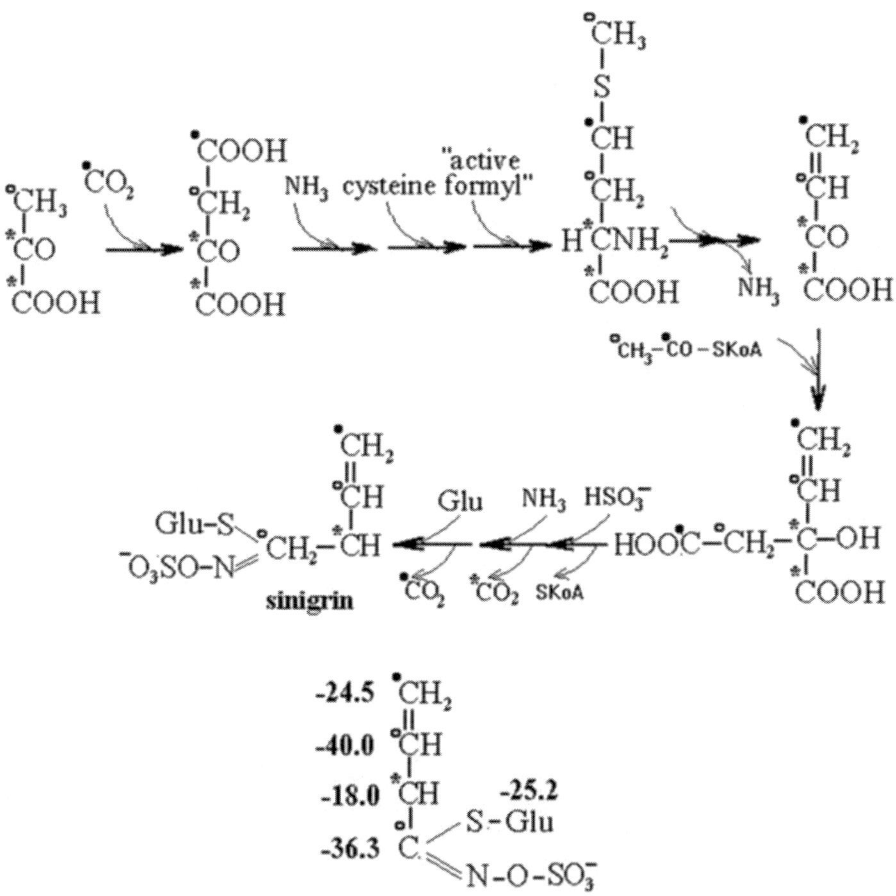

Figure 12. *Angelic acid synthesis.* Metabolic biosynthesis pathway of sinigrin and the ^{13}C distribution obtained from the model and by means of direct measurements (Schmidt et al., 1985).

Comparing the carbon isotope ratio of different fatty acids, it is possible to establish the sequence of their syntheses. In this way by analyzing the carbon isotope ratio of different metabolites of post-photosynthetic metabolism and the metabolic pathways of their syntheses, we have found that in photosynthesizing organisms, isoprenoids and carotenoids are usually produced before fatty acids; the fatty acids syntheses occur prior to the Krebs cycle function (Ivlev, 2008).

Moreover by examining carbon isotope distribution within the molecule, it is possible to trace the duration of its synthesis (Figure 13). In fact, taking into consideration that the known metabolic pathways of plant monoterpenes biosyntheses occur via C_2-fragments, and analyzing their carbon isotope distribution (the carbon isotope composition of carboxyl groups), it is possible to establish what part of the molecule is built first. The intriguing problem is to investigate the impact of environmental factors on the temporal organization of cells.

Metabolic pool studies. One more attractive possible area of isotope heterogeneity application is in metabolic pool studies seeking to establish the source of carbon for metabolites synthesis. The basis for such studies is the existence of carbohydrate pools with different carbon isotope compositions appearing in photosynthetic oscillations.

It has already been shown that fragments of plant glucosilonates have two isotopically different carbon sources (Figure 7). In particular, the aglycon pieces of the molecules use, as a source of carbon for the synthesis, carbohydrates formed in the carboxylase phase. At the same time, the carbohydrate fragments of the molecules are supplied from the carbohydrate pool formed in the oxygenase phase. It should be noted that one of the studied glucosilonates has a serine fragment that is also inherited from photorespiration.

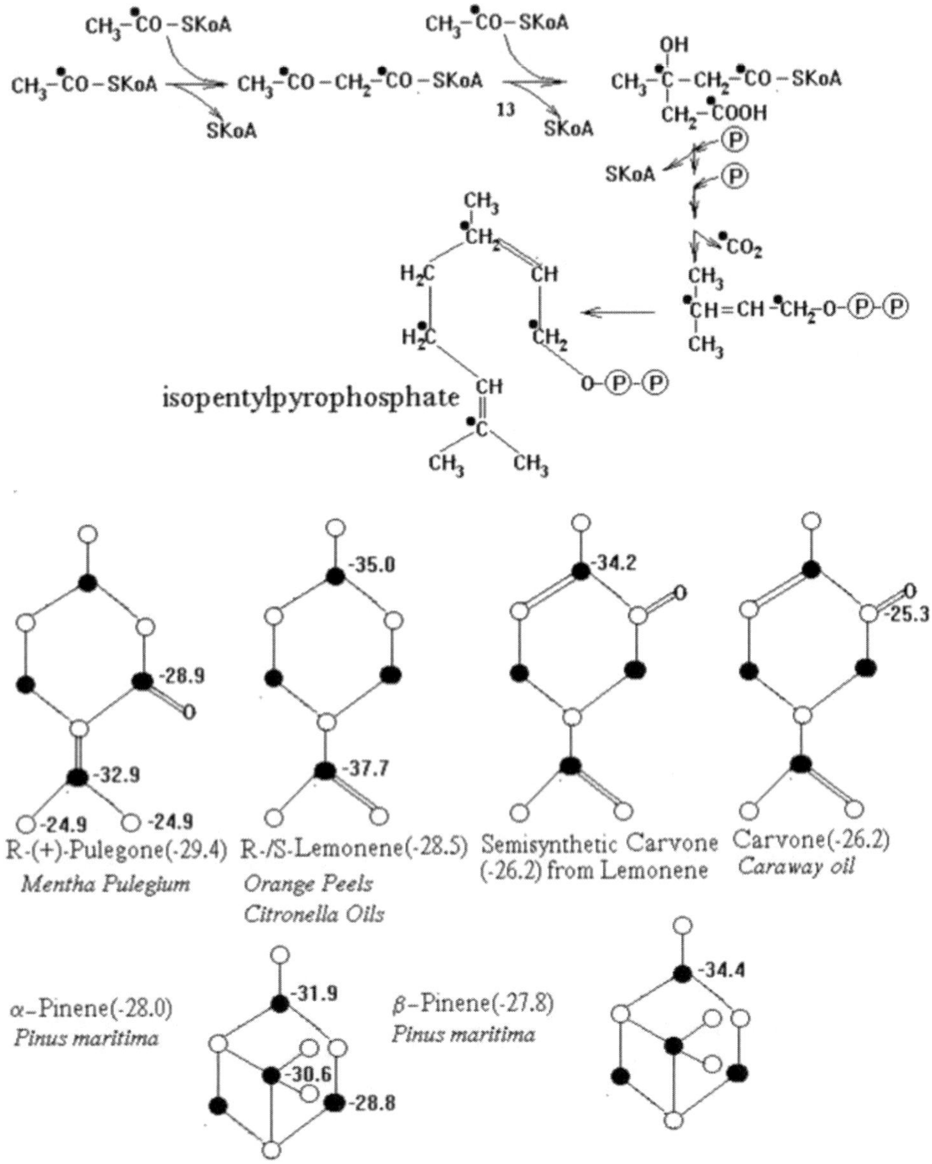

Figure 13. Metabolic biosynthesis pathway for sinigrin and the corresponding ^{13}C distribution of the molecule obtained from the model and by means of direct measurements (Schmidt et al., 1985).

As early as 1961, Abelson and Hoering found that among the amino acids isolated from the protein fraction of different photosynthesizing organisms, amino acids from the triose family are mostly enriched in ^{13}C. The reason for this was unclear. The present model explains this fact and allows the conclusion that these amino acids (glycine, serine, and alanine) are tightly linked with photorespiration, much like the serine fragment in the above glucosilonate.

Other examples of applications for metabolic carbon isotope heterogeneity in different fields can be found in our previous works. In the work of Voronin et al. (2012), model presentations were used to assimilate transport studies. In the work of Ivlev (2013), model presentations were used for the evolution of problem studies.

CONCLUSION

The analysis of carbon isotopic data allows the conclusion that all the carbon metabolic processes in photosynthesizing cells, including photosynthesis and post-photosynthetic metabolism, have an oscillatory character. The photosynthetic oscillations, consisting in the alternation of CO_2 assimilation and photorespiration processes, are short-term events with a period of a few seconds. The post-photosynthetic oscillations (oscillations in the glycolytic chain), consisting of two phases: glycolysis and gluconeogenesis are long-term events with a period ranging up to 12 hours. Photosynthetic oscillations occur in the light. The glycolysis phase of glycolytic oscillations corresponds to the dark period; the gluconeogenesis phase of oscillations corresponds to the light period.

The isotopic data analysis results in the conclusion that the carbon isotope fractionation in cell processes is followed by a Raleigh effect. This means that the effect-generating reaction proceeds with substrate pool depletion. The Raleigh effect was found in the RuBP carboxylation reaction in CO_2 photo-assimilation, in the glycine decarboxylation reaction in photorespiration, and in the pyruvate decarboxylation reaction in post-photosynthetic metabolism. The Raleigh effect's existence allows us to conclude that photosynthesizing cells work during the filling/depleting regime. This regime correlates well with the assumed oscillations.

The Raleigh effect translates to the existence of the strict temporal organization of cell processes and evidences that oscillations are synchronous and that the synthesis of each component occurs within certain limits of the cycle.

The combination of the kinetic nature of the isotopic effects with the Rayleigh effects overlap on a variety of metabolic pathways in the cell at a given functional state of the organism, which leads to a variety of manifestations of isotope heterogeneity. These are displayed as the differences in the carbon isotope composition of biochemical fractions and of individual metabolites and fragments of the biomolecules; they are also expressed as different isotopic patterns of the metabolites.

The above combination leads to regularities in inter- and intramolecular ^{13}C distributions. It should be stressed that certain regularities are displayed for a given functional state of the organism, which is determined by the environmental conditions, by internal factors, and by the ontogenetic phase.

Knowing the nature of carbon isotope heterogeneity and having the corresponding model describing carbon isotope fractionation in a cell, it is possible to use the observed regularities of ^{13}C distributions in a biomass to examine a living organism. In particular, this model can be applied to study metabolic pathways, the interrelation of metabolite pools, the temporal organization of cell processes and the biological clock, and other problems.

REFERENCES

Abelson, P. H., Hoering, T. C., 1961. Carbon isotope fractionation in formation of amino acids by photosynthetic organisms. *Proc. Nat. Acad. Sci. USA* 47, 623–629.

Blair, N., Lea, A., Munez, E., Olsen, J., Kwong, E., Des-Marais, D., 1985. Carbon isotopic fractionation in heterotrophic microbial metabolism. *Appl. Environ. Microbial.* 50, 996–1002.

Bender, M. M., 1971. Variations in the $^{13}C/^{12}C$ ratios in the relation to the pathway of photosynthetic dioxide fixation *Phytochemistry* 10, 1239–1244.

Borland, A. M., Griffiths, H., Broadmeadow, M. S., Fordham, M. C., Maxwell, C., 1994. Carbon isotope composition of biochemical fractions and the regulation of carbon balance in leaves of the C_3-Crassulenean acid metabolism intermediate *Clusia minor* L. growing in Trinidad. *Plant Physiol.* 105, 493–501.

Buchachenko, A. L., 2003. Are chemical transformations in an enzyme-substrate complex reversible? The experience of the fractionation of isotopes in enzymatic reactions. *Russ. J. Phys. Chem.* 77, 1298–1302.

Degens, E. T., Gullard, K. R. L., Sackett, W. M., Hullebust, J. A., 1968. Metabolic fractionation of carbon isotopes in marine plankton. II. Data on samples collected off the coasts of Peru and Ecuador. *Deep Sea Res.* 15. 11–20.

Deleens, E., Garnier-Dardart, J.,1977. Carbon isotope composition of biochemical fractions isolated from leaves of *Bryophyllumdaigramontianum* Berger, a plant with CAM: some biological aspects related to CO_2 dark respiration. *Planta* 135. 241–248.

DeNiro, M. J., Epstein, S., 1977. Mechanism of carbon isotope fractionation associated with lipid synthesis. *Science* 197, 261–267.

Dubinsky, A. Yu., Ivlev, A. A., 2011. Computational analysis of the possibility of the oscillatory dynamics in the processes of CO_2 assimilation and photorespiration. *BioSystems* 103, 285–290.

Dynnik, V. V., Sel'kov, E. E., 1975. Generator of oscillations in lower part of glycolytic chain. *Biofizika (Russ.)* 20, 288–292.

Galimov, E. M., 1985. *The biological fractionation of isotopes*. Academic Press, London.

Hayes, J. M., Monson, K. D., 1982. Biosynthetic control of the natural abundance of carbon - 13 at specific positions within fatty acids in *Saccharomyces cerevisiae. J. Biol. Chem.* 257, 5568–5575.

Hayes, J. M., 2001. Fractionation of carbon and hydrogen isotopes in biosynthetic processes In: Review in mineralogy and geochemistry V.43 Stable isotope geochemistry. Valley, J. W., Cole, D. R. (eds.). Mineralogical Society of America. Washington, DC.

Ivanov, M. V., Zyakun, A. M., Gogotova, G. I., Bondar, V. A., 1978. Fractionation of carbon isotopes by photosynthesizing bacteria, grown on bicarbonate enriched in carbon –13. *Dokl. Akad. Nauk SSSR* (Russ.). 242**,** 1417-1420.

Igamberdiev, A. U., Ivlev, A. A., Bykova, N. V., Threlkeld, Ch., Lea, P. J., Gardestrom, P., 2001. Decarboxylation of glycine contributes to carbon isotope fractionation in photosynthetic organisms. *Photosynthesis Research*. 67, 177–184.

Igamberdiev, A. U., Mikkelsen, T. N., Ambus, P., Bauwe, H., Lea, P. J., Gardeström, P., 2004. Photorespiration contributes to stomatal regulation and carbon isotope fractionation: A study with barley, potato and *Arabidopsis* plants deficient in glycine decarboxylase. *Photosynthesis Research* 81, 139–152.

Ivlev, A. A., 1993. On the flows of light and heavy carbon during photosynthesis and photorespiration coupling. *Russ. J. Plant Physiol.* 40, 752–758.

Ivlev, A. A. 2011. Oscillatory character of carbon metabolism in photosynthesizing cell according to data on carbon isotope composition. *Uspekhi Sovr. Biology* (Russ.). 131, 178–193.

Ivlev, A. A., 2008. Isotope fractionation and cell mechanisms of carbon metabolism in photosynthesizing cell. RGAY–MSKhA of Timiryazev. Moscow.

Ivlev, A. A., 2010. Oscillatory character of carbon metabolism in photosynthesis. Arguments and facts. *Biol. Bull.* 37, 211–220.

Ivlev, A. A., Dubinski, Yu. A., 2011. On the nature of the light-induced component of dark respiration of plants. *Biophysics* 56, 679–686.

Ivlev, A. A., 2012. Oscillatory nature of metabolism and carbon isotope distribution in photosynthesizing cells Oscillatory nature of metabolism and carbon isotope distribution in photosynthesizing cells. In: Photosynthesis: fundamental aspects Najafpour, M. M. (ed.). Intech Publishers. Croatia. p. 341–366.

Ivlev, A. A., Kalinkina, L. G., 2001. Experimental evidence for the isotope effect in photorespiration. *Russ. J. Plant. Physiology* 48, 400–412.

Ivlev, A. A., Shnoll, S. E., 1978. About concept of "thermodynamic" carbon isotope distribution in biogenic compounds. *Geochmiya* (Russ.) 10, 1561–1569.

Ivlev, A. A., Bykova, N. V., Igamberdiev, A. U., 1996. Fractionation of carbon ($^{13}C/^{12}C$) isotopes in glycine decarboxylase reaction. *FEBS Letters*, 386, 174–176.

Ivlev, A. A., Igamberdiev, A. U., Threlkeld, Ch., Bykova, N. V., 1999. Carbon isotope effects in glycine decarboxylase reaction *in vitro* on mitochondria from pea and spinach. *Russ. J. Plant Physiol.*46, 653–660.

Ivlev A. A., Lapin, A. V., Brizanova, L. Ya., 1987. Distribution of carbon isotopes ($^{13}C/^{12}C$) in glucose of maize starch. *Fiziol. Rast.*34, 493–498.

Ivlev, A. A., Esikov, A. D., Knyazev, D. A., Matyushenko, I. N., 1990. Correlation between distribution of carbon isotopes in lysine and the pathways of its biosynthesis by Brevibacterium flavum and Corynebacterium glutamicum. *Microbiologya* (Russ.) 59, 373–381.

Ivlev, A. A., Knyazev, D. A., Matyushenko, I. N., Esikov, A. D., Balitskaya, R. M., Kuznetsova, N. N., 1991. The relationship carbon isotope distribution in glutamic acid and specific features of metabolism of *Corinebacterium glutamicum*. *Biofizika* 36, 365–372.

Kalinkina, L. G., Udel'nova, T. M., 1990. Effect of photorespiration on the stable carbon isotope fractionation in marine *Chlorella*. *Fiziol. Rast.* (Russ.) 37, 96–104.

Laing, W. A., Orgen, W. L., Hageman, R. H., 1974. Regulation of soybean net photosynthetic CO_2 fixation by interaction of CO_2, O_2 and ribulose-1,5-bisphosphate carboxylase. *Plant Physiology* 54, 678–685.

McFadden, B. A., 1974. The oxygenase activity of ribulose-1,5-bisphosphate carboxylase from *Rhodospirillum rubrum. Biochem. Biophys. Research Comm.* 60, 312–314.

Melander, L., Saunders, W. H., 1983. *Reaction rates of isotopic molecules.* Wiley Interscience Publ., New York.

Monson, K. D., Hayes, J. M., 1982. Carbon isotopic fractionation in the biosynthesis of bacterial fatty acids. Ozonolysis of unsaturated fatty acids as a means of determining the intramolecular distribution of carbon isotopes. *Geochim. et cosmochim. Acta.* 46, 139–149.

O'Leary, M. H., Yapp, C. J., 1978. Equilibrium carbon isotope effect on a decarboxylation reaction. *Biochem. Biophys. Res. Comm.* 80, 155–160.

Park, R., Epstein, S., 1960. Carbon isotope fractionation during photosynthesis. *Geochim et Cosmochim. Acta,* 21, 110–119.

Park, R., Epstein, S., 1961. Metabolic fractionation of $^{13}C/^{12}C$ in plants. *Plant. Physiol.* 36, 133–139.

Pye, K., 1969. Biochemical mechanisms underlying the metabolic oscillations in yeast. *Canad. J. Bot.* 47, 271–285.

Raven, J. A., Griffiths, H., Glidewell, S. M., Preston, T., 1982. The mechanism oxalate biosynthesis in higher plants: investigations with the stable isotopes oxygen-18 and carbon-13. *Proc. R. Soc. (London),* 216, 87–101.

Rinaldi, G. G., Meinschein, W. G., Hayes, J. M., 1974. Intramolecular carbon isotopic distribution in biologically produced acetoin acid. *Biomed mass-Spectrometry* 1, 415–417.

Rivera, E. R., Smith, B. N., 1979. Crystal morphology and $^{13}C/^{12}C$ composition of solid oxalate in Cacti. *Plant Physiol.* 64, 966–970.

Rossmann, A., Schmidt, H. L., Reniero, F., Verzini, G., Moussa, I., Merle, M. H., 1996. Stable carbon isotope content in ethanol of EC data bank wines from Italy, France and Germany. *Z. Lebensm.-Unters. Forsch.* 208, 293–301.

Rossmann, F., Butzenlechner, M., Schmidt, H. L., 1991. Evidence for nonstatistical carbon isotope distribution in natural glucose. *Plant Physiology.* 96, 609–614.

Roussel, M. R., Igamberdiev, A. Y., 2011. Dynamics and mechanisms of oscillatory photosynthesis. *BioSystems* 103, 230–238.

Roussel, M. R., Ivlev, A. A., Igamberdiev, A. Y., 2007. Oscillations of the internal CO_2 concentration in tobacco leaves transferred to low CO_2. *J. Plant Physiol.* 164. 1188–1196.

Sanadze, G. A., Black, C. C., Tevzadze, I. T., Tarkhnishvili, G. M., 1978. A change in the $^{13}CO_2/^{12}CO_2$ isotope ratio during photosynthesis by C_3 and C_4-plants. *Fiziol. Rast.* (Russ.) 25, 171–172.

Schmidt, H. L., 2003. Fundamentals and systematics of the non-statistical distributions of isotopes in natural compounds. *Naturwissenschaften,* 90, 537–552.

Schmidt, H. L., Kexel, H., Butzenlechner, M., Schwarz, S., Gleixner, G., Thimet S., Werner, R.A., Gensler, 1995. M.2. *Non-statistical isotope distribution in natural compounds: mirror of their biosynthesis and key for their origin assignment.* Wada, E., Yoneyama, T., Minagawa, M. Ando, T., Fry, B.D. (eds.), Kyoto: Kyoto University Press, pp. 17–35.

Sel'kov, E. E., 1978. Temporal organization of energy metabolism and cell clock. In: *Regulation of energy exchange and physiological state of organism.* Kondrashova (ed.), Nauka, Moscow, pp. 15–32,

Sel'kov, E. E., 1975. Stabilization of energy charge, oscillations and multiplicity of stationary states in energy metabolism as a result of purely stochoimetric regulation. *Europ J Biochem* 59, 151–157.

Sel'kov, E. E., Shevelev, E. A., 1989. The key role of fructose-1,6,bishosphate in temporal organization of carbohydrate metabolism. Autovibrational mathematical model. *Biofizika* (Russ.) 34, 797–803.

Sel'kov, E. E., Betz, A., 1973. On the mechanism of single-frequency glycolytic oscillations. In: Biological and Biochemical Oscillators. Chance, B. (ed.) New York. Acad. Press, p. 197–220.

Smith, B. N., Epstein, S., 1971. Two categories of $^{13}C/^{12}C$ ratios for higher plants. *Plant Physiol.* 47, 380–384.

Varshavsky, Ya. M., 1988. On distribution of carbon heavy isotope (^{13}C) in biological systems. *Biofizika* (Russ.) 33, 351–355.

Voglar, E. A., Hayes, J. M., 1979. Carbon isotopic composition of carboxyl groups of biosynthesized fatty acids. In: Advances in Organic Geochemistry. Pergamon Press, New York, p. 697–704.

Voznesenskii, V. L., Glagoleva, T. A., Zubkova, E. K., Mamushina, N. S., Filippova, L. A., Chulanovskaya, M. V., 1982. Metabolism of ^{14}C in *Chlorella* during prolonged cultivation in the presence of $^{14}CO_2$. *Fiziol. Rast.* (Russ.). 29, 564–571.

Weber, D., Kexel, H., Schmidt, H. L., 1997. ^{13}C-pattern of natural glycerol: origin and practical importance. *J. Agric. and Food Chem.* 45, 2042–2046.

In: Photosynthesis
Editor: Nafees Khan

ISBN: 978-1-63463-304-8

Chapter 5

PHOTOSYNTHESIS AND PHOTOINHIBITION IN AQUATIC PLANTS WITH DIFFERENT CARBON UPTAKE PHYSIOLOGY

Søren Laurentius Nielsen*
Department of Environmental, Social and Spatial Change
Roskilde University, Denmark

ABSTRACT

There is ample evidence in existing scientific literature that photorespiration acts towards reducing photoinhibition in plants. The internal generation of CO_2, resulting from the oxidation of ribulose bisphosphate by the enzyme RUBISCO acting as an oxygenase, is often sufficient to prevent photoinhibition partly or even completely, whereas photoinhibition is observed, when photorespiration is prevented. To prevent photorespiration in terrestrial plants it is, however, necessary to create a highly artificial composition of air that is replete in CO_2 and/or deplete in oxygen. The use of highly artificial and manipulated atmospheres makes it difficult to assess the environmental significance of photorespiratory alleviation of photoinhibition.

In contrast, aquatic plants naturally grow in an environment where carbon availability as well as ambient oxygen concentration varies from location to location as a result of varying carbonate content in the underlying soil as well as diurnally in the same location as a result of biological processes (photosynthesis and respiration). Probably as a result of the availability of free, dissolved, CO_2, often limiting for photosynthesis due to the low diffusion rate of CO_2 in water, aquatic plants have evolved a number of different physiological carbon accumulation strategies, including bicarbonate use, inducible C_4-type photosynthesis and CAM photosynthesis. This makes aquatic plants an obvious tool for the study of photorespiration/photoinhibition interactions.

In this chapter, I am presenting data from a study on photoinhibition in three aquatic plants with different carbon uptake strategies: *Callitriche cophocarpa*, a plant reportedly relying entirely on dissolved CO_2 in the water for photosynthesis, *Elodea canadensis*, a bicarbonate user, and *Crassula helmsii*, an aquatic CAM-plant, taking up CO_2 during the night, storing it as malate for photosynthetic use the following day. The plants were

* Corresponding author address: nielsen@ruc.dk

grown under varying carbon-availability, and were compared based on their photosynthetic oxygen evolution, degree of photoinhibition (evaluated using PAM-fluorometry), and their activity of antioxidative enzymes.

Photosynthetic oxygen evolution decreased with decreasing carbon availability in *Elodea* and *Callitriche*, but not in the aquatic CAM plant *Crassula*, where the internal malate pool partly compensates for low ambient carbon availability. Photoinhibition increased with increasing carbon availability (decreasing photorespiration) in both *Callitriche* and *Elodea*, but not in *Crassula*, where the internal malate pool may prevent any significant photorespiration as well as any significant photoinhibition. Photoinhibition was higher in the bicarbonate user *Elodea*, compared to *Callitriche*, as bicarbonate uptake functions as a carbon concentration mechanism, reducing photorespiration. The antioxidative enzyme ascorbate peroxidase evinced higher activities with lower carbon availability, especially in *Callitriche*, to a lesser degree in *Elodea* and not at all in *Crassula*.

The results confirm the previously described interaction between carbon uptake physiology and photoinhibition. Interestingly, CAM-photosynthesis may prevent both photorespiration and photoinhibition, but further studies are needed.

Keywords: Aquatic plants; Carbon uptake strategies; Photoinhibition; Photorespiration

INTRODUCTION

Photorespiration is intrinsically linked to the dual function of the central enzyme of the Calvin cyclus, RUBISCO, as both a carboxylase and an oxygenase. In leaf cells of terrestrial plants in a normal atmosphere of 380 ppm CO_2 and 21 % O_2, a large part of the assimilated carbon is channeled through the photorespiratory pathway in C_3 plants (Krause, Kirk, Heber, & Osmond, 1978). The lower the available CO_2 concentration at the active site of the RUBISCO enzyme, the more carbon is diverted to photorespiration as a consequence of the kinetic properties of RUBISCO. At the CO_2 compensation point, the endogenous generation of CO_2 by photorespiration balances the CO_2 fixation in photosynthesis. However, it has become evident that because the photosynthetic aparatus in plant leaves has little control over the influx of light, the supply of CO_2 and O_2 and temperature, the option to function inefficiently may be very important to plants (Osmond & Grace, 1995). It has thus been proposed that photorespiration protects leaves against photoinhibition by acting as a sink for reducing equivalents from the light process of photosynthesis as well as by preventing over-reduction of the electron carriers between photosystems II and I (Brestic, Cornic, Fryer, & Baker, 1995; Katona, Neimanis, Schonknecht, & Heber, 1992).

Photoinhibition has been defined as “the debilitating effect of high intensities of visible light on the photosynthetic capability of green organisms” (Kok, 1956). Today, we distinguish between chronic photoinhibition, which is the slowly reversible decline of quantum yield associated with loss of photosystem II activity, and dynamic photoinhibition, the rapidly reversibly decline in photosystem II quantum yield at light saturation, which is under physiological control (C.B. Osmond, 1994). Both types of photoinhibition are linked to the absorption of excessive photons by the photosynthetic apparatus of the leaf. Photorespiration, on the other hand, introduces adaptive inefficiencies in the Calvin cycle reactions of photosynthesis by stimulating non-asimilatory electron transport (Schreiber, Bilger, & Neubauer, 1994), meaning electron flow that does not result in CO_2 assimilation.

The recovery processes of photorespiration thus cause significant non-assimilatory electron flow that potentially mitigates photoinhibition (Osmond, 1981).

Although, as stated above, photorespiration does divert a significant amount of fixed carbon in C_3 plants even under a normal atmospheric composition, photorespiraton increases in significance under conditions of drought in terrestrial plants because the plants under these conditions reduce stomata opening, whch in turn lead to lower CO_2 conditions at the active site of RUBISCO (Noctor, Veljovic-Jovanovic, Driscoll, Novitskaya, & Foyer, 2002). In terrestrial plants, alternative photosynthetic pathways, C_4 and CAM photosynthesis, have evolved, which reduce or eliminate the effect of photorespiration. These plants are primarily found in drier and warmer climates, compared to normal C_3 plants (Ehleringer & Monson, 1993).

Aquatic plants are, for obvious reasons, never water stressed in their natural habitats. However, the aquatic environment does have a number of properties that often leads to the characterization of the aquatic habitat as inferior to plant growth, compared to the terrestrial habitat (Maberly & Spence, 1989).

One important factor is the diffusion coefficients for gasses, which are much lower in water than in air. The diffusion coefficient for CO_2 is thus 10^4 times lower in water compared to air (Proctor, 1982), and although the CO_2 concentration in air and in water in equilibrium with air is approximately the same (approx. 14 μM), CO_2 is virtually always a limiting factor for photosyntesis at atmosphere equlibrium. The effect of the lower diffusion rates of gasses in water are exacerbated by the higher viscosity of water compared to air, which means that boundary layers around aquatic plant leaves are much thicker than boundary layers around the leaves of terrestrial plants (Proctor, 1982). Fortunately for aquatic plants, the availability of free (dissolved) CO_2 in many aquatic habitats is much higher than it would be at atmospheric equlibrium. This is due to the influx of CO_2 rich ground water in streams (Rebsdorf, Thyssen, & Erlandsen, 1991), and the prevalence of respiratory processes in the bottom layer of water in lakes (Maberly, 1985; Nygaard & Sand-Jensen, 1981).

In addition, many species of aquatic plants have physiological adaptations, enhancing their uptake of inorganic carbon in the aquatic environment. These adaptations include inducible C_4 photosynthesis (Salvucci & Bowes, 1982), CAM photosynthesis (Baattrup-Pedersen & Madsen, 1999; Pedersen, Rich, Pulido, Cawthray, & Colmer, 2011) and, perhaps most importantly, use of bicarbonate (HCO_3^-) as a substrate in photosynthesis (Maberly & Spence, 1983). Bicarbonate constitutes 30% of the available dissolved inorganic carbon at pH 6.0 and 95% at pH 9.0, and the utilization of bicarbonate in photosynthesis therefore has great significance for aquatic plants in bicarbonate rich lakes and streams.

It has been documented that the presence of alternative carbon outake and utilization mechanisms reduce the occurrence of photorespiration in aquatic plants (Maberly & Spence, 1989; Pedersen et al., 2011; Salvucci & Bowes, 1981), but the interaction between carbon uptake physiology, photorespiration and photoinhibition is not well studied in these organisms. Given that photorespiration mitigates photoinhibition, as described above, it can be hypothesized that aquatic plants that posses alternative carbon uptake mechanisms, reducing photorespiration, will be more susceptible to photoinhibition than species not possessing such mechanisms. In this chapter, that hypothesis will be tested by studying the susceptibility to photoinhibition in three species of aquatic plants with different carbon uptake strategies: The C_3 plant *Callitriche cophocarpa* Sendtner, which has been shown to rely entirely on the uptake of free dissolved CO_2 (Madsen, 1991), *Elodea canadensis* L. C. Rich., a known bicarbonate user (Madsen & Sand-Jensen, 1987), and *Crassula helmsii* (Kirk) Cockayne, an aquatic CAM plant (Newman & Raven, 1995).

METHODS

Plant Material

Callitriche cophocarpa was collected from the stream Himmelev Bæk, immediately adjacent to Roskilde University (55° 39.383 N, 12° 8.196 E). *Elodea canadensis* was collected from the stream Pøleåen, 38 km north of the university (55° 59.508 N, 12° 13.771 E). *Crassula helmsii,* which is an invasive species in Denmark, was collected from a location on the island of Møn, 76 km south of the university (54° 56.758 N, 12° 17.129 E).

Growth Conditions

In the laboratory, the plants were grown in aquaria under an irradiance of 250 μmol m^{-2} s^{-1} (photosynthtically active radiation) in a 16 h : 8 h light/dark cycle at 15°C. Plant shoots were planted in mineral wool, that in itself is inert with respect to nutrient content. To ensure sufficient nutrient supply, both macro- and micronutrients were added to the rooting medium at regular intervals, using commercially available slow-release fertilizers (ASB-Grünland, Vallensbæk Strand, Denmark).

The water in the aquaria was a 50 : 50 mixture of tap water and demineralized water, which was replaced twice per week to keep the aquaria free of algal growth. The plants were grown for 14 days under two diffeent carbon-availablity treatments – high and low availability of free dissolved CO_2 – after which the measuring program was initiated. The availability of free CO_2 in the aquaria was controlled by adjusting the pH to 6.2 and 7.7, respectively, with CO_2 concentrations of 1.5 mM and 40 μM, as measured by Gran-titrations (Stumm & Morgan, 1996). The total concentration of dissolved inorganic carbon in the growth medium was approximately 2.2 mM. The pH, and thus the CO_2 concentration, were maintained by aerating the water with CO_2 from a gas cylinder controlled by an aquarium CO_2-fertilizing device (JBL Proflora vario 500; JBL, Neuhofen, Germany). Three plants of each of the three species were planted in each of three replicate aquarias of the two treatments.

Carbon Affinity

The carbon affinity of *Elodea* and *Callitriche* was measured on detached leaves in closed bottles mounted on a rotating wheel in a thermostatic incubator (15°C) with an irradiance of 390 μmol m^{-2} s^{-1}. The experimental water was aerated with atmospheric air before the experiment to bring it to equilibrium with atmospheric O_2 and CO_2. Carbon affinity was determined by measuring alkalinity and total carbon concentration in the bottles before and after at least 24 h of continous illumination. The carbon affinity is expressed as the ratio of the final concentration of total dissolved inorganic carbon (C_T) to initial alkalinity (Alk) (Maberly & Spence, 1983).

Crassula Malate Pools

The night time malate accumulation of the CAM-plant *Crassula* was determined by harvesting plant material in the morning. Approximately 0.8 g leaf material (fresh weight) was boiled in 10 ml demineralized water for 10 min. A subsample of a known volume was then titrated with 0.01 M NaOH to pH 8.3 (Klavsen & Maberly, 2009).

Photosynthetic Oxygen Production

To measure photosynthesis, individual leaves were incubated at 15°C and at eight incident light intensities, ranging from 0 to 500 μmol m^{-2} s^{-1}, provided by a Hansatech (King's Lynn, UK) LS2 high-intensity light source fitted with precalibrated neutral filters. Photosyntesis was measured as O_2 evolution in a stirred Hansatech DW3 leaf chamber fitted with a Clark-type O_2 electrode. The leaf chamber was temperature adjusted to 15°C with a circulating water jacket and stirred with a Hansatech magnetic follower and stirrer unit to reduce boundary layer effects. The leaf was held in place perpendicular to the direction of light inside the chamber by a piece of mesh inside the chamber on the shared side of the leaf. The water in the O_2 chamber was the same as used in the growth aquaria, so that photosynthesis was measured under conditions equivalent to those for growth. Photosynthesis was calculated as mmol O_2 produced m^{-2} leaf surface.

Chlorophyll Fluorescence (PAM) Measurements

Chlorophyll fluorescence parameters were measured with a Hansatech FMS-1 system. The leaves were placed in a dark-adaptation leaf-clip system at fixed working distance from the FMS-1 measuring fiber optics at a temperature of 15°C. Dark adaptation was obtained by application of a 5.5 s far-red 735 nm light pulse at 6 μmol m^{-2} s^{-1}, which ensured the full relaxation of PSII. F_0 was measured and immediately followed by a saturating light pulse provided by a broad spectrum halogen lamp (intensity 2400 μmol m^{-2} s^{-1}, duration 0.8 s) and simultaneous measurement of F_m. The maximum efficiency of dark-adapted PSII (F_v/F_m) was calculated as:

$$\frac{F_v}{F_m} = \frac{F_m - F_0}{F_m}$$ (Maxwell & Johnson, 2000).

Scavenging Enzymes

Superoxide dismutase (SOD) was analyzed using a SOD assay kit from Sigma-Aldrich (Sigma-Aldrich product no. 19160). The analysis is an indirect assay method based on xanthine oxidase and a water-soluble tetrazolium salt as color reagent. The color reagent is oxidized with superoxide in a reaction inhibited by SOD, the activity of which consequently can be determined colorometrically.

The guidelines enclosed with the asaay kit were followed without modifications. Ascorbate peroxidase (APX) was analysed colorimetrically in an assay in which the oxidation of ascorbate was followed by a decrease in absorbance at 290 nm (Chen & Asada, 1989; Nakano & Asada, 1981).

For both SOD and APX assays, 0.2 – 0.3 g leaf material (fresh weight) was homogenized in liquid nitrogen and extracted in the relevent buffers. The activity of both enzymes was expressed per dry mass, using a fresh weight / dry weight relationship determined individually for the three plant species.

Statistical Analysis

Due to the experimental setup, data analysis was performed as a nested analysis of variance (ANOVA) design (Sokal & Rohlf, 1995) with Tukey's test as a posthoc test (Sokal & Rohlf, 1995). The significance level was set to P = 0.05 in all cases. Data were tested for normal distribution using the Kolmogorov-Smirnov test (Sokal & Rohlf, 1995), and homogeneity of variance was analysed using Bartlett's test (Sokal & Rohlf, 1995). Curves were fitted to the photosynthesis – irradiance curves using tangent hyperbolar functions:

$$P = P_{\max} \tanh \frac{\Phi I}{P_{\max}},$$

where P is the photosynthesis, P_{max} the maximal achievable photosynthesis, Φ is the initial slope of the photosynthesis – irradiance curve, and I is the irradiance (Jassby & Platt, 1976). The fitted parameters are evaluated statistically based on their mean values ± 95% confidence intervals (CI).

RESULTS

Carbon Affinity and Malate Pools

Carbon affinity in *Elodea* clearly evince a large degree of plasticity (Figure 1). Under high availability of free CO_2, the C_T/Alk is in the range 0.89 – 0.92, which is indicative of CO_2 use only, with no contribution of bicarbonate. On the other hand, under low availability of free CO_2, the C_T/Alk is in the range 0.36 – 0.39, which indicates the uptake of bicarbonate under these conditions. The difference in C_T/Alk between the two treatments is highly significant (ANOVA, $p < 0.00001$). In *Callitriche*, there are no statistically significant differences in C_T/Alk between the two CO_2 – treatments (ANOVA, $p > 0.05$), so no plasticity in carbon uptake strategy as a consequence of CO_2 availability is seen. Regardless of free CO_2 availability, the C_T/Alk falls in the range 0.50 – 0.72 (Figure 1). Interestingly, this indicates either a very high affinity for free CO_2 or some contribution of bicarbonate for photosynthesis.

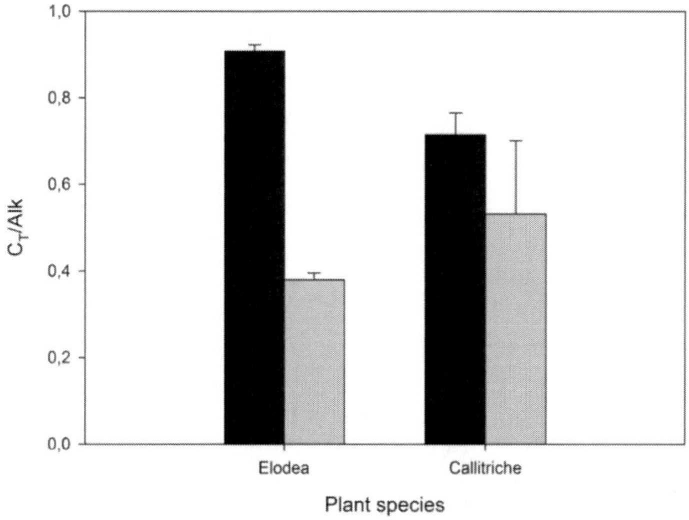

Figure 1. Ratio of the final concentration of total dissolved inorganic carbon (C_T) to initial alkalinity (Alk) as a measure of the carbon affinity of experimental plants. Black bars: High CO_2 availability, grey bars: low CO_2 availability. Error bars represent 1 SD.

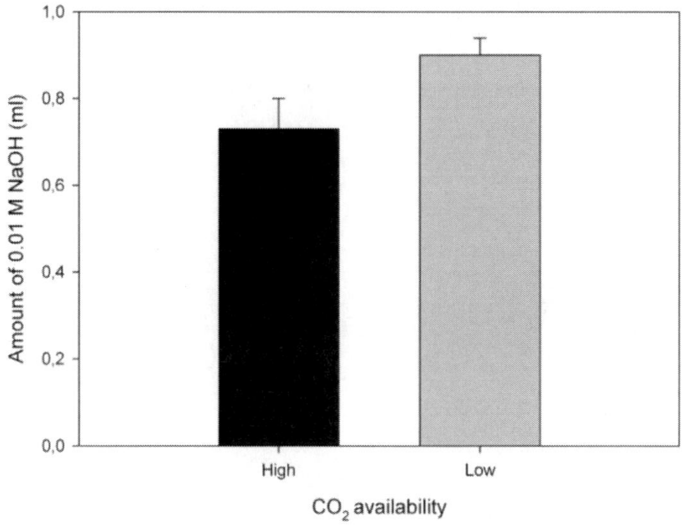

Figure 2. Malate pools in Crassula helmsii measured as the amount of 0.01 M NaOH added per 8 ml plant extract to achieve a pH of 8.3. Black bar: high free CO_2 availability, grey bar: low free CO_2 availability. Error bars represent 1 SD.

The malate pools built up in *Crassula* during night-time CO_2 uptake are relatively large, confirming it as a CAM-plant (Newman & Raven, 1995) (Figure 2). However, at high availability of free CO_2, the plants tended to accumulate less malate (corresponding to 0.73 ± 0.07 ml 0.01 M NaOH) than they did if the availability of free CO_2 was low, where malate accumulation corresponded to 0.90 ± 0.04 ml 0.01 M NaOH (Figure 2). This difference between treatments is statistically significant (ANOVA, $p = 0.023$).

Photosynthetic Oxygen Production

The photosynthesis – irradiance curves for the three species (Figure 3) clearly show that *Elodea,* under high availability of free CO_2, achieves a much higher maximal photosynthetic rate (P_{max}) compared to the same species under low availability of free CO_2 (Figure 3, top panel) as well as compared to the two other species, regardless of free CO_2 availability (Figure 3, middel and bottom panels).

The same difference in P_{max}, but less pronounced, is seen in *Callitriche* that also achieves a significantly higher P_{max} under high availability of free CO_2, compared to the P_{max} achieved under a low free CO_2 availability. In *Crassula* there is no significant difference between P_{max} as a consequence of CO_2 treatment. There are no differences between the initial slopes of the photosynthesis – irradiance curves in any of the three species as a consequence of varying availability of free CO_2. This indicates that the quantum yield of photosynthetic oxygen production (Φ) is independent of free CO_2 availability in all three species under the present experimental conditions. In addition, it can be seen that Φ does not differ significantly among the three species. The fitted photosynthetic parameters are summarized in Table 1.

Table 1. Fitted photosynthetic parameters of the three species investigated in this chapter. The parameters refer to the photosynthesis – irradiance curves in Figure 3

Species	Treatment	P_{max} (μmol O_2 m^{-2} s^{-1}) (mean ± 95 % CI)	Φ (μmol O_2 μmol^{-1} photons) (mean ± 95 % CI)
Elodea canadensis	High CO_2	4.77 ± 0.49	0.050 ± 0.013
	Low CO_2	2.74 ± 0.75	0.038 ± 0.029
Callitriche cophocarpa	High CO_2	2.50 ± 0.22	0.028 ± 0.006
	Low CO_2	1.92 ± 0.29	0.025 ± 0.011
Crassula helmsii	High CO_2	2.18 ± 0.36	0.021 ± 0.008
	Low CO_2	2.14 ± 0.53	0.023 ± 0.015

Maximum Quantum Yield (F_v/F_m)

The maximum quantum yields of the three plant species, measured using PAM chlorophyll fluorometry, are shown in Figure 4. A two-way ANOVA analysis revealed statistically significant effects of both plant species and CO_2 treatment ($p < 0.001$ in both cases).

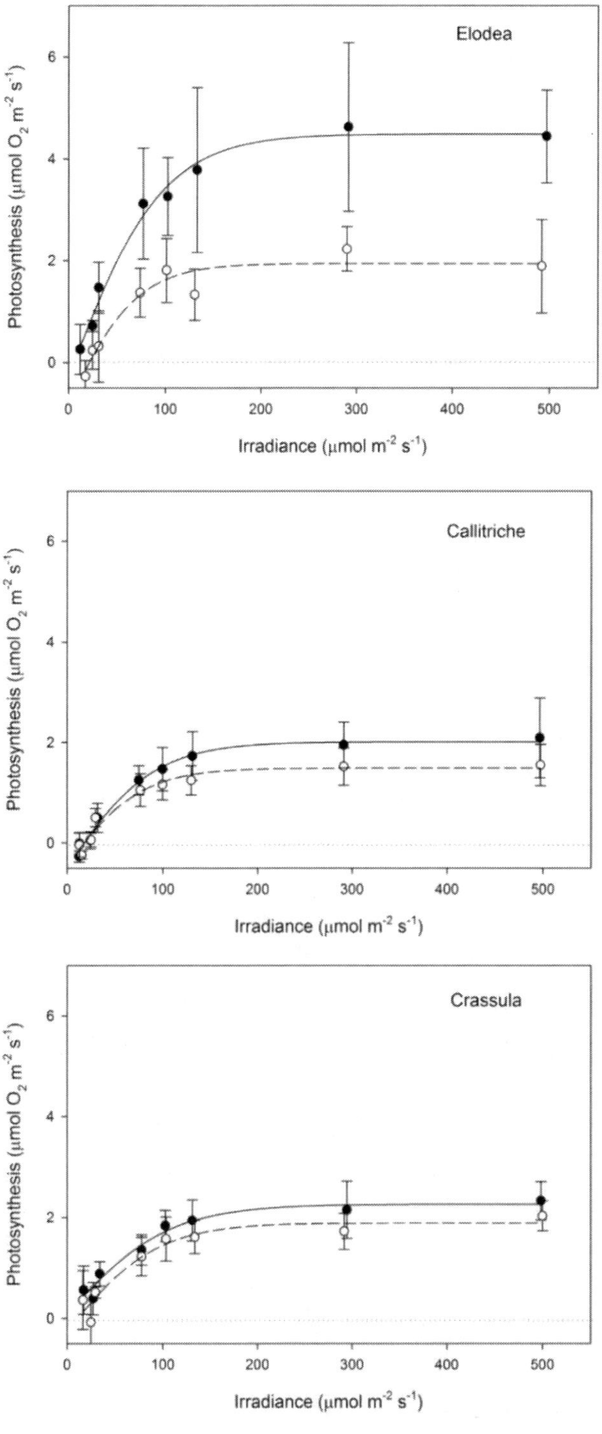

Figure 3. Photosynthesis – irradiance curves for the three plant species. Top panel: *Elodea canadensis*, middle panel: *Callitriche cophocarpa*, bottom panel: *Crassula helmsii*. Filled circles, solid lines: High CO_2 availability, open circles, dashed lines: low CO_2 availability. Error bars represent ± 1 SD. Fitted lines are tangent hyperbolas (Jassby & Platt, 1976).

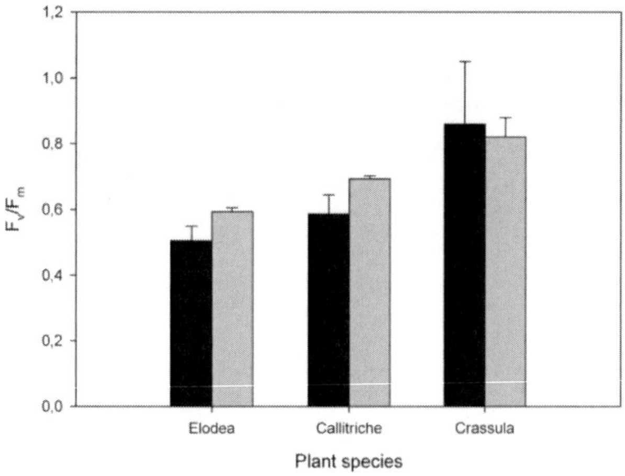

Figure 4. Maximum quantum yield of PSII in the three plant species under different availabilities of free CO_2. Black bars: High availability of free CO_2, grey bars: low availability of free CO_2. Error bars represent 1 SD.

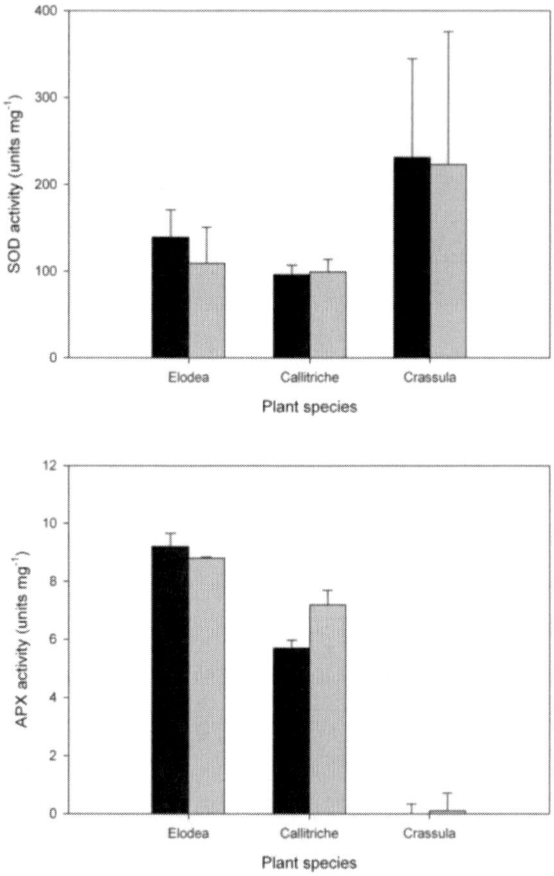

Figure 5. Content of scavenging (antioxidative) enzymes in the plants. Upper panel superoxide dismutase (SOD), lower panel ascorbate peroxidase (APX). Black bars: High availability of free CO_2, grey bars: low availability of free CO_2. Error bars represent 1 SD.

The ensuing Tukey's posthoc test showed that *Crassula* achieved significantly higher maximum quantum yields than the two other species (Figure 4, $p < 0.001$), but also that the maximum quantum yield in *Crassula* was independent of CO_2 availability (Figure 4, $p > 0.05$), with high values in the range 0.76 – 1.00. Comparisons among the other two species showed that the maximum quantum yield of *Elodea,* grown at a high availability of free CO_2, was the significantly lowest maximum quantum yield (Figure 4, $p < 0.001$), with values averaging 0.50, and that maximum quantum yield of *Callitriche* grown at low free CO_2 availability was significantly higher ($p < 0.001$), and the highest maximum quantum yield achieved by the two non-CAM-plants, with values averaging 0.69. Maximum quantum yields of *Elodea* grown under low free CO_2 availability, and of *Callitriche* grown at high free CO_2 availability, were intermediate, both averaging 0.59, were not significantly different from each other ($p > 0.05$) and also not significantly different from either values for *Elodea* grown at high free CO_2 availability or values for *Callitriche* grown at low free CO_2 availability ($p > 0.05$ in both cases).

Scavenging Enzymes

There is a non-significant trend for activities of superoxide dismutase (SOD) to be lower in *Callitriche* and higher in *Crassula* (Figure 5, upper panel), but the ANOVA analysis reveals no statistically significant differences among the species or as a consequence of availability of free CO_2 within species. It must be noted that the variability in SOD activity especially in *Crassula* is quite high. Values of SOD activity found in the plant material range from an average of 98 units mg^{-1} in *Callitriche* to 227 units mg^{-1} in *Crassula* (figure 5, upper panel).

Ascorbate peroxidase (APX) activity shows a completely different pattern (figure 5, lower panel). A two-way ANOVA analysis reveals significant effects of both plant species and free CO_2 availability. The Tukey's posthoc test indicated that *Elodea* contains significantly higher activities of APX (averaging 9.0 units mg^{-1}) than the two other species ($p < 0.001$), but there was no statistically significant difference within this species as a consequence of availability of free CO_2 ($p > 0.05$) (Figure 5, lower panel). *Callitriche* contain generally lower activities of APX than *Elodea* (Figure 5, lower panel, Tukey's test, $p < 0.001$), regardless of free CO_2 availability, but the Tukey's test also revealed that *Callitriche* grown under low availability of free CO_2 contained higher activities of APX (averaging 7.2 units mg^{-1}) than plants grown under high free CO_2 availability (averaging 5.7 units mg^{-1}) ($p < 0.001$). In *Crassula* measured activities of APX are not significally different from 0 in either free CO_2 treatment ($p > 0.05$, Figure 5, lower panel).

Conclusion

The bicarbonate user *Elodea* reveals a large plasticity in inorganic carbon affinity (Figure 1). Its affinity for inorganic dissolved carbon at high levels of free CO_2 is relatively low and does not show any indications of bicarbonate uptake under these conditions. This is in contrast to the situation at low levels of free CO_2, where the C_T/Alk values are much lower and indicative of bicarbonate uptake (Maberly & Spence, 1983). Similar plasticity in inorganic carbon affinity has previously been observed in this species (Madsen & Sand-Jensen, 1987). *Callitriche* shows a similar, but in this species non-significant, pattern since even in this species there is a tendency for a higher affinity for inorganic carbon under a low availability of free CO_2. *Callitriche* shows a surprisingly high affinity for inorganic carbon regardless of the level of free CO_2 availability during growth, which indicates either a very high affinity for free CO_2 or a low level of bicarbonate uptake, which has not been previously shown in this species (Olesen & Madsen, 2000). The data, howver, remain consistent with *Elodea* shifting from only free CO_2 uptake at high concentations of this carbon species to bicarbonate use at low concentrations of free CO_2, while *Callitriche* retains the same carbon uptake physiology regardless of carbon availability.

The CAM – physiology of photosynthesis previously found in *Crassula* (Newman & Raven, 1995) is confirmed in this study, as this species accumulates significant amounts of malate during night-time carbon uptake. The night-time malate pools of *Crassula* are obviously dependent on CO_2 availability (Figure 2), with higher pools building up in plants grown under low availability of free CO_2, where the likelihood of CO_2 becoming limiting for photosynthesis during the day is largest. This could indicate that *Crassula* is operating a form of facultative CAM photosynthesis (Neales, 1976).

Both *Elodea* and *Callitriche* attain higher maximum photosynthetic rates (P_{max}) and higher light saturation points at high availability of free CO_2 than they do at low availability (Figure 3). However, this significant tendency is much more pronounced in *Elodea*, where the P_{max} doubles as a consequence of higher CO_2 availability, than it does in *Callitriche*, where the increase is only about 25%. This large difference between these two species is surprising, given that *Elodea* switches to bicarbonate uptake under low free CO_2 availability, but apparently this is not sufficient to compensate for the lower availability of free CO_2. In contrast to these two species, *Crassula* have the same P_{max} and the same light saturation point regardless of free CO_2 concentration in the growth medium (Figure 3). This is probably an effect of the malate pools accumulated during night-time carbon uptake, which enables this species to compensate for varying carbon availability in the growth medium, giving it a competitive advantage under low availabilities of inorganic dissolved carbon (Klavsen & Maberly, 2009).

In both *Elodea* and *Callitriche* the expected relationship between carbon availability and photoinhibition is seen: At low availability of free CO_2, presumably corresponding to a higher photorespiration, the photoinhibition, measured as maximal PSII quantum yield, F_v/F_m, is lower (Figure 4). This relationship, as previously mentioned, presumably arises because photorespiration protects leaves against photoinhibition by acting as a sink for reducing equivalents from the light process of photosynthesis as well as by preventing over-reduction of the electron carriers between photosystems II and I (Brestic et al., 1995; Katona et al., 1992). The capability to utilize bicarbonate in photosynthesis apparently does not enable *Elodea* to compensate for a low availability of free CO_2, which is in accordance with the pattern displayed by photosynthetic oxygen production (Figure 3). Interestingly, the lowest photoinhibition (highest F_v/F_m, Figure 4) is seen in *Crassula*, which could have been expected to have the lowest photorespiration and thus the highest photoinhibition due to the accumulation of malate pools during the night and its CAM photosynthesis characteristics. Either the accumulated malate pools are too small to prevent significant photorespiration during the day time photosynthesis, which, however, does not seem to be in accordance with the data on photosynthetic oxygen production (Figure 3), or other mechanisms influence the relationship between photorespiration and photoinhibition in CAM plants, so that the CAM characteristics of photosynthesis protect the plants against photoinhibition, which has been observed in some terrestrial CAM plants (Herrera, 2009). The relationship between photorespiration and photoinhibition in aquatic CAM plants warrant further study.

Interestingly, the activity of the important antioxidative enzyme superoxide dismutase (SOD) does not show any relationship to carbon availability or any species related differences (Figure 5). In the litterature, an inverse relationship between the activity of antioxdative enzymes in plants and photoinhibition is often found (Nielsen & Simonsen, 2011) as the antioxidative enzymes are degraded as a consequence of a high fomation of reactive oxygen species resulting from photoinhibition (Hegedüs, Erdei, & Horváth, 2001). This phenomenon can be seen for ascorbate peroxidase (APX) activity in *Callitriche*, while interestingly APX in *Elodea* shows the opposite pattern (Figure 5). *Crassula* show very low activities of APX or none at all (Figure 5).

In conclusion, the expected relationships between carbon availability and photosynthesis physiology, acting through photorespiration, on one hand and photoinhibition on the other has been found in *Elodea* and *Callitriche*, confirming that photorespiration does protect plants against photoinhibition, also in aquatic plants, and that the significance of this protection is modified by the carbon uptake physiology of the plant. The results for *Crassula* strongly indicate that the ecological significance of CAM photosynthesis in aquatic plants warrant further study.

REFERENCES

Brestic, M., Cornic, G., Fryer, M.J., Baker, N.R., 1995. Does photorespiration protect the photosynthetic apparatus in French bean leaves from photoinhibition during drought stress? *Planta,* 196(3), 450-457.

Baattrup-Pedersen, A., Madsen, T.V., 1999. Interdependence of CO_2 and inorganic nitrogen on crassulacean acid metabolism and efficiency of nitrogen use by *Littorella uniflora* (L.) Aschers. *Plant, Cell and Environment,* 22, 535-542.

Chen, G.X., Asada, K., 1989. Ascorbate peroxidase in tea leaves: Occurence of two isozymes and the differences in their enzymatic and molecular properties. *Plant & Cell Physiology,* 30(7), 987-998.

Ehleringer, J.R., Monson, R.K., 1993. Evolutionary and ecological aspects of photosynthetic pathway variation. *Annu. Rev. Ecol. Syst.,* 24, 411-439.

Hegedüs, A., Erdei, S., Horváth, G., 2001. Comparative studies of H_2O_2 detoxifying enzymes in green and greening barley seedlings under cadmium stress. *Plant Science,* 160, 1085-1093.

Herrera, A., 2009. Crassulacean acid metabolism and fitness under water deficit stress: if not for carbon gain, what is facultative CAM good for? *Annals of Botany,* 103(4), 645-653.

Jassby, A.D., Platt, T., 1976. Mathematical formulation of the relationship between photosynthesis and light for phytoplankton. *Limnol. Oceanogr.,* 21, 540-547.

Katona, E., Neimanis, S., Schonknecht, G., Heber, U., 1992. Photosystem I dependent cyclic electron transport is important in controlling photosystem II activity in leaves under conditions of water stress. *Photosynthesis Research,* 34(3), 449-464.

Klavsen, S.K., Maberly, S.C., 2009. Crassulacean acid metabolism contributes significantly to the in situ carbon budget in a population of the invasive aquatic macrophyte Crassula helmsii. *Freshwater Biology,* 54(1), 105-118.

Kok, B., 1956. On the inhibition of photosynthesis by intense light. *Biochimica Et Biophysica Acta,* 21(2), 234-244.

Krause, G.H., Kirk, M., Heber, U., Osmond, C.B., 1978). O_2 - dependent inhibition of photosynthetic capacity in intact isolated chloroplasts and isolated cells from spinach leaves illuminated in absence of CO_2. *Planta,* 142(2), 229-233.

Maberly, S.C., 1985. Photosynthesis by *Fontinalis antipyretica*, II: Assessment of environmental factors limiting photosynthesis and production. *New Phytol.,* 100, 141-155.

Maberly, S.C., Spence, D.H.N., 1983. Photosynthetic inorganic carbon use by freshwater plants. *J. Ecol.,* 71, 705-724.

Maberly, S.C., Spence, D.H.N., 1989. Photosynthesis and photorespiration in freshwater organisms: Amphibious plants. *Aquat. Bot.,* 34, 267-286.

Madsen, T.V., 1991. Inorganic carbon uptake kinetics of the stream macrophyte *Callitriche cophocarpa* Sendt. *Aquat. Bot.,* 40, 321-332.

Madsen, T.V., Sand-Jensen, K., 1987. Photosynthetic capacity, bicarbonate affinity and growth of *Elodea canadensis* Michaux. exposed to different concentrations of inorganic carbon. *Oikos,* 50(2), 176-182.

Maxwell, K., Johnson, G.N., 2000. Chlorophyll fluorescence - a practical guide. *Journal of Experimental Botany,* 51(345), 659-668.

Nakano, Y., Asada, K., 1981. Hydrogen peroxide is scavenged by ascorbate-specific peroxidase in spinach chloroplasts. *Plant & Cell Physiology,* 22(5), 867-880.

Neales, T.F., 1976. The gas exchange patterns of CAM plants. In R. H. Burris & C. C. Black (Eds.), *CO_2 metabolism and plant productivity* (pp. 299-310). Baltimore: University Park Press.

Newman, J.R., Raven, J.A., 1995. Photosynthetic carbon assimilation by *Crassula helmsii. Oecologia,* 101, 494-499.

Nielsen, S. L., Simonsen, A.M., 2011. Photosynthesis and photoinhibition in two differently coloured varieties of Oxalis triangularis - the effect of anthocyanin content. *Photosynthetica,* 49(3), 346-352.

Noctor, G., Veljovic-Jovanovic, S., Driscoll, S., Novitskaya, L., Foyer, C.H., 2002. Drought and oxidative load in the leaves of C-3 plants: a predominant role for photorespiration? *Annals of Botany,* 89, 841-850.

Nygaard, G., Sand-Jensen, K., 1981. Light climate and metabolism of *Nitella flexilis* (L.) Ag. in the bottom waters of oligotrophic Lake Grane Langsø, Denmark. *Int. Revue ges. Hydrobiol.,* 66(5), 685-699.

Olesen, B., Madsen, T.V., 2000. Growth and physiological acclimation to temperature and inorganic carbon availability by two submerged aquatic macrophyte species, *Callitriche cophocarpa* and *Elodea canadensis. Functional Ecology,* 14, 252-260.

Osmond, C.B., 1981. Photorespiration and photoinhibition. Some implications for the energetics of photosynthesis. *Biochimica Et Biophysica Acta,* 639(2), 77-98.

Osmond, C.B., 1994. What is photoinhibition? Some insights from comparisons of shade and sun plants. In N. R. Baker & J. R. Bowyer (Eds.), *Photoinhibition of photosynthesis from molecular mechanisms to the field* (pp. 1-24). Oxford: BIOS Scientific Publishers.

Osmond, C.B., Grace, S.C., 1995. Perspectives on photoinhibition and photorespiration in the field: quintessential inefficiencies of the light and dark reactions of photosynthesis? *Journal of Experimental Botany,* 46(Special Issue), 1351-1362.

Pedersen, O., Rich, S.M., Pulido, C., Cawthray, G.R., Colmer, T.D., 2011. Crassulacean acid metabolism enhances underwater photosynthesis and diminishes photorespiration in the aquatic plant Isoetes australis. *New Phytologist,* 190(2), 332-339.

Proctor, M.C.F., 1982. Physiological Ecology: Water relations, light and temperature responses, carbon balance. In: Smith, A.J.E. (Ed.), *Bryophyte ecology* (pp. 333-381). London: Chapman & Hall.

Rebsdorf, A., Thyssen, N., Erlandsen, M., 1991. Regional and temporal variation in pH, alkalinity and carbon dioxide in Danish streams, related to soil type and land use. *Freshwat. Biol.,* 25, 419-435.

Salvucci, M.E., Bowes, G., 1981. Induction of reduced photorespiratory activity in submersed and amphibious aquatic macrophytes. *Plant Physiol.,* 67, 335-340.

Salvucci, M.E., Bowes, G., 1982. Photosynthetic and photorespiratory response of the aerial and submerged leaves of *Myriophyllum brasiliense*. *Aquat. Bot.,* 13, 147-164.

Schreiber, U., Bilger, W., Neubauer, C., 1994. Chlorophyll fluorescence as a non-intrusive indicator for rapid assessment of *in vivo*photosynthesis. In: Schulze, E.D., Caldwell, M.M. (Eds.), *Ecophysiology of photosynthesis* (pp. 49-70). Berlin: Springer-Verlag.

Sokal, R.R., Rohlf, F.J., 1995. *Biometry* (3ed.). New York: W.H. Freeman and Company.

Stumm, W., Morgan, J.J., 1996. *Aquatic chemistry* (3ed.). New York: Wiley Interscience.

In: Photosynthesis
Editor: Nafees Khan
ISBN: 978-1-63463-304-8

Chapter 6

STOMATAL AND PHOTOSYNTHETIC RESPONSES TO FLOODING

Mohammad Mobin*, Mohammad Nasir Khan and Zahid bin Khorshid Abdukhaliq Nourabbas
Department of Biology, Faculty of Science, University of Tabuk, Tabuk, Saudi Arabia

ABSTRACT

Geological records indicate that plant life on the land has aquatic origins, started approximately 460 million years ago. However, nearly all of the land-dwelling plants are very sensitive to the inundation. Nonetheless, the most basic attributes of flooding tolerance seemingly involved an adaptive *return* to aquatic competence by terrestrial plants. The negative impact of flooding on terrestrial plant life is a consequence of oxygen deprivation. Flooding often disrupts plant water relations, gas exchange and stomatal conductance. Under the diffusional limitation of carbon uptake imposed by stomatal closure, the photosynthetic capacity of flood-sensitive plants would be significantly declined and can exhibit non-stomatal inhibition of photosynthesis with alterations in plant hormone concentrations induced by tissue anoxia. A tight regulation of the plant water status is thus required to maintain photosynthesis, nutrient uptake and growth in flooded environments. Here, we review the importance of stomatal attributes and photosynthesis for the survival of flooded land-dwelling plants.

INTRODUCTION

Flooding or waterlogging can be defined as the presence of excess water in the root environment of land plants. About 12% of the world's soils have excess water problems (Dudal 1976). Heavy rainfalls in imperfectly drained soils produce transient flooding in many areas of seasonal rainfed agriculture in the tropics (Sandhu et al. 1986). According to Van

* Corresponding Author, Email: mohammad_mobin@hotmail.com.

Toai et al. (2001), the flooding can be classified into two groups: (i) stream flooding, characterized by the overflow of rivers or creeks into a flood plain and (ii) lowland flooding, characterized by inadequate surface drainage and slow soil permeability of areas with depressions.

Roots; the hidden half of the plants has been considered as not only the organ for absorption of minerals and water but also as important sensors that perceive signals and regulate whole plant growth (Aiken and Smucker 1996). Since submergence takes place mainly at root-level, alterations in oxygen availability have a huge impact on root biochemistry and metabolism. Therefore, signals to the plant shoot, leaf, flower and fruit can change in response to excess water and govern whole-plant processes (Drew 1997).

The most significant effect of soil flooding is a massive slow-down of the exchange of oxygen, as gas diffusion rates are 10,000 times slower in water than in air (Jackson 1985). Under aerobic conditions, the plant cells cycles 1 mol of glucose to yield a maximum of 38 mol of ATP via glycolysis, Krebs' cycle, and mitochondrial electron transport. However, when soil becomes waterlogged, the molecular oxygen is unavailable (i.e., O_2 partial pressure is the limiting factor), and generation of ATP by the cytochrome chain of mitochondria is depleted. The state of such debilitated capacity of ATP generation is known as hypoxia (Drew 1997). As hypoxia proceeds further, fermentation takes place in response to NADH accumulation which suppresses the Krebs' cycle. Consequently, anoxia occurs; which is a point at which the ATP produced by aerobic glycolysis becomes negligible in comparison to the ATP produced by fermentation.

Flood events can damage the germinating seedlings through several processes including substrate saturation and root hypoxia, inhibition of photosynthetic carbohydrate generation, erosion of sediments or plants, or burial by sediments. Physiological damage during inundation includes suppressed shoot and root growth, delayed leaf formation, accelerated senescence, and root decay (Kozlowski 2002).

In this way, flood associated anoxic environment triggers an array of alterations that include modifications of physiological events along with anatomical and morphological arrangements linked to their survival under anaerobic conditions. In this article we have focused on the impact of flooding on soil conditions, flood tolerance strategies and photosynthetic responses.

FLOOD-INDUCED CHANGES IN THE SOIL

When soil is flooded, gas-filled soil pores are replaced with water, resulting in the elimination of soil oxygen, except within a few millimeters at the soil-water interface (Kozlowski 1997). In non-flooded soils, oxygen and carbon dioxide, as well as other gasses, are maintained in the soil through gas diffusion. Upon flooding, this diffusion process ceases and gas transport is replaced by molecular diffusion from the air through the flood water into the soil profile. This type of diffusion is 10^4 times slower and less efficient than through air, resulting in a drop of oxygen flow into the soil (Ponnamperuma 1984). Molecular diffusion rates are further decreased with increased water depth and/or reduced rates of water flow. As oxygen from the air is depleted, it is also consumed by plant roots and soil microorganisms which depend on aerobic respiration. Subsequently, aerobic soil microbes are rapidly replaced in flooded soils by anaerobic bacteria, which scavenge essentially all available soil oxygen (Kozlowski 2002). During flooding, the by-products of anaerobic respiration such as methane, acetic acid butyric acid, and ethylene may turn toxic within the soil profile. The concentration of these by-products relies on different factors such as soil temperature and the availability of carbon and other specific mineral substrates required for anaerobic respiration (Drew 1991).

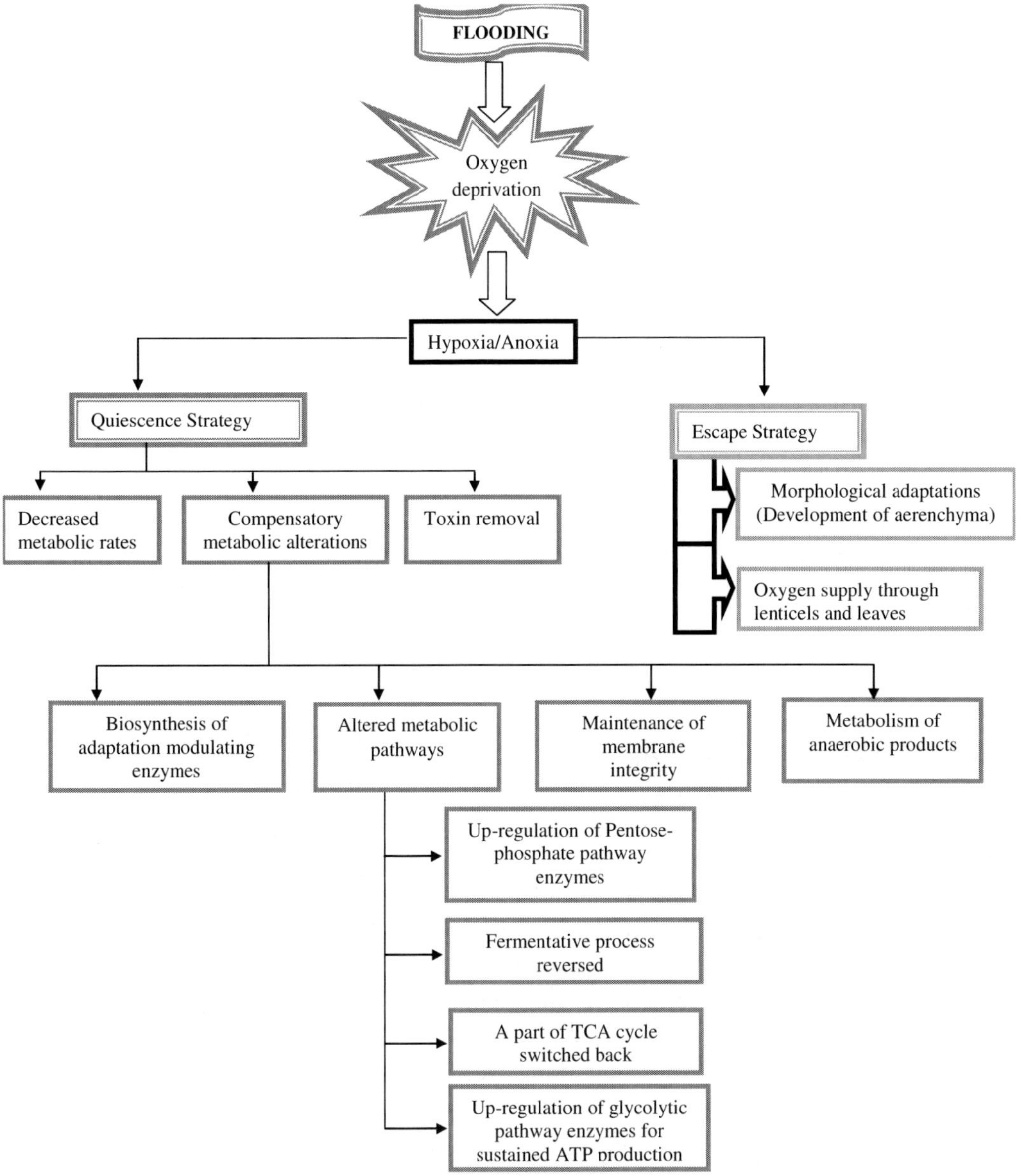

Figure 1. Flood tolerance strategies: Flood tolerant plant species demonstrated either quiescence strategy characterized by restrained growth and cellular metabolism, or escape strategy facilitated by the development of aerenchyma and adventitious roots, alterations in leaf cell ultrastructure and chloroplast positioning to enhance photosynthesis underwater and the maintenance of gas films on submerged leaves.

The effects of flooding on soil conditions can be quantified by using soil redox potential (Eh), an indicator of the severity of the flooding on a number of soil processes. Redox potential is a measure of the tendency to accept or donate electrons in a soil. It is normally measured in millivolts and standard values can range from -300 to +700 mV. Well aerated soils are > 400 mV, while non-aerated soils are < 350 mV (Pezeshki 2001). Eh values ranges from -250 to -300 mV in waterlogged soils. Specific chemical reactions can be predicted based on specific redox levels. As soil pH increases, Eh values decreases. For example, the reduction of FeOOH to Fe^{2+} occurs at 65 mV in a soil with pH4, while the same reaction would shift to 25 mV in a soil at pH 6 (Brady and Weil 2002). Reduced soil Eh values are significantly correlated with both reduced plant gas exchange *gs* and reduced net photosynthetic rates in flooded oak seedlings (Anderson and Pezeshki 1999).

In flooded soil, the conversion of organic nitrogen to inorganic forms (ammonification) terminates with ammonium (NH_4^+) due to a lack of oxygen. Any nitrate that is present within the soil at the onset of flooding will be highly unstable and may ultimately lost as either NO or N_2through de-nitrification. The rate of de-nitrification is dependent on the flooding interval, soil organic matter content and soil temperature. High soil temperatures, alternating flooding patterns and high levels of soil organic matter increases the denitrification rates (Ponnamperuma 1984). For ammonium, some will be immobilized due to chemical fixation within the flooded soil, depending on pH and soil texture. In addition, some ammonium will also be immobilized by anaerobic soil bacteria, especially at the onset of flooding. Additionally, nitrogen can be lost from a flooded soil through volatilization. Finally, some N_2 is fixed in flooded soils in the form of ammonium by blue-green algae and/or anaerobic bacteria, depending upon the soil pH and organic matter content (Ponnamperuma 1984).

Flood Tolerance Strategies

In the last three decades considerable progress has been made in our understanding of the mechanisms that enable certain plant species and cultivars to withstand periods with excess soil water, or even complete submergence. Partial to complete submergence of aerial organs may deplete the oxygen levels, diminish the rates of photosynthesis and promote ethylene entrapment in plants. Under partially or completely inundated conditions, the flood tolerant plant species showed either quiescence or escape strategy. (i) Quiescence strategy: ability of plant to endure submergence events for prolonged periods of time; characterized by restrained growth and cellular metabolism (Bailey-Serres and Voesenek 2008). (ii) Escape strategy: ability of plants to adapt to partial submergence, facilitated by the development of aerenchyma and adventitious roots (Voesenek et al. 2006), alterations in leaf cell ultrastructure and chloroplast positioning to enhance photosynthesis underwater (Mommer et al. 2005; Voesenek et al. 2006), and the maintenance of gas films on submerged leaves (Pedersen et al. 2009; Winkel et al. 2011).

The underlying advantage of these adaptive strategies is the better maintenance of oxygen in submerged cells or tissues during submergence. The activation of both the quiescence and escape responses largely depend upon the gaseous plant hormone ethylene, which is entrapped in cells by submergence and often further produced during the inundation. Ethylene regulates or interplays with the plant hormones abscisic acid (ABA), gibberellin (GA), and auxin. Through complex regulatory mechanisms these compounds control the metabolic, morphological and anatomical changes manifested during submergence.

Leaf Responses to Flooding

Leaf responses to flooding include the closing of stomata, epinasty, senescence, abscission, ability or inability to maintain leaf canopy, reduction of stomatal conductance (gs) and net CO_2 assimilation (A), reduction in leaf and stem water potential (ψl and ψs), nutrient deficiencies, and buildup of leaf carbohydrate concentrations (Kozlowski 1997). Some of these responses are due to a decrease in water uptake from the roots, and a reduced ability to translocate photosynthate from source (leaf) to sink (roots or other storage structures). Additionally, electrolyte leakage has also been noticed in the needles of black spruce (*Picea mariana*) and tamarack (*Larix laricina*) when exposed to flooding indicated that flood stress can damage cell membrane function in the leaves (Islam et al. 2003). Moreover, the closure of stomata seems to be an early response involved in the reduction of net photosynthetic rate as a result of flooding. For mango, avocado, banana, citrus, and *Annona* spp., A, gs, and transpiration (E) significantly decreased as early as 1 to 3 d after onset of flooding, and even at some instances internal leaf CO_2 concentration (Ci) values was increased (Larson et al. 1991a; 1991c; Nuñez-Elisea et al. 1998).

When roots are submerged, the synthesis and translocation of cytokinin (CK) from roots to leaves is reduced (Reid and Railton 1974). Cytokinin is known to help keep stomata open.

Abscisic acid (ABA), which is synthesized in the root and leaf is known to be involved in the stomatal closure and the ratio of ABA: CK provides a critical signal to the leaf stomata (Else et al. 1996). However, under flood conditions, studies have shown that it may not be an increase in root production of ABA, which determines the stomatal closure, but rather the ABA present in the apoplast adjacent to the guard cells regulates the stomatal movement (Else et al. 1996). But in flooded avocado trees, Gil et al. (2009) did not find any relation between alterations in root or leaf ABA activity and stomatal closure.

Effects of Flooding on Water Potential

Water potential is the measure of free energy found in a given volume of water, generally measured in pascals, a pressure unit (Taiz and Zeiger 2002). Its components are solute potential, pressure potential, and matric potential. Understanding water potential can be a useful tool for understanding the soil-plant-atmosphere continuum, and water status/health of the plant. For example, as water evaporates into the mesophyll, the air spaces of the leaf leading to transpiration via the stomata, more water is drawn by cohesion through the leaf xylem, thus drawing more water up through the petiole, stem, and trunk xylem. Besides bulk flow, roots are also capable of absorbing ions from the soil solution, building up solute or osmotic potential inside the root xylem, and causing the movement of water into the roots. The path of water into the root involves either the symplastic or apoplastic routes. Measurements of stem water potential may indicate the ability of the plant to take up water. If the stem water potential remains relatively high, then the roots are probably still functioning sufficiently to allow water to be drawn up by the plant. If the stem water potential drops, then it is likely that the root function has been negatively impacted by hypoxic or anoxic conditions, compromising the root cells' ability to function properly. If stem water potential declines very rapidly, then it is likely that the leaves will undergo necrosis, wilt, turn brown and dry, as opposed to undergoing the physiological processes involved with senescence, chlorosis, and abscission. If the stem water potential does not drop too rapidly, the plant may have sufficient time to respond to flooding conditions via hormonal and other signaling mechanisms. It has been noted that stem water potential after 9 d of flooding in citrus remain unaltered (García-Sánchez et al. 2007). Flooding of mango in containers for 14 d did not have any impact on the leaf water potential of flooded plants in comparison to the non-flooded plants, although A did decrease within 7 d (Larson 1991a).

Flooding Responses to Chlorophyll Content

According to Boru et al. (2001) the leaf chlorosis could be used as a measure of waterlogging tolerance. Zhou and Lin (1995) reported that root flooding was responsible for decreases in chlorophyll content at various growth stages and that the most significant effects were observed in the leaves at the base. Furthermore, the degradation of chlorophyll proceeds more intensively in those leaves that is closer to the flooded roots. Losses of chlorophyll and leaf chlorosis are stress symptoms observed in tropical or temperate tree species subjected to soil flooding (Gardiner and Krauss 2001). Leaf chlorosis as a result of waterlogging was observed in pea and maize (Przywara and Stepniewski 1999) and lucerne (Smethurst and Shabala 2003). Decreases in chlorophyll content of 17% and 35% were reported by Gardiner and Krauss (2001) for *Quercus pagoda* seedlings subjected to soil flooding under partial (27%) and full sunlight, respectively.

Stomatal Limitation of Photosynthesis in Response to Flooding

Stomatal limitations on photosynthesis is executed through a decrease in the internal CO_2 concentration (Ci) in leaf (Farquhar and Sharkey 1982). The *Ci/Ca* ratio is considered to be an appropriate indicator for the stomatal limitation of photosynthesis. The smaller the *Ci/Ca* ratio, the greater is the stomatal limitation of photosynthesis. A reduction in CO_2 availability decreased the electron sink in chloroplasts, which may result in excess available light energy (Allen 1995). Efficiency of photosynthesis can be assessed by following the fate of absorbed light energy in plant. This is particularly important in conditions where CO_2 fixation is reduced relative to light harvesting as is the case with flood induced stomatal closure (Demmig-Adams and Adams 1992). As is evident in flood induced stomatal closure where decline in photosynthetic rate could be attributed to lower concentrations of CO_2 present in the air spaces of the leaf mesophyll and restriction on the replacements of CO_2 from ambient air through closed stomata. This phenomenon has been evidenced as an early plant response to flooding in numerous fruit crops such as mango, avocado, banana, citrus, and *Annona* spp. (Larson et al. 1991a; 1991c; Núñez-Elisea et al. 1998). However, this does not explain all observed leaf gas exchange responses, particularly over extended periods of flooding.

Non-stomatal Limitations of Photosynthesis in Response to Flooding

Non-stomatal limitations of photosynthesis include down-regulation of photosynthetic enzymes (Soares-Cordeiro et al. 2009), low foliar nitrogen concentrations (Heckathorn et al. 1997) or photodynamic damage by excess light, reduced carboxylation efficiency and decreased RuBP regeneration. As discussed above, ATP and NADPH is required for the reduction reaction, and the regeneration of RuBP. Thus, if any part of the thylakoid reactions is disrupted, ATP regeneration would be hampered due to the curtailed proton gradient across the thylakoid membrane. Predominantly due to depleted ATP levels, RuBP regeneration would thus be reduced (Lawlor 2002). Reduced relative water content in the leaves have important consequences leading to the depleted ATP synthesis, due to increased relative concentrations of Mg^{2+} in the chloroplast as relative water content decreases (Lawlor 2002). In seedlings of the non-tolerant species *Nothofagus menziesii* and *Nothofagus solandri*, A, gs and non-stomatal conductance, i.e., mesophyll conductance plus carboxylation efficiency (CE) decreased markedly in response to waterlogging (Sun et al. 1995). However, in the flood-tolerant tropical species, *Pouteria orinocoensis*, experimental flooding of seedlings to complete submergence decreased CE by 70% and Asat by 61% relative to drained seedlings (Ferna´ndez 2006).

In general survival to flooding includes one or more of the following adaptive responses; re-opening of stomata, recovery of stomatal conductance, independence of photosynthesis and stomatal conductance from flooding, ability to transport oxygen internally to submerged tissues, a high capacity to respire anaerobically, maintain glycolytic metabolism and the capacity to avoid oxidative damage from excess light energy not used in carbon fixation.

REFERENCES

Aiken, R. M., Smucker, A.J.M., 1996. Root system regulation of whole plant growth. *Annu. Rev. Phytopathol.* 34, 325-346.

Allen, R.D., 1995. Dissection of oxidative stress tolerance using transgenic plants. *Plant Physiol.* 107, 1049-1054.

Anderson, P.H., Pezeshki SR., 1999. The effects of intermittent flooding on seedlings of three forest species. *Photosynthetica* 37, 543-552.

Bailey-Serres, J., Voesenek, L.A.C.J., 2008. Flooding stress: acclimations and genetic diversity. *Ann. Rev. Plant Biol.* 59, 313–339.

Boru, G., van Ginkel, M., Kronstad, W.E., Boersma, L., (2001) Expression and inheritance of tolerance to waterlogging stress in wheat. *Euphytica* 117, 91-98.

Brady, N.C., Weil, R.R., 2002. The nature and properties of soils. 13th ed. Prentice Hall, Inc. Upper Saddle River, NJ.

Demmig, A.B., Adams, W.W., 1992. Photoprotection and other responses of plants to high light stress. *Ann. Rev. Plant Physiol. Plant Mol. Biol.* 43, 599-626.

Drew, M.C., Stolzy L.H., 1991. Growth under oxygen stress, In Ed. Waisel, Y., Eshel, A., Kafkafi, U. Plant roots; the hidden half. *Marcel Dekker, New York* 331-350.

Drew, M.C., 1997. Oxygen deficiency and root metabolism: Injury and acclimation under hypoxia and anoxia. *Annu. Rev. Plant Physiol. Plant Mol. Biol.* 48, 223-250.

Dudal, R., 1976. Inventory of the major soils of the world with special reference to mineral stress hazards. Ed. M.J. Wright. Plant adaptation to mineral stress in problem soils. *Cornell University Press, Ithaca, New York*. 3-13.

Else, M.A., Tiekstra, A.E., Croker, S.J., Davies, W.J., Jackson, M.B., 1996. Stomatal closure in flooded tomato plants involves abscisic acid and a chemically unidentified anti-transpirant in xylem sap. *Plant Physiol.* 112, 239-247.

Farquhar, G.D., Sharkey, T.D., 1982. Stomatal conductance and photosynthesis. *Annu. Rev. Plant Physiol.* 33, 317-345.

Fernández, M.D., 2006. Changes in photosynthesis and fluorescence in response to flooding in emerged and submerged leaves of *Pouteria orinocoensis*. *Photosynthetica* 44, 32-38.

García-Sánchez, F., Syvertsen, J.P., Gimeno, V., Botía, P., Perez-Perez, J.G., 2007. Responses to flooding and drought stress by two citrus rootstock seedlings with different water-use efficiency. *Physiol. Plant.* 130, 532-542.

Gardiner, E.S., Krauss, K.W., 2001. Photosynthetic light response of flooded cherrybark oak (*Quercus pagoda*) seedlings grown in two light regimes. *Tree Physiol.* 21, 1103-1111.

Gil, P.M., Gurovich, L., Schaffer, B., García, N., Iturriaga, R., 2009. Electrical signaling, stomatal conductance, ABA, and ethylene content in avocado trees in response to root hypoxia. *Plant Signal. Behav.* 4, 100-108.

Heckathorn, S.A., DeLucia, E.H., Zielinski, R.E., 1997. The contribution of drought related decreases in foliar nitrogen concentration to decreases in photosynthetic capacity during and after drought in prairie grasses. *Physiol. Plant.* 101, 173-182.

Islam, M.A., MacDonald, S.E., Zwiazek, J.J., 2003. Responses of black spruce (Picea *mariana*) and tamarack (*Larix laricina*) to flooding and ethylene. *Tree Physiol.* 23, 545-552.

Jackson, M.B., 1985. Ethylene and responses of plants to soil waterlogging and submergence. *Ann. Rev. Plant Physiol.* 36, 145-174.

Kozlowski, T.T., 2002. Physiological-ecological impacts of flooding on riparian forest ecosystems. *Wetlands* 22, 550-561.

Kozlowski, T.T., 1997. Responses of woody plants to flooding and salinity. *Tree Physiol.* 1, 1-29

Larson, K.D., Davies, F.S., Schaffer, B., 1991a. Floodwater, temperature, and stem lenticels hypertrophy in *Mangifera indica* (Anacardiaceae). *Amer. J. Bot.* 78, 1397-1403.

Larson, K.D., Schaffer, B., Davies, F.S., 1991c. Mango responses to flooding in limestone soil. *Proc. Fla. State Hort. Soc.* 104, 33-39.

Lawlor, D., 2002. Limitation to photosynthesis in water-stressed leaves: stomata vs. metabolism and the role of ATP. *Ann. Bot.* 89, 871-885

Mommer, L., Pons, T.L., Wolters-Arts, M., Venema, J.H., Visser, E.J.W., 2005. Submergence-induced morphological, anatomical, and biochemical responses in a terrestrial species affects gas diffusion resistance and photosynthetic performance. *Plant Physiol.* 139, 497-508.

Núñez-Elisea, R., Schaffer, B., Crane, J.H., Colls, A.M., 1998. Impact of flooding on *Annona* species. *Proc. Fla. State Hort. Soc.* 111, 317-319.

Pedersen, O., Rich, S.M., Colmer, T.D., 2009. Surviving floods: leaf gas films improve O_2 and CO_2 exchange, root aeration, and growth of completely submerged rice. *Plant J.* 58, 147-156.

Pezeshki, S.R., 2001. Wetland plant responses to soil flooding. *Environ. Expt. Bot.* 46, 299-312.

Ponnamperuma, F.N., 1984. Effects of flooding on soil. In: Flooding and plant growth. Ed. T.T. Kozlowski *Acad. Press, Inc., Orlando,* FL. 9-45.

Przywara, G., Stepniewski, W., 1999. The influence of waterlogging at different temperatures on penetration of roots and on stomatal diffusive resistance of pea and maize seedlings. *Acta. Physiol. Plant.* 21, 405-411.

Reid, D.M., Railton, I.D., 1974. Effect of flooding on the growth of tomato plants: Involvement of cytokinins and gibberellins. Eds. In R.L. Bieleski, A.R. Ferguson, and M.M. Cresswell. Mechanisms of regulation of plant growth. Bul. 12. *The Royal Society of New Zealand, Wellington.* 789-792.

Sandhu, B.S., Singh, B., Singh, B., Khera, K.L., 1986. Maize response to intermittent submergence, straw mulching and supplemental N-fertilization in subtropical region. *Plant Soil* 96, 45-56.

Smethurst, C.F., Shabala, S., 2003. Screening methods for waterlogging tolerance in lucerne: comparative analysis of waterlogging effects on chlorophyll fluorescence, photosynthesis, biomass and chlorophyll content. *Funct. Plant Biol.* 30, 335-343.

Soares-Cordeiro, A.S., Carno-Silva, A.E., da Silva, A., da Silva, J., Keys, A.J., Arrabaca, M.C., 2009. Effects of rapidly imposed water deficit on photosynthetic parameters of three C4 grasses. *Photosynthetica* 47, 304-308.

Sun, O.J., Sweet, G.B., Whitehead. D., Buchan, G.D., 1995. Physiological responses to water stress and waterlogging in Nothofagus species. *Tree Physiol.* 15, 629-638.

Taiz, L., Zeiger, E., 2002. Plant Physiology. Third Ed. Sinauer Associates, Inc., Publishers. Sunderland, Mass.

VanToai, T.T., St. Martin, S.K., Chase, K., Boru, G., Schnipke, V., Schmitthennner, A.F., Lark, K.G., 2001. Identification of a QTL associated with tolerance of soybean to soil waterlogging. *Crop Sci.* 41, 1247-1252.

Voesenek, L.A.C.J., Colmer, T.D., Pierik, R., Millenaar, F.F., Peeters, A.J.M., 2006. How plants cope with complete submergence. *New Phytol.* 170, 213-226.

Winkel, A., Colmer, T.D., Pedersen, O., 2011. Leaf gas films of *Spartina anglica* enhance rhizome and root oxygen during tidal submergence. *Plant Cell Environ.* 34, 2083–2092.

Zhou, W.J., Lin, X.Q., 1995. Effects of waterlogging at different growth stages on physiological characteristics and seed yield of winter rape. *Field Crops Res.* 44, 103-110.

In: Photosynthesis
Editor: Nafees Khan
ISBN: 978-1-63463-304-8

Chapter 7

PHOTOSYNTHESIS IN CHANGING ENVIRONMENT: ANALYZING THE ROLE OF PHYTOHORMONES IN MODULATION

M. Iqbal R. Khan*•, *Nafees A. Khan**, *Mohd. Asgher*, *Mehar Fatma*, *Shabina Syeed*, *Asim Masood*, *Tasir S. Per* and *Wasim Kaiser
Department of Botany, Aligarh Muslim University,
Aligarh, India

ABSTRACT

Photosynthetic organisms sustain life by converting light energy into chemical energy. Photosynthesis is a key phenomenon contributing substantially to plant growth and development. It is a well regulated and multistep process to produce assimilatory power NADPH and ATP to synthesize carbohydrate. The changing environmental conditions have forced plant scientists to better understand photosynthetic process more importantly for agricultural sustainability. The abiotic stresses such as salt stress, heavy metal stress, water stress, heat stress, chilling stress, nutrient stress etc. adversely affect photosynthesis because of induced oxidative stress. Various mechanisms may provide protection to the photosynthetic apparatus under abiotic stress, but phytohormones have potential to modify biochemical, physiological processes and molecular mechanisms in plants to optimize photosynthesis under optimal and stressful environments. They are a strong tool to protect and alleviate stress by modulating metabolic processes in plants. The present chapter gives an insight into the role of phytohormones in modifying biochemical, physiological and molecular responses to adapt plants in stressful environments.

• E-mail: amu.iqbal@gmail.com.
* E-mail: naf9@lycos.com.

Keywords: Abiotic stress, Climate change, Photosynthesis, Phytohormones, Tolerance

INTRODUCTION

The exploitation of soil, air and water by man for extracting maximum benefit has resulted in the degradation of the environment leading to conditions such as salinity stress, elevated levels of heavy metals, nutrients limitations, drought, and temperature stress (low and high) together with global warming and extinction of the biodiversity (Thomas et al. 2004; Khan and Khan 2013). The resulting climate change is now being recognized as the biggest and potential threat for agriculture all over the world. It affects the physiological, biochemical and molecular mechanisms in plants from early stage of seed germination to maturity of plants, subsequently leading to reduction in economic yield of crops. It has been estimated that increasing salinization of arable land will result in 30% land loss by the end of 2028 and 50% by mid of the 21st century (Wang et al. 2003). The International Panel on Climate Change (IPCC) (2007) has suggested that the concentration of CO_2 may rise to 500-1000 ppm by the year 2100 which may increase the global mean temperature by approximately 1°C to 3°C above the present value by 2025 and 2100, respectively (Houghton et al. 1990). The increased release of toxic heavy metals (HMs) in to soil and water l by industries has also become a major concern to the environment and agriculture. A report released by U.S. environmental action group says that the health of more than 10 million people in eight countries is threatened due to heavy metals toxicity. These countries include Russia, China, Dominican Republic, India, Kyrgyzstan, Peru, Ukraine and Zambia (Environmental News Service 2006). These climate changes activate the possibility of extinction of about 15-37% flora and fauna (Thomas et al. 2004).

Among various processes affected by the changing climatic conditions, photosynthesis is the first and primary process inhibited under stressful environments. The inhibition of photosynthesis could significantly lead to reduction in plant growth and also loss in yield and quality of crop plants. The adverse effects of abiotic stresses on photosynthesis at whole plant to cellular level are imposed either by diffusion limitations through the stomata or mesophyll or by induced oxidative stress causing reduction in photosynthetic capacity of plants (Chaves et al. 2009). Phytohormones are the signaling molecules with important roles in regulation of developmental processes through their different signaling networks in plants under abiotic stresses. Recent studies have shown the potential of phytohormones in reducing the negative effects of abiotic stress on photosynthesis (Hao et al. 2012; Khan and Khan 2014; Khan et al. 2013a; Khan et al. 2012a; Nazar et al. 2011; Yoon et al. 2009). They played a significant role in increasing the ability of plants to grow successfully under both optimal and stressful conditions with the improvement and/or protection of plant growth and yield. Phytohormones can directly affect photosynthesis by modifying enzymes of photosynthesis or by interacting with other phytohormones. Auxin, cytokinin (CK), gibberellins (GAs), abscisic acid (ABA), ethylene, salicylic acid (SA), jasmonic acid (JA), nitric oxide (NO), brassinosteorids (BRs) are considered as the potential phytohormones for improvement of photosynthesis under optimal and stressful environments. The present chapter focuses on the involvement of phytohormones in protection of photosynthesis and in abiotic stress tolerance through regulation of various metabolism in plants.

HISTORICAL PERSPECTIVE OF PHOTOSYNTHESIS

- The term 'photosynthesis' was coined by Charles Reid Barnes and Conway MacMillan
- 1727: Stephen Hales suggested that plants used air as nutrients and light participated somehow in this process
- 1771: Joseph Priestley discovered oxygen evolution in the process
- 1779: Jan Ingenhousz reported essential role of light in photosynthesis
- 1818: Pierre Joseph Pelletier and Joseph Bienaimé Caventou isolated the green pigment in plants that acted as principal light-absorbing molecule in the process of photosynthesis and named it as chlorophyll
- 1837: René-Joachim-Henri Dutrochet recognized the essential role of chlorophyll in photosynthesis
- 1837: Hugo von Mohl first described the 'chlorophyllkörnern' or chlorophyll granules in detailed
- 1845: Mayer showed that plants transformed energy of sunlight into chemical energy
- 1881: Theodor Wilhelm Engelmann demonstrated that red and blue colors are most effective in photosynthesis and in oxygen evolution
- 1882: Englemann characterized the action spectrum of photosynthesis and relative effectiveness of different wavelengths of light in promoting photosynthesis and used rate of oxygen production as a measure of rate of photosynthesis
- 1883: A.F.W. Schimper introduced the term 'plastid'
- 1883: A. Meyer coined the term 'grana'
- 1885: Schimper described the 'stroma' as the dense, dot-like structures**1915:** Richard Martin Wilstätter was awarded Nobel prize for his significant contributions on plant pigments, especially chlorophyll
- 1920: Otto Warburg observed that O_2 inhibits photosynthesis. This phenomenon, originally known as the ''Warburg effect'' was later recognized as the light-dependent release of CO_2 by photosynthetic organisms
- 1927: C. Zirkle demonstrated that chloroplasts were derived from colorless precursor organelles called 'primordia' (proplastids)
- 1930: C. B. Van Niel discovered that O_2 produced during green plant photosynthesis results from splitting of water and not carbon dioxide and concluded that photosynthesis is a redox (reduction–oxidation) process
- 1930: Hans Fischer was awarded Nobel Prize for investigating chlorophylls and hemes and also for giving the complete chemical structure of chlorophyll in 1940
- 1932: Robert Emerson and William Arnold reported that light is absorbed by clusters of chlorophylls and accessory pigment molecules called 'photosystems'
- 1937 and 1939: Hill showed oxygen evolution by chloroplasts suspensions using ferric oxalate, ferricyanide as electron acceptors. This reaction is known as Hill reaction
- 1941: Ruben and Kamen studied the breakdown of water and resulting liberation of O_2 during photosynthesis. They also discovered the radioactive isotopes ^{14}C which was successfully used by later workers to study photosynthetic CO_2 assimilation

- 1943: Emerson and Lewis discovered 'red drop' phenomenon
- 1950: Robins Hill showed that light energy is used by plants to generate reducing power. The reduction of carbon dioxide into organic molecules in the light-independent reactions is called carbon fixation
- 1954: Arnon, Allen and Whatley reported that photophosphorylation occurred in isolated chloroplasts with simultaneous assimilation of $^{14}CO_2$ and O_2 evolution
- 1954: Daniel Arnon reported the formation of ATP from ADP and Pi in chloroplasts
- 1960: Emerson and Rabinowitch postulated the existence of two light reactions in photosynthesis
- 1960: Hill and Bendall put forward the 'Z scheme' of photosynthesis involving the operations of two photosystems in photosynthetic electron transport and phosphorylation
- 1961: Calvin and co-workers studied the incorporation of $^{14}CO_2$ into photosynthesizing algae. They studied the reduction of CO_2 to sugars by dark enzymic reactions and also found that two molecules of $NADPH_2$ and three molecules of ATP were required for the reduction of every molecule of CO_2. For the explanation of pathway of photosynthetic carbon reduction, Calvin was awarded Nobel Prize in 1961
- 1962 and 1990: Wilhelm Menke introduced the term 'thylakoids'
- In 1963: Ruth Sager and Masahiro R. Ishida provided strong evidence for the presence of unique DNA in chloroplasts
- 1965: Kortschak and co-workers in 1965 and also Hatch and Slack in 1966 discovered C_4 Pathway of CO_2 fixation
- 1975: Alan Stemler and Richard Radmer reported that the source of O_2 in photosynthesis was H_2O
- 1988: Harmut Michel, Robert Huber, and Johann Deisenhofer were awarded Nobel prize "for the determination of the three-dimensional structure of a photosynthetic reaction centre "
- 1989: Hartmut Michel, Johann Deisenhofer, Robert Huber and co-workers determined three-dimensional structure of the reaction center from the purple photosynthetic bacterium *Rhodopseudomonas viridis*
- 1996: Zhang et al. characterized and crystalized the lumen side domain of chloroplast Rieske iron-sulfur protein
- Staehelin (2003) and Gunning et al. (2006) fully discussed the structure of choloroplasts
- Horst Witt (2004), Petra Fromme and Paul Mathis (2004) presented the X-ray structures of photosystem II and I

ROLE OF PHYTOHORMONES UNDER DIFFERENT ENVIRONMENTAL CONDITIONS

Plant growth and yield are important agronomic traits in the agricultural crops. Plants are repeatedly subjected to varied environmental stress factors such as salinity , heavy metals, unavailability of nutrients, drought and temperature that significantly influence plant productivity and plant distribution. Plant species differ in their tolerance to abiotic stress factors owing to different adaptive mechanisms.

Salinity Stress and Photosynthesis

Salinity is one of the major environmental stress factors that adversely affects plant growth and development (Dash and Panda 2001), enzymes activity (Seckin et al. 2009; Khan et al. 2014), photosynthesis (Khan et al. 2012a; b) and cell division (Tabur and Demir 2010). Salinity stress decreases nutrients and water availability, value of arable lands, and also alters the structure of ecological communities. Salinity-induced osmotic stress, the physiological equivalent of drought stress, typically reduces growth and photosynthesis in plants (Nazar et al. 2011). The role of phytohormones in the alleviation of the harmful effects of salinity stress has been documented by many researchers (Kabar 1987; Mutlu and Bozcuk 2000; Khan et al. 2010; Nazar et al. 2011; Khan et al. 2012a; Khan et al. 2014). The salt treated plants exhibited a significant reduction in chlorophyll a and chlorophyll b and increased chlorophyll a/b ratio (Al-Whaibi et al. 2012). This decreased photosynthesis pigments by salt stress was shown due to instability of protein complexes and destruction of chlorophyll by the enhanced activity of chlorophyllase, a chlorophyll degrading enzyme (Reddy and Vora 1986) resulting in reduction of photosynthesis (Khan et al. 2012a).

Cytokinins play a central role in plant growth and development including cell expansion and differentiation, seed germination and leaf senescence (Choi and Hwang 2007). Cytokinins have been shown to control plant adaptation to salinity stress (Wu et al. 2012). In plants, salinity stress changed the content of CK and suppressed growth parameters (Albacete et al. 2008). Under stress condition, changes in the content of individual CK forms occur. It was found that maize roots subjected to NaCl show increased content of zeatin riboside (type of CK; ZR) with the decreased content of zeatin (type of CK; Z), whereas in pea plants, a pronounced accumulation of isopentenyl (type of CK; iP) was observed (Atanasova et al. 1999). In tomato leaves, salinity stress induced a decrease in the total CK content, most expressed in the case of ZR level (Albacete et al. 2008). Salinity reduced the contents of Z and iP in roots and leaves of the common ice plants (Shevyakova et al. 2013). Since CK activates plant growth and increases photosynthesis (Kulaeva and Kusnetsov 2002), the observed partial inhibition of Z and ZR synthesis at salinity seems quite logical. Under water deficit experienced by plants at salinity stress, cell metabolism is usually not activated but slightly suppressed, and this reduces the degree of injury and favours organism survival. In contrast, the total content of ZR in plants increased (Shevyakova et al. 2013) which indicates a difference in the stress dependent regulation of the level of different CK forms (Kulaeva and Kusnetsov 2002; Hirose et al. 2008). Seed enhancement (seed priming) with CK has been reported to increase plant salt tolerance (Iqbal et al. 2006). The quantitative and qualitative responses of plants to different CKs may differ considerably under saline conditions. Pre sowing treatment with different CKs, kinetin (type of CK; Kin) and benzylaminopurine (type of CK; BAP), alleviated the inhibitory effect of salt stress on wheat cultivars and caused maximum increase in shoot growth and improved grain yield. This increase in growth and grain yield was found to be associated with increased net photosynthesis and improved water use efficiency under saline condition (Iqbal and Ashraf 2005). Gurmani et al. (2011) have shown that benzyl adenine (type of CK; BA) caused a significant ameliorative effect on three spring wheat cultivars under salt stress. Further, they suggested that BA pre-treatment increased K^+ and Ca^{2+} with reduced Na^+ and Cl^- concentrations in shoot of all the cultivars.

Gibberellins are the phytohormones that can control plant growth and development under salt stress. Exogenous GA treatment of salt-stressed wheat plants resulted in an increased photosynthetic capacity which was a major aspect of greater dry matter production (Iqbal and Ashraf 2013b). Reports are available on the influence of GA on the increased net photosynthesis (Khan 1996; Khan et al. 2002; Yuan and Xu 2001) or the decreased net photosynthesis (Dijkstra et al. 1990). The use of GA biosynthesis inhibitors has also confirmed the increase (Sairam et al. 1991; Thetford et al. 1995) or decrease of net photosynthesis (Bode and Wild 1984; Heide et al. 1985). These conflicting findings may be due to various causes, including measurement and expression of photosynthesis, and the timing of application of GA or GA inhibitors (Nagel and Lambers 2002). Gibberellins-mediated photosynthetic responses may also include the involvement of other hormones, as perception of one plant hormone affects the concentrations, perception, and responses to the others (Khan et al. 2003). Gibberellic acid has also been shown to alleviate the effects of salt stress on pigment content, Hill activity (Aldesuquy and Gaber 1993) and water use efficiency (Shah 2007). Gibberellins were involved in the adjustment of plants under limited environmental conditions and maintenance of source-sink relationship (Iqbal et al. 2011). Under salinity stress, an increase in sucrose in source leaves has been noted due to reduced sink enzymes activities because of decrease in photosynthesis rate by feedback inhibition responses (Poljakoff-Mayber and Lerner 1994). Iqbal and Ashraf (2013b) have shown that GA priming increases wheat grain yield which was attributed to the GA priming-induced modulation of ion uptake, partitioning within shoots and roots and hormone homeostasis under salinity. Hamayun et al. (2010) reported that GA application significantly promoted plant length and fresh and dry biomass of plants which were markedly hindered by NaCl induced salt stress in soybean plants. Alonso et al. (2009) have shown that seed germination and seedling establishment of the *FsGASA4* transgenic plants overexpressing a GA-responsive gene from beechnut, coding for a member of the GASA family (*FsGASA4*) were resistant to high salt stress. Achard et al. (2006) found a correlation between survival of salt toxicity and the function of DELLA proteins. The role of the DELLA proteins as potential integrators of phytohormone in the regulation of germination, cell redox state, growth, and stress responses is also worth mentioning (Grant and Jones 2009). DELLA proteins are repressors of GA signaling, modulate ROS levels and are involved in growth-regulatory mechanisms (Achard et al. 2008). These results show that the salt-inducible DDF1 gene is involved in growth responses under high salinity conditions through altering GA levels (Kaya et al. 2009).

Abscisic acid synthesis in roots and its xylem transport are well correlated to the stomatal reactions under salt stress. Stomatal closure is one of the most rapid plant responses to stress. It is directly related to ABA involvement in the inter organ water deficit signal (Kholodova et al. 2006). Jae-Ung and Youngsook, (2001) reported that ABA as a signal for stomatal closure induced rapid depolymerization of cortical actin filaments and the slower formation of a new type actin which was randomly oriented throughout the cell. ABA treatment accelerated plant transition to CAM photosynthesis (Kuznetso et al. 2003). It was also demonstrated that treatment with ABA of Arabidopsis leaves (Morillon et al. 2001), rice seedlings (Li et al. 2010) or bean roots (Khadri et al. 2006) increased plant tolerance to salt stress. Moreover,

Fricke et al. (2004) reported that salinity transiently increased ABA concentration in the barley growth zones.

Ethylene has long been identified as a phytohormone that directs plant reactions under salt stress. Salinity affects the number of plant traits such as the rate of photosynthesis or leaf morphology, affects the growth of a plant with elevated levels of ethylene production (Cao et al. 2007; Achard et al. 2006; Khan et al. 2012b). Ethylene constitutively sensitive mutant, *ctr1-1* largely improves plant salt tolerance with respect to biomass accumulation and photosystem function. Moreover, the exogenous supply of ethephon (an ethylene releasing compound) to salt stressed *Col-0* plants improved *Arabidopsis* salt tolerance (Yang et al. 2010). However, a large amount of information is available describing the role of ethylene in photosynthesis. Ethylene concentrations enhanced the rate of photosynthesis (Subrahmanyam and Rathore 1992; Khan 2004), possibly as a consequence of an increased stomatal conductance (Khan 2004), although the ethylene-induced effects on stomatal regulation and photosynthesis may differ between species. Ethylene plays a role in the regulation of photosynthesis via its role in sugar sensing. Paul and Pellny (2003) have shown that the control of photosynthetic activity and gene expression are the result of down regulation of the Calvin-Benson cycle enzymes by carbohydrate end-products. Further, Danny et al. (2006) found that ethylene-insensitive tobacco plants have a lower photosynthetic capacity as a result of a lower Rubisco expression. Ethylene production may be significant in promoting growth under elevated carbon dioxide levels (Seneweera et al. 2003) and to overcome the harmful effects of high endogenous glucose concentrations on photosynthesis. Contrarily, the rate of ethylene production increased in salt-stressed tomato and cucumber, promoting senescence in these plants (Feng and Barker 1992; Helmy et al. 1994). Loss of chlorophyll under salt stress has been reported due to ethylene induced reduction in intermediates of chlorophyll biosynthesis (Khan 2003). Furthermore, ethylene interacts with other hormones such as GA, ABA etc. Ethylene-induced change in ABA biosynthesis may cause alteration in photosynthesis through the established role of ABA in stomatal movement and gas exchange. Ethylene accumulation in the submerged tissue led to decrease in ABA levels (Kende and Zeevaart 1997; Grossmann and Huesen 2001) and an increase in the responsiveness of the tissue to GA.

Application of SA helped in restoration of altered chlorophyll concentration induced by salt stress (Karlidag et al. 2009; Noreen et al., 2012). The increased chlorophyll a/b ratio was linked with the changes in pigment composition of the photosynthetic apparatus which possessed a lower level of light harvesting chlorophyll proteins (LHCPs) (Loggini et al. 1999). The reduction in LHCPs content is an adaptive defense mechanism of chloroplast, leaves and plants, which allows them to endure the adverse conditions (Asada et al., 1998). The increased chlorophyll a/b ratio indicates the availability of nutrients to the plant leaf under stress condition. SA could be responsible for improving the plant tolerance to salt stress by increasing chlorophyll contents and decreasing chlorophyll a/b ratio (Fahad and Bano 2012). Carbonic anhydrase (CA) activity decreased due to salinity stress in wheat plants (Al-Whaibi et al. 2012). Under salt stress, the inactivation of Rubisco (Soussi et al. 1998) reduces the net photosynthetic carbon metabolism and photosynthetic efficiency (Seeman and Critchley 1985). In contrast, SA exposed plants exhibited improved CA activity because of increased reversible hydration of CO_2 and its constant supply to Rubisco and improved the tolerance of plants to salt stress. The effect of endogenous SA variation in signaling salt tolerance might have been due to differences in net photosynthesis and growth changes in all the tested *Arabidopsis* mutants. Maximum PS II efficiency (Fv/Fm) represents the efficiency of energy capture of open PS II reaction centers. No major change of Fv/Fm was detected in *nahG* and *npr1-1* plants exposed to 100 and 200 mM NaCl, but the values decreased at 300 mM NaCl. In contrast, a significant decline of Fv/Fm occurred in *snc1* plants exposed to 100 mM NaCl, and in wild and *snc1/nahG* plants at 200 mM or higher NaCl. This indicated that the PS II activity was more resistant to NaCl stress in *nahG* and *npr1-1* plants than in other plants (Hao et al. 2012). In fact, Fv/Fm value always exhibits more tolerance to salt stress than other chlorophyll fluorescence parameters (Tezara et al. 2003). Hao et al. (2012) reported that important chlorophyll fluorescence parameter, quantum efficiency of PS II (ΦPS II) was more sensitive to salinity than Fv/Fm as indicated by significant decrease occurred at lower level of NaCl in the most tested plants of Arabidopsis. Taken together, the values of net photosynthesis, Fv/Fm and Φ PS II in the tested plants suggested that the high accumulation of SA aggravated NaCl induced impairment in photosynthetic process of *snc1* plants, and this was effectively ameliorated by the decrease of SA level. Furthermore, *nahG* and *npr1-1* plants always exhibited more resistance in the photosynthetic process than wild plants under 100, 200 and 300 mM concentrations of NaCl. Recently, Khan et al. (2014) have shown that application of SA under salt stress alleviated salt-induced photosynthesis and growth inhibition in mungbean (*Vigna radiata* L.). They suggested that interaction of SA and ethylene was responsible for tolerance in plants as application of SA resulted in the formation of optimal ethylene by suppressing stress ethylene under salt stress and regulated stomatal conductance necessary for increased carboxylation efficiency and water use efficiency.

The alleviation of salt stress effects on photosynthesis by jasmonates has been demonstrated in broccoli and barley (Walia et al. 2007). In tomato, JA was found as requirement for salt-induced gene expression (Dombrowski 2003) and activation of JA signaling can promote salt tolerance (Capiati et al. 2006). It has been observed that barley seedlings treated with JA along with 50-100 mM NaCl have an increased activity of CA, with a more pronounced effect on the cytoplasmic localized enzyme. It has been suggested that these enzymes are involved in the plant's adaptation to osmotic stress as a possible biochemical mechanism (Popova et al. 1991). Maslenkova et al. (1989; 1990) have established that JA application to the growth medium of barley seedlings leads simultaneously to changes in photosynthetic light reactions connected with chloroplast thylakoid membranes. The effect of NaCl salinity on photosynthetic capacity is also reflected by the inhibition of Rubisco activity. The seedlings when exposed to NaCl concentration and pretreated with JA before salinization showed restored Rubisco activity (Tsonev et al. 1998). Barley seedlings treated with JA or grown with 100 mM NaCl showed an increase in the activities of PEPC and CA (Tsonev et al. 1998). The increase in PEPC activity was due to the increase in the quantity of the enzyme protein (Chu et al. 1990).

Brassinosteroids have been found to protect photosynthetic apparatus under saline condition and exhibit higher net photosynthesis. Reduction in photosynthetic performances under stress was correlated with the inhibition in Rubisco activation state (Sharkey 2005; Yamori et al. 2006) and activity of Rubisco and phosphoenol pyruvate carboxylase (Lefebvre et al. 2005).

Yu et al. (2004) have reported increase in photosynthesis by increasing Rubisco activation state and carboxylation efficiency (Ogweno et al. 2008) by BR application. Similarly, Fv/Fm has also been found to be protected with the application of BRs under salt stress in wheat plants. Application of BRs helped in the protection of PS II against over excitation under salt stress avoiding damage to the thylakoid membrane. It is recognized that the thylakoid membranes are very sensitive to stress (Haldimann and Feller 2005). Application of BR alone or as a follow up treatment to NaCl stressed plants enhanced CA activity possibly by accelerating the assimilatory rate of CO_2 through the expression of specific genes (Khripach et al. 1999; Swamy and Rao 2006). Further, Dalio et al. (2011) reported that salt stress affected values of gas exchange parameters (net photosynthesis, transpiration rate, stomatal conductance) along with water use efficiency and photosynthetic pigment content in *Cajanus cajan*. However, these effects were ameliorated by the application of BRs in the studied parameters.

Nitric oxide triggers redox-regulated defense related gene expression directly or indirectly to establish a plant stress tolerance (Sung and Hong 2010). Garcia-Mata and Lamattina (2002) showed that exogenous NO application reduced transpiration and induced stomatal closure in several species such as *Vicia faba*, *Salpichroa* and *Tradescantia* and indicated to be a component of ABA signaling pathways in ABA-induced stomatal closure. ABA promotes NO formation through the reaction of nitrate reductase in guard cells (Desikan et al. 2004). NO stimulated mitogen activated protein kinase (MAPK) activity and cGMP production (Sokolovski et al. 2005; Neill et al. 2008), and phospholipase C (PLC) and phospholipase D (PLD)-derived phosphatidic acid (PA) (Distefano et al. 2008) during stomatal closure. Phosphoenolpyruvate carboxylase (PEPCase: EC 4.1.1.31), the enzyme responsible for primary carboxylation in C4 photosynthesis is phosphorylated by a light-dependent phosphoenolpyruvate carboxylase-kinase (PEPCase-k). Monreal et al. (2013) reported that NO is involved in the complex control of PEPCase-k activity in sorghum leaves, and it mediated some of plants responses to salinity that are related to this enzyme. Additionally, light is essential for plant growth and development, but the excess light can be harmful because it causes the accumulation of ROS (Asada, 2006). Xu et al. (2010) elucidated the role of applied exogenous NO in alleviating light induced oxidative damage in leaves of two varieties of tall fescue (Arid3 and Houndog5). Plants when exposed to a photosynthetic photo flux density (PPFD) higher than that required for CO_2 fixation leads to increased ROS generation. Wu et al. (2013) reported that the application of NO (in the form of sodium nitroprusside; SNP) to salt-stressed eggplant seedlings resulted in enhanced photosynthetic parameters such as the net photosynthesis, stomatal conductance, transpiration rate, and intercellular CO_2 concentration, F_v/F_m, efficiency of excitation capture of open PSII centers, quantum yield of PSII and photochemical quenching indicating the protective function of exogenous NO against salt stress. NO has also been found to inhibit photosynthetic electron transport and modulate reaction-center-associated non-photochemical quenching in plants (Wodala et al. 2008). Wodala et al. (2005) found that photochemical quenching (qP) and the effective quantum efficiency (ΦPSII) decreased while non-photochemical quenching (NPQ) remained unchanged with the treatment of NO donor, SNP. NO activates different biochemical pathways via directly or indirectly interaction with proteins to produce metal proteins with heavy metals, S-nitrosylation with sulfuhydrl groups and nitro groups in the process of nitration to provide resistance against salt stress (Leterrier et al. 2011; 2012).

Thus, understanding the mechanisms of salt tolerance in plants by exogenous NO will make it possible for plant breeder and plant physiologists to develop specific techniques to reduce salinity toxicity in soils. The summary of phytohormones involvement in photosynthetic attributes under salt stress is shown in Table (1).

Table 1. Mediation of phytohormones in protection of photosynthetic attributes under salt stress

	Plant	Photosynthetic attributes	Response	Treatment of salt	Reference
Auxin	*Zea mays*	Stomatal CO_2 concentration, net assimilation rate, stomatal conductance, transpiration rate, chlorophyll a, chlorophyll b, carotenoids	+	100 mM	Khalid et al. 2013
Auxin	*Triticumaestivum*	CO_2 assimilation rate	+	150 mmol L^{-1}	Iqbal and Ashraf et al. 2013a
CK	*Solanum melongena*	Photosynthetic gas exchange, chlorophyll fluorescence	+	90 mM	Wu et al. 2012
GA	*Triticum aestivum*	Net CO_2 assimilation rate, transpiration, stomatal conductance	+	15 dSm^{-1}	Iqbal and Ashraf 2013b
GA	*Hibiscus sabdariffa*	Chlorophyll a, b content	+	40, 80, 120 mM	Ali et al. 2011
ABA	*Triticum aestivum*	Chlorophyll content, net assimilation, transpiration rate	+	100 mM	Gurmani et al. 2013
Ethylene	*Brassica juncea*	Chlorophyll content, quantum yield, Rubisco activity, stomatal conductance, intercellular CO_2, net photosynthesis	+	100 mM	Nazar et al. 2014
SA	*Helianthus annuus*	Chlorophyll content	+	150 mM	Nooreen et al. 2012
BR	*Solanum melongena*	Chlorophyll content, chlorophyll fluorescence	+	90 mM	Wu et al. 2012
NO	*Gossypium hirsutum*	Photosystem II, net photosynthetic rate, transpiration rate	+	100 mM	Liu et al. 2013
NO	*Solanum lycopersicum*	Chlorophyll content, chlorophyll fluorescence, gas exchange parameters	+	100 mM	Wu et al. 2010
PA	Cucumber	Chlorophyll content, photochemical quenching, non-photochemical quenching, actual PSII efficiency	+	75 mmol L^{-1} NaCl	Shu et al. 2012
PA	Sugar beet	Chlorophyll fluorescence, CO_2 assimilation, transpiration rate, net photosynthesis, stomatal conductance	+	25, 50, 150 mM	Hajiboland et al. 2012

Heavy Metals and Phytohormones

Heavy metals such as cadmium (Cd), lead (Pb), nickel (Ni), mercury (Hg), copper (Cu), cobalt (Co), arsenic (As) etc. are toxic to plants at low concentrations. They can be accumulated to high levels in leaves and affect photosynthetic process adversely by disrupting the electron transport chain, aggregation of pigment protein complexes of the photosystems and inhibiting the activities of enzymes of the Calvin–Benson cycle (Mobin and Khan 2007;

Parmar et al. 2013; Khan and Khan 2014). In order to avoid deleterious effects of heavy metals toxicity, plants have evolved a number of adaptive strategies to cope with the stresses by initiating various appropriate physiological and cellular responses. One such response in plants is the phytohormones actions that play an important role in enhancing the crop productivity by reducing the oxidative stress. Reduction in plant growth under stress conditions could result from an altered hormonal balance or hormone supplementation provides an attractive approach to cope with the stress. The central metal atom of chlorophyll, Mg can be replaced with Cu, Zn, Cd, Pb, Hg or Ni. This substitution prevents light harvesting by the affected chlorophyll molecules. Heavy metals decreased net photosynthesis, efficiency of PS II, photosynthetic pigments, impaired net CO_2 assimilation rate and decreased Rubisco activity (Dias et al. 2013; Khan and Khan 2014).

Heavy metals inhibited the production of chlorophyll by affecting both the synthesis of 5-aminolevulinic acid and protochlorophyllide reductase and net photosynthesis. In Cd treated plants, level and distribution of auxin were both greatly altered in primary root tips and cotyledons. The results also showed that after Cd treatment, the Indole-3-acetic acid (IAA) content significantly decreased which was accompanied by an increase in the activity of IAA oxidase and alteration in the expression of several putative auxin biosynthetic and catabolic genes, as IAA oxidase is responsible for the degradation of IAA. Cd stress also altered the expression of PINs and AUX1 in Arabidopsis roots implying that the auxin transport pathway is required for lateral root (LR) formation (Hu et al. 2013). Application of 100 μM IAA alleviates the toxic effects of 80 μM Cu in roots by enhancing greater root length and root hair formation. Photosynthetic pigments significantly decreased under Cu stress along with increased net photosynthesis and water use efficiency in sunflower (Ouzounidou and Ilias 2005).

Moreover, the increment of chlorophyll content by Kin may be due to the effects of Kin on chlorophyll synthesis and increased net photosynthesis. Foliar application of Kin exhibited a favorable effect on the accumulation of some ions and ameliorated the inhibitory effect by increasing chlorophyll a, chlorophyll b and net photosynthesis under Cd stress (Al-Hakimi 2007). Thidiazuron (TDZ) a synthetic CK significantly enhanced seedling growth indices, decreased lipid peroxidation rate in the seedlings treated with heavy metals (Sazanova et al. 2012). Kinetin can alleviate the harmful effects of Cu and Zn on the growth of *Lupinus termis* plants through stimulation of Cu and Zn incorporation into metal-binding proteins and caused reduction of chlorophyll degradation and enhanced photosynthetic activities (Gadallah and El-Enany 1999). Plants treated with CK antagonist, PI-55 and INCYDE (2-chloro-6-(3-methoxyphenyl) amonopurine) showed improved plant growth in the presence of Cd. It is suggested that modulating CK status with inhibitors of CK perception or degradation may be useful in protecting plants against the adverse effects of high cadmium levels (Gemrotova et al. 2013). Cytokinin treatment reduces the negative effects of water deficits on chlorophyll and carotenoid content which are important protective factors against oxidative stress caused by heavy metal uptake and on photosynthetic activity (Metwally et al. 1998).

It has been proposed that GA has a crucial role in protecting plants against stress. Application of GA protected *Chlorella vulgaris* against stress of Pb and Cd at low concentrations of 10^{-7} and 10^{-6} M and induced an increase in photosynthetic pigments. However, at higher doses (10^{-4} M) of Pb and Cd, photosynthetic inhibition was reported (Falkowska et al. 2011). GA could be a potential modulator of Cd stress alleviation through an increase in sulfur-use efficiency (SUE) of crops. The increase in SUE by GA through increase in growth, CO_2 exchange rate and nitrogen use-efficiency has been reported by Khan et al. (2005). Ni-stressed seedlings exposed to GA exhibited an improved growth. Application of GA decreased Ni uptake by roots with less reduction in growth. Further, the presence of GA reduced lipoxygenase (LOX) activity and MDA production in Ni-treated plants. Based on this study, it can be concluded that GA certainly provides plant protection against Ni-induced oxidative stress and alleviate Ni toxicity to a certain extent (Saeidi-Sar et al. 2007).

Ethylene is a plant growth regulator involved in various aspects of growth and photosynthesis. Ethylene increased the stomatal conductance of plants, thereby increasing the diffusion of CO_2 resulting in increased net photosynthesis in mustard (Iqbal et al. 2011). Masood et al. (2012) have shown that ethylene was involved in S-mediated alleviation of photosynthetic inhibition by Cd. Further, they showed that exogenous application of ethephon resulted in increased activity ATPS and content of Cys and GSH which accounted for enhanced S assimilation, and increased net photosynthesis and growth under Cd stress. Recently, Khan and Khan (2014) reported that exogenously sourced ethylene (200 $\mu L\ L^{-1}$ ethephon) protected photosynthesis against Ni and Zn-accrued stress in mustard. Under the metal stress, ethylene protected photosynthetic potential by efficient PS II activity and through increased activity of Rubisco and photosynthetic nitrogen use efficiency.

Salicylic acid application decreased the uptake and the transport of Cd, alleviated the Cd-induced inhibition of nutrient absorption and led to a significant increase of chlorophyll and carotenoid content (Saidia et al. 2013). Recent evidence also suggests that SA is an important regulator of photosynthesis, and the activity of enzymes such as Rubisco and CA (Pancheva and Popova 1998; Al-Whaibi et al. 2012). Studies carried out in the presence of SA have shown that this hormone partially protects chlorophyll loss under Cd stress (Noriega et al. 2012).

Nitric oxide has also been reported to be involved in defense responses to heavy metal stress. It has been reported that NO exerts a protective effect in response to heavy metal stress. Application of NO (SNP) to *Brassica napus* plants enhanced the chlorophyll content and the Hill reaction activity of chloroplasts under Cd stress (Jhanji et al. 2012). Sodium nitroprusside application decreased ion leakage and lipid peroxidation levels and altered the level of antioxidant metabolites by reducing the ascorbate redox ratio and glutathione redox ratio and protected chickpea plants from Cd toxicity (Kumari et al. 2010). Sodium nitroprusside reduced Cu toxicity and NH_4^+ accumulation in rice leaves, and the protective effect of SNP on the toxicity and NH_4^+ accumulation was found reversed by c-PTIO, a NO scavenger, suggesting that the protective effect of SNP is attributable to NO released. These findings suggested that the reductions of Cu-induced toxicity and NH_4^+ accumulation by SNP are most likely mediated through its ability to scavenge active oxygen species (Yu et al. 2005).

Polyamines, such as spermine, spermidine, and putrescine are produced to combat abiotic stress in plants and the modulation of their biosynthetic pathway confers tolerance to heavy metal. Choudhary et al. (2012) reported that application of spermidine (Spd) protected photosynthetic capacity inhibition by increasing the content of chlorophyll a, b and carotenoids and by effectively improving the Cr-stress management capacity of radish seedlings.

It has been shown that exogenously applied, JAs elicit several physiological responses to stress and therefore increase plant resistance. Keramat et al. (2009) have shown that exposure of soybean plants to methyl-jasmonate (MeJA) alleviated Cd-induced damages by the reduction of MDA and H_2O_2 content and enhanced activities of antioxidant enzymes. In Arabidopsis plants, JA regulates genes involved in GSH and phytochelatins synthesis under Cd treatment (Xiang and Olivier 1998). However, contradictory result was obtained by Piotrowska-Niczyporuk et al. (2012) and they showed that JA enhanced heavy metal toxicity leading to an increase in the metal biosorption and ROS generation, decrease in cell number, chlorophylls, carotenoids, soluble proteins, ascorbate and GSH content as well as antioxidant enzyme activity.

Brassionosteriods have also been extensively used in research to ameliorate the impact of abiotic stress in plants including heavy metal toxicity. BRs application alleviated the toxic effect of heavy metals and increased the percentage of seed germination and seedling growth. The activities of antioxidant enzymes CAT, SOD, APX and GPX were increased in the seedlings of *Raphanus sativus* treatments with Cd along with BRs. Application of brassinolide alleviated the toxic effect of Cu applied through shotgun approach in sunflower. BRs application improved assimilation of carbon and nitrogen by the stabilization of membrane structures in the stressful conditions and also improved general growth and plant photosynthesis (Yu et al. 2004; Zhang et al. 2008). The ameliorative effects of BRs have been shown in Cu grown sunflower plants (Filová et al. 2013). Choudhary et al. (2012) studied the effect of application of 24-EBL on antioxidative defense system of chromium metal-stressed *Raphanus sativus* plants. Presoaking treatment with 24-EBL modulated the activities of antioxidative enzymes considerably. The summary of phytohormones involvement in photosynthetic attributes under heavy metals stress is presented in Table (2).

Water Stress and Photosynthesis

Drought stress limits crop productivity and the effects of drought stress are continuously growing in dimension of severity in many regions of the world (Shukla et al. 2012). It affects growth parameters, membrane integrity, total protein quantity and quality and causes a broad range of biochemical and physiological changes that negatively affect plant development and productivity. Plant growth is most sensitive to water deficit largely due to an inhibition of cell elongation. Due to water stress plants are shorter and have a smaller leaf area which reduces the amount of photosynthetically active radiation absorbed by leaves, the rate of photosynthesis and yield. Water stress in plants enhances oxidative stress (Apel and Hirt 2004) and induces stomatal closure which limits CO_2 uptake by leaves and reduces the potential activity of the Calvin cycle enzymes, particularly Rubisco due to limited availability of substrate i.e., CO_2 (Chaves et al. 2003). This decrease in the photosynthetic carbon fixation disturbs the utilization of NADPH resulting in its increased level while the level of $NADP^+$ decreased.

As a result transport of electrons occurs from PS I to molecular oxygen instead of $NADP^+$, thereby generating ROS (Lawlor and Tezara 2009). Oxygen-evolving complex and PSII are also damaged in water-stressed plants (Rensen and Curwiel 2000). The observations of Georgieva et al. (2007) suggest that the Calvin-cycle enzymes are more affected by dehydration than membrane bound electron transport reactions. Under stress conditions, Rubisco protein undergoes down-regulation or degradation (Feller et al. 2008). Transcripts encoding the small subunit of Rubisco were reported to be down-regulated in response to water stress (Bernacchia et al. 1996).

Table 2. Mediation of phytohormones in protection of photosynthetic attributes under heavy metal Stress

	Plant	Photosynthetic attributes	Response	Treatment of Heavy metal	Reference
Auxin	*Helianthus annuus*	Net photosynthesis , water use efficiency	+	Cu 80 μM	Ouzounidou and Ilias 2005
CK	*Pisumsativum*	Chlorophyll a, b content, net photosynthesis		Cd 25 or 50 μM	Al-Hakimi 2007
GA	*Chlorella vulgaris*	Photosynthetic pigments	+	10^{-7} M and 10^{-6} M (Pb and Cd)	Falkowska et al. 2011
ABA	*Oryzasativa*	Chlorophyll content	-	0.5 mM Cd	Hsu and Kao 2003
Ethylene	*Brassica juncea*	Gas exchange parameters, Rubisco, PS II activity, chlorophyll, water use efficiency	+	200 mg Ni and Zn	Khan and Khan 2014
Ethylene	*Brassica juncea*	Rubisco, net photosynthetic rate	+	Cd 200mg	Masood et al. 2012
SA	*Phaseolus vulgaris*	Chlorophyll, carotenoid content	+	Cd 20μM	Saidia et al. 2013
	Pisumsativum	Chlorophyll,CO_2 assimilation, stomatal conductance, water use efficiency, net photosynthesis	+	Cd 0.5, 1,2, 5 μM	Popova et al. 2008
JA	*Glycine max*	Chlorophyll content	+	Cd 50-500 μM	Noriega et al. 2012
JA	*Cajanus cajan*	Photosynthetic pigments	+	Cu	Sharma et al. 2013
BR	*Solanumlycopersicum*	Chlorophyll content, chlorophyll fluorescence, CO_2 assimilation rate, stomatal conductance, intercellular CO_2 concentration	+	Cd 100 μM	Ahammed et al. 2013
NO	*Brassica napus*	Chlorophyll content and Hill reaction activity	+	Cd 1,2 or 4 mM	Jhanji et al. 2012
PA	*Zea mays*	Chlorophyll content, gas exchange parameters	+	As 2 or 5 mg dm^{-3}	Stoeva et al. 2005

Pigments like Chl-a, Chl-b, and carotenoids are also significantly affected by drought stress (Ashraf et al. 2002) together with fluorescence of PS II. At severe drought stress, Fv/Fm strongly decreases indicating that there is a strong photo-inhibition, thus reducing the quantum efficiency and net photosynthesis (dos Santos et al. 2013). Different types of signals are utilized by pants in order to respond to soil water deficit and dry air. Under drought conditions these responses are triggered by primary osmotic stress signals (Chaves et al. 2003) or by secondary signal metabolites. The latter include hormones (e.g., ABA, ethylene, CK), ROS and intracellular secondary messengers (e.g., phospholipids, sugars, NO, etc.).

Phytohormones are one of the most important stress signals involved in the mechanisms of susceptibility or tolerance in photosynthetic inhibition of various plant species. Modulation in the concentration of auxins, GAs, CKs, ABA, JA, ethylene, and SA is often observed in drought stressed plant. Drought stress decreased IAA concentration in leaves of wheat (Xie et al. 2003; Bano and Yasmeen 2010) and corn (Wang et al. 2008). However, some studies have indicated that the adaptation to drought is a result of enhancement of IAA concentration (Pustovoitova et al. 2004). Similarly, in wheat and corn plants, drought stress maximally decreased concentrations of ZT and GA (Bano and Yasmeen 2010, Wang et al. 2008). It has been reported that several genes involved in the GA signaling pathway were down or up-regulated on exposure to water stress (Achard et al. 2008). GA20-oxidase, a key enzyme in the GA biosynthesis pathway, was down-regulated under drought conditions in maize and Arabidopsis (Qin et al. 2008; Zheng et al. 2010) which was probably responsible for the decrease in the endogenous GA content when plants were grown in drought stress environment. In contrast, recent study has shown that exogenous GA counteracts the inhibitory effects of abiotic stress (Alonso-Ramírez et al. 2009). Studies have revealed that ABA signaling components were activated in guard cells. A decrease in concentration of auxins, GAs and CKs and an increase of ABA, JA, ethylene and SA levels are often observed in drought stressed plants. ABA decreases the H^+ concentration (high pH) in the guard cell cytosol by acting on the H^+ transporters, which enhance K^+ out channel activity and long-term efflux of anions from guard cells that contribute to the loss of guard cell turgor leading to stomatal closing (Schroeder et al. 2001). In addition, some workers have suggested that stomatal functions under conditions of water deficit were regulated by JA, ethylene and SA and the complex of intracellular signaling events led to stomatal closure (Acharya and Assmann 2009; Siddiqui 2011). The other hormones particularly, CK, SA, ethylene and JA also play significant roles, direct or indirect, in plants under stress conditions. Cytokinins increase the life span of green leaves by regulating cell division. They induce the biosynthesis of many proteins particularly important under the condition of water stress. It has also been suggested that JA may interact with ABA synthesis under water stress conditions (Bandurska et al. 2003; Adie et al. 2007) and that ABA and JA could regulate stomatal closure (Acharya and Assmann 2009). However, our understanding about changes in SA, ethylene and JA concentrations in plants under drought stress is poor.

Auxin plays a positive regulatory role in stomatal opening, although a high exogenous application of auxin has been shown to inhibit stomatal opening (Lohse and Hedrich 1992). Antagonistic stomatal regulation has been observed between auxin and ABA. Auxin represses stomatal closure in response to ABA in epidermal peels of Arabidopsis (Tanaka et al. 2006) under water stress conditions. The change in IAA level under drought stress is very inconsistent. Drought stress decreased IAA concentration in the leaves of wheat (Xie et al. 2003; Bano and Yasmeen 2010) or corn (Wang et al. 2008). However, some studies have indicated that the adaptation to drought could enhance the IAA concentration (Pustovoitova et al. 2004).

Exogenously applied CKs alleviated the negative effects of water stress on chlorophyll and carotenoid contents, photochemical activities of photosystem I and II and content and activity of Rubisco and phosphoenolpyruvate carboxylase (Metwally et al. 1998; Chernyad'ev and Monakhova 1998; Pandey et al. 2000; Singh et al. 2001). However, it is not known whether improved photosynthesis is due to specific effect of CK on photosynthetic enzyme activity or due to delayed senescence caused by CK treatment. In isolated systems (detached epidermis and leaves), IAA and some CKs antagonize ABA-induced stomatal closure. In intact cotton plants, foliar sprays of 50 μM IAA and GA_3 partially counteracted the water deficit on net photosynthesis, stomatal conductance and transpiration rate (Kumar et al. 2001).

The principle effect of ABA on photosynthesis is due to stomatal closure with decreased intercellular CO_2 concentration (Meyer and Genty 1998). However, it has been observed that there was ABA-induced reduction in net photosynthesis at constant internal CO_2 concentration (Raschke and Hedrich 1985) suggesting a possible mechanism of ABA in carboxylation capacity. The possible mechanisms by which ABA reduces photosynthesis are decreased activity of Rubisco, decreased generation of ribulose-1, 5-bisphosphate or inhibition of ATP-synthase. Moreover, the treatment of ABA suppresses the increase in electrolyte leakage and decline in Fv/Fm ratio, which suggest the role of ABA in reducing drought injury by protecting cell membrane and photosynthetic apparatus. Rajasekaran and Blake (1999) treated old jack pine seedlings with ABA and found higher photosynthetic rate together with lower electrolyte leakage than untreated plants. ABA application resulted in the decline of Fv/Fm ratio during drought stress in Kentucky bluegrass (Wang et al. 2003). It has been found that GA_3 promotes the formation of stomata in hypocotyl cells of *Arabidopsis thaliana*. In addition, auxin and ethylene enhanced GA-mediated formation of stomatal complexes in this organ (Saibo et al. 2003).

The role of SA in the limitation of plant dehydration is associated with its regulatory role in promoting stomatal closure (Acharaya and Assmann 2009). Prior to exposure of plants to water stress SA reduced the damaging action of water deficit on cell membranes and increased the ABA content in barley leaves (Bandurska and Stroinski 2005). SA-treated wheat seedlings under drought stress displayed higher moisture content, Rubisco activity, stomatal conductance and total chlorophyll content. SA treatment also protected nitrate reductase activity and thereby assisted in maintaining normal levels of various proteins in leaves (Singh and Usha 2003). Exogenous spraying of SA ameliorated leaf chlorophyll and carotenoid loss of barley plants grown under water stress (Fayez and Bazaid 2013).

Jasmonic acid is an important player in stomatal closure during drought stress (Creelman and Mullet 1997). Its accumulation plays positive regulatory role in stomatal closure (Munemasa et al. 2007). A rapid JA accumulation was observed in water-deficient soybean leaves (Creelman and Mullet 1995). It has been shown that JA regulates the accumulation of soybean vegetative storage protein (VSP) (Anderson 1991; Mason and Mullet 1990). Once the water deficit is removed, amino acids in VSP are remobilized and used for leaf and stem growth. JA induces the loss of chlorophyll that decreases the amount of energy absorbed by photosynthetic apparatus. These changes help to ameliorate the photochemical damage when water deficit conditions persist (Galli and Levi 1982; Pietrini et al. 2002).

Ethylene plays an important role in guard channel signaling under drought stress. Ethylene interacts with ABA and affects stomatal movement (Tanaka et al. 2005). However, detailed studies are required to unravel the interactive play between ethylene and ABA in stomatal movement and photosynthetic functions under drought stress. Polyamines applied at 10 μM increased net photosynthesis, water-use efficiency and leaf water content (Farooq et al. 2009). Similarly, BR has been reported to induce the expression of stress-related genes, leading to the maintenance of photosynthesis activity, the activation of antioxidant enzymes, the accumulation of osmoprotectants and the induction of other hormone responses (Divi and Krishna 2009). Pustovoitova et al. (2001) observed that cucumber plants sprayed with a synthetic BR, 24- EBR, improved resistance to dehydration. BR treatment has also been reported to increase germination and seedling growth of *sorghum vulgare* under osmotic stress (Vardhini and Rao 2003) and enhances drought tolerance in *Phaseolus Vulgaris* (Upreti and Murti 2004). In addition, it has been shown that treatment with EBR increased the survival rate of *Arabidopsis thaliana* and *Brassica napus* seedlings subjected to drought stress (Krishna 2003; Kagale et al. 2007).

Table 3. Mediation of phytohormones in protection of photosynthetic attributes under water stress

	Plant /transgenic	Photosynthetic attributes	Response	Treatment of water stress	Reference
Auxin	*Triticumaestivum*	Photosynthesis, water deficiency	+	Withholding of irrigation	Abdoli et al. 2013
CK	*Prunus armeniaca*	Net photosynthesis, stomatal conductance	+	by increasing VPD and limiting watering frequency	Arzani et al. 2005
GA	*Helianthus annus*	Chlorophyll content, relative water content	+	By maintaining the soil moisture level 50% pot capacity whereas well watered maintained in full pot capacity	Kaya et al. 2006
ABA	*Phaseolus vulgaris, Nicotianatabacum, Beta vulgaris, Zea mays*	Pigment contents, chlorophyll fluorescence	+	by cessation of watering	Haisel et al. 2006
SA	*Myric rubra*	Chlorophyll content, net photosynthesis, stomatal conductance	+	Watered and maintained at 20–25 % field capacity	Ying et al. 2013
SA	*Triticumaestivum*	Chlorophyll content, gas exchange parameters, water use efficiency	+	PEG–grown in C + PEG 6000 (−0.75 MPa) causing osmotic stress	Marcinska et al. 2013
Auxin + JA	*Allium sativum*	Chlorophyll content, carotenoids content	+	by reducing available water supply by 40% via irrigation tapes	Bideshki et al. 2013
BR	*Oryzasativa*	Gas exchange parameters, transpiration rate	+	Drought was maintained at 50 % field capacity by applying water on alternate days or whenever needed	Farooq et al. 2009
BR	*Solanum lycopersicum*	Gas exchange parameters	+	Withholding water	Yuan et al. 2010
NO	*Tageteserecta*	Chlorophyll content, chlorophyll fluorescence	+	by cessation of watering	Liao et al. 2012
PAs	*Oryzasativa*	Net photosynthesis, water-use efficiency and leaf water content	+	Drought was maintained as 50% of field capacity by curtailing the water supply on alternate days	Farooq et al. 2009

Schilling et al. (1991) reported that homobrassinolide treatment increased tap-root weight, sucrose content, and yield of sugar beets grown under drought stress. Sairam et al. (1996) found that homobrassinolide application increased water uptake and membrane stability as well as reduced ion leakage in two wheat varieties. The summary of phytohormones involvement in photosynthetic attributes under water stress is presented in Table (3).

Nutrient Stress and Photosynthesis

The practice of addition of mineral nutrients to soils for improving plant growth is known in agriculture for many years. It has been shown that limitation in nutrient status in agricultural system brings nutrient stress in plants and also disturbance in metabolism and crop productivity. Inadequate supply of the essential mineral elements results in a nutritional disorder and characteristic deficiency symptoms. Nutrient deficiency may result in alteration in photosynthates or phytohormones levels. Similarly, excess of nutrient elements may also cause stress and limit the other nutrients uptake and show deficiency symptoms. Studies have shown that nitrogen stress reduces photosynthetic capacity of C_3 and C_4 plants and alters the partitioning of carboxylase activity between Rubisco and PEPC. In addition to the overall expected decline in photosynthesis, N stress substantially reduces the intrinsic efficiency of the CO_2 concentrating system associated with C_4 pathway and reduces quantum yields in sugarcane (Meinzer and Zhu 1998). Low NO_3^-, PO_4^{2-}, or K^+ concentration decreased the net rate of CO_2 uptake and total chlorophyll content due to lowering of stomatal conductance in *Gossypium hirsutum* L. (Longstreth and Nobel 1980). To avoid the nutrient limitations, plants induce certain mechanisms of adaptation including control of phytohormones biosynthesis. The interaction between nutrients and phytohormones has been reported. Nitrogen application causes an increase in CK content in xylem sap and also in roots and shoots of maize (Takei et al. 2001), indicating that CK functions as a root-to-shoot long-distance signal of nitrogen supplement. In *Arabidopsis thaliana*, trans- ZR-59-monophosphate and trans-ZR were accumulated in roots in response to nitrate resupply suggesting CK as root-to-shoot signal communicator (Takei et al. 2002). Auxin plays an important role in mediating nitrogen signals from shoot to root because auxin is transported basipetally and enhances lateral root initiation and development (Forde 2002; Fukaki and Tasaka 2009). Increased nitrogen supply stimulates plant growth and productivity as well as photosynthetic capacity of leaves through increased amounts of stromal and thylakoid proteins in leaves. Nitrogen influences the quantity, structure, and composition of the photosynthetic apparatus and hence plays a crucial role in determining the photosynthetic capacity of the plant by increasing Rubisco activity in *Brassica juncea* (Iqbal et al. 2012). Under nutrient limitation, growth of wild-type tobacco cultures was reduced and carbon use efficiency decreased by 42–63%. In wild-type mitochondria, the capacity of the cyt pathway increased slightly by growth under low P while it was significantly reduced at low N (Sieger et al., 2005). Moniuszko et al. (2013) have shown that the regulation of ethylene signaling is a part of plants' response to S deficiency. Excess-S induces oxidative stress that leads to high ethylene evolution which is potentially harmful to plants, while sufficient-S produces ethylene in optimal range and enhances photosynthetic capacity in *Brassica juncea* (Masood et al. 2012). Anjum et al. (2011) reported that S deprivation caused acute depressions in net photosynthesis and its related variables, such as intercellular CO_2, chlorophyll fluorescence and chlorophyll content in mustard. The decrease in photosynthetic attributes under S deprivation induced photo-damage to PS II reaction centers and reduced the maximum quantum efficiency of PS II photochemistry. The deficiency of various nutrients such as phosphorus, potassium, calcium, magnesium, nitrogen and iron influenced ethylene production together with altered agricultural productivity. It was, thus suggested that the nutrients shortage can be maintained by targeting ethylene biosynthesis genes for increased nutrients-use-efficiency (Iqbal et al. 2013). Phosphorus deficient plants showed decreased CO_2 assimilation and stomatal conductance, but increased intercellular CO_2 concentration. The other photosynthetic parameters, Rubisco activity, contents of chlorophyll and total soluble protein in P-deficient leaves decreased to a lesser extent than CO_2 assimilation in *Camellia sinensis* (Lin et al. 2009). Ethylene is also involved in the control of K starvation responses. It has been reported that low K supply induced the expression of several genes controlling biogenesis of ethylene which is correlated with increased ethylene biosynthesis (Jung et al. 2009). This increase in ethylene levels could account for the reduced number of lateral roots found in K-deprived Arabidopsis plants (Rubio et al. 2009). However, still more research should be focused on the involvement of phytohormones in the regulation of photosynthetic attributes under nutrient stress.

Heat Stress and Photosynthesis

Heat stress has potential to disturb cellular homeostasis and can lead to severe reduction in growth and development of plants (Khan et al. 2013b). Alterations in photosynthetic attributes under heat stress are the prime indicators of the plant that are correlated with growth of the plants. Wise et al. (2004) have suggested that photochemical reactions in thylakoid lamellae and carbon metabolism in the stroma of chloroplast are primary sites of injury at high temperatures. Studies have established that moderately high temperature (35-40°C) reduces net photosynthesis (Santarius and Müller 1979; Berry and Björkman 1980; Enami et al. 1994; Crafts-Brandner and Salvucci 2000; Khan et al. 2013a). Heat stress has also potential to reduce PS II activity (Bukhov et al. 1999; Camejo et al. 2005) due to the properties of thylakoid membranes where PS II is located (Mcdonald and Paulsen 1997). Similarly, high temperature dissociates oxygen-evolving complex (OEC), resulting in an imbalance between the electron flow from OEC towards the acceptor side of PSII in the direction of the PSI reaction center (De Ronde et al. 2004). The optimum temperature is necessary for photosynthesis and for kinetic properties of Rubisco (Monson et al. 1982; Jordan and Ogren 1984; Sage and Sharkey 1987). High temperature inhibits Rubisco activity in several plants including spinach (*Spinacia oleracea*), wheat (*Triticum aestivum*), rice (*Oryza sativa*) and soybean (*Glycine max*) (Weis 1981; Kobza and Edwards 1987; Eckhardt and Portis 1997; Eckhardt et al. 1997; Salvucci et al. 2001; Vu et al. 2001; Gesch et al. 2003; Kim and Portis 2005; Yamori et al. 2006). Recently, Djanaguiraman et al. (2013) have shown that heat stress decreased quantum yield of PS II, electron transport rate and transcript levels of Rubisco activase. As heat stress has been considered as one of the potential threats for agricultural productivity, strategies are needed to reduce the adverse effects of heat stress in plants.

Studies on the influence of phytohormones such as ethylene (Biddington and Robinson 1993; Karagezov, 2002), CK (Arnison et al. 1990; Sayed, 1999) and SA (Khan et al. 2013a) have shown their role in alleviating the detrimental effect of heat stress. Salicylic acid has been the focus of studies because of its ability to induce thermo tolerance in mustard (Dat et al. 1998), wheat (Khan et al. 2013a), potato (Lopez-Delgado et al. 1998), tomato (Senaratna et al. 2000), grape (Wang and Li 2006) and *Arabidopsis* (Larkindale and Knight 2002; Clarke et al. 2004). Khan et al. (2013a) have shown that application of 0.5 mM SA induced temperature tolerance by reducing oxidative stress and maintained photosynthesis and growth of wheat plants. Dobra et al. (2010) reported that heat stress (40°C) caused a transient increase in the level of CKs with a transient decrease of ABA level along with increased content of protective xanthophyll cycle pigments (violaxanthin, antheraxanthin and zeaxanthin) during the heat stress. In another study, Cvikrová et al. (2012) reported that heat stress increased the levels of free and conjugated putrescine and also the levels of free spermidine, norspermidine and spermine after a 2-h lag phase. These changes were correlated with the stimulation of the activity of the corresponding biosynthetic enzymes.

Table 4. Mediation of phytohormones in protection of photosynthetic attributes under heat stress

	Plant	Photosynthetic attributes	Response	Treatment of temperature	Reference
CK	*Creeping bentgrass*	Chlorophyll, chlorophyll fluorescence	+	35/30°C	Veerasamy et al. 2007
GA	*Helianthus annuus*	Chlorophyll	+	51°C	Sayed and Gadallah. 2013
ABA	*Solanumlycopersicum*	Net CO_2 assimilation, stomatal conductance	+	sub-optimal day/night T (17/14∘C), optimalday/ night T levels (22/18∘C)	Ntatsi et al. 2014
Ethylene	*Quercus ilex*	Chlorophyll fluorescence	+	50°C	Munne-bosch et al. 2004
SA	*Triticum aestivum*	Chlorophyll, Rubisco Fv/Fm, water use efficiency, gas exchange parameters	+	40°C	Khan et al. 2013a
SA	*Vitisvinifera*	Gas exchange parameters, PS II activity	+	43°C	Wang et al. 2010
BR	*Solanummelongena*	Chlorophyll, gas exchange parameters, PS II activity	+	high temperature	Wu et al. 2014
NO	*Chrysanthemummorifolium*	Photosynthetic pigment, net photosynthetic rate	+	45°C	Yang et al. 2011
NO	*Oryzasativa*	Chlorophyll fluorescence, net photosynthesis, photochemical quenching	+	40, 45, 50°C	Song et al. 2013

Khan et al. (2013a) reported that proline production protected plants under heat stress. In heat sensed plants, proline protected OEC dissociation and helped in feeding electrons to PSII, maintaining an acceptable NADPH level and preventing further damage (De Ronde et al. 2004). Examination of transcriptome data for heat and MeJA responsive genes (Zimmermann et al. 2004) has indicated a preponderance of genes associated with protein translation and photosynthetic electron transport which could represent features associated with cellular recovery following heat stress (Clarke et al. 2009).

Further, they suggested that JA interacts with SA and confers thermo tolerance, while ethylene may promote cell death under heat stress. Vettakkorumakankav et al. (1999) reported that changes in chlorophyll fluorescence were associated with GA. The Fv/Fm ratio decreased after heat stress in the normal and dwarf barley plants treated with GA_3. This decline did not occur in the untreated-dwarf and in the normal plant treated with paclobutrazol (antagonist of the gibberellin).

The Fv/Fo ratio denotes the efficiency of electron transfer from photosystem II to the quinone pool. This ratio was relatively unchanged when measured 24 h after stress in the case of paclobutrazol treated normal and untreated dwarf plants, whereas it decreased significantly in the normal and in the GA_3-treated-dwarf plants providing further evidence for a stress-induced injury to the plants. They concluded that treatment of normal plants with paclobutrazol induced stress protection whereas treatment of dwarf plants with GA_3 enhanced susceptibility under heat stress. The summary of phytohormones involvement in photosynthetic attributes under heat stress is presented in Table (4).

Conclusion and Future Prospects

The reduction in agricultural productivity due to the changing environmental conditions has initiated multidirectional research to protect the photosynthetic apparatus. The protection of photosynthetic process involving different strategies is the only option for agricultural sustainability. Attempts have been made to induce tolerance in a range of plant species using different approaches. These include exogenous applications of phytohormones on plants or modulation of phytohormones biosynthesis. Phytohormones have been found very potential in enhancing photosynthetic mechanisms under optimal conditions and also in alleviation of adverse effects of abiotic stress on photosynthetic functions. The involvement of phytohormones in regulation of physiological and biochemical processes and molecular mechanisms accounts for the protection of photosynthesis under stress. Regulatory points of photosynthesis need to be identified to target photosynthesis using phytohormones that could improve tolerance in plants.

However, phytohormones regulate photosynthesis by initiating the complex network of physiological and molecular mechanisms, and the action of phytohormones depends on specific hormone combinations. The crosstalk of various phytohormones in protection of photosynthesis in the era of climate change might also help. The role of different phytohormones in regulation of photosynthetic processes under various abiotic stress has been discussed in this chapter, but more studies need to be focused on enhancing our understanding on the interaction between different phytohormones cross talk in regulating the photosynthesis processes.

ACKNOWLEDGMENT

Authors are thankful to University Grants Commission, Department of Science and Technology, Department of Biotechnology, Council of Scientific and Industrial Research, New Delhi for providing research financial assistance in the laboratory of Prof. Nafees A Khan.

REFERENCES

Abdoli, M., Saeidi, M., Jalali-Honarmand, S., Azhand, M., (2013). The effect of foliar application of Indole-3-Acetic Acid (IAA) and roles of ear photosynthesis on grain yield production of two wheat cultivars (*Triticum aestivum* L.) under post anthesis water deficit. *Int. Res. J. App. Basic Sci.,* 4, 1406-1413.

Achard, P., Cheng, H., De Grauwe, L., Decat, J., Schoutteten, H., Moritz, T., Van Der Straeten, D., Peng, J., Harberd, N. P., (2006). Integration of plant responses to environmentally activated phytohormonal signals. *Science,* 311, 91-94.

Achard, P., Renou, J. P., Berthomé, R., Harberd, N. P., Genschik, P., (2008). Plant DELLAs restrain growth and promote survival of adversity by reducing the levels of reactive oxygen species. *Curr. Biol.,* 18, 656-660.

Acharya, B. R., Assmann, S. M., (2009). Hormone interactions in stomatal function. *Plant Mol. Biol.,* 69, 451-462.

Adie, B. A., Perez-Perez, J., Perez-Perez, M. M., Godoy, M., Sanchez-Serrano, J. J., Schmelz, E. A., Solano, R., (2007). ABA is an essential signal for plant resistance to pathogens affecting JA biosynthesis and the activation of defences in Arabidopsis. *Plant Cell,* 19, 1665-1681.

Ahammed, G. J., Choudhary, S. P., Chen, S., Xia, X., Shi, K., Zhou, Y., Yu, J., (2013). Role of brassinosteroids in alleviation of phenanthrene cadmium co-contamination–induced photosynthetic inhibition and oxidative stress in tomato. *J. Exp. Bot.,* 64, 199-213.

Albacete, A., Ghanem, M. E., MartínezAndujar, C., Acosta, M., SánchezBravo, J., Martínez, V., Lutts, S., Dodd, I. C., and PérezAlfocea, F., (2008). Hormonal changes in relation to biomass partitioning and shoot growth impairment in salinized tomato (*Solanum lycopersicum* L.) plants. *J. Exp. Bot.,* 59, 4119-4131.

Aldesuquy, H. S., Ibrahim, A. H., (2001). Water relations, abscisic acid and yield of wheat plants in relation to the interactive effect of seawater and growth bioregulators. *J. Agron. Crop Sci.,* 187, 97-104.

Aldesuquy, H. S., Gaber, A. M., (1993). Effect of growth regulators on *Vicia faba* plants irrigated by seawater, leaf area, pigment content and photosynthetic activity. *Biol. Plant,* 35, 519-527.

Al-Hakimi, A. M. A., (2007). Modification of cadmium toxicity in pea seedlings by kinetin. *Plant Soil Environ.,* 53, 129-135.

Ali, H. M., Siddiqui, M. H., Basalah, M. O., Al-Whaibi, M. H., Sakran, A. M., Al-Amri, A., (2011). Effects of gibberellic acid on growth and photosynthetic pigments of *Hibiscus sabdariffa* L. under salt stress. *Afr. J. Biotechnol.,* 11, 800-804.

Alonso-Ramírez, A., Rodríguez, D., Reyes, D., Jiménez, J., Nicolás, G., López_Climent, M., Gómez-Cadenas, A., Nicolás, C., (2009). Evidence for a role of gibberellins in salicylic acid modulated early plant responses to abiotic stress in Arabidopsis seeds. *Plant Physiol.,* 150, 1335-1344.

Al-Whaibi, M. H., Siddiqui, M. H., Basalah, M. O., (2012). Salicylic acid and calcium-induced protection of wheat against salinity. *Protoplasma,* 249, 769-778.

Anderson, J. M., (1991). Jasmonic acid dependent increase in vegetative storage protein in soybean tissue cultures. *J. Plant Growth Regul.,* 10, 5-10.

Anjum, N. A., Umar, S., Iqbal, M., Ahmad, I., Pereira, M. E., Khan, N. A., (2011). Protection of growth and photosynthesis of *Brassica juncea* genotype with dual type sulfur transport system against sulfur deprivation by coordinate changes in the activities of sulfur metabolism enzymes and cysteine and glutathione production. *Russ. J. Plant Physiol.,* 58, 892-898.

Anuradha, S., Rao, S. S. R., (2001). Effect of brassinosteroids on salinity stress induced inhibition of seed germination and seedling growth of rice (*Oryza sativa* L.). *Plant Growth Regul.,* 33, 151-153.

Apel, K., Hirt, H., (2004). Reactive oxygen species: metabolism, oxidative stress, and signal transduction. *Annu. Rev. Plant Biol.,* 55, 373-379.

Arnison, P. G., Donaldson, P., Jackson, A., Semple, C., Keller, W., (1990). Genotype specific response of cultured boroccoli (*Brassica oleraceavar.* Italica) anthers to cytokinins. *Plant Cell Tissue Organ Cult.,* 20, 217–222.

Arzani, K., Wood, D. E. S., Lawes, G. S., (2005). Influence of water stress and exogenous cytokinin application on growth and gas exchange of young 'trevatt' apricot grown under aeroponic system and controlled environment conditions. International Conference and Exhibition on Soilless Culture. ISHS Acta Horti. 742.

Asada, K., (2006). Production and scavenging of reactive oxygen species in chloroplasts and their functions. *Plant Physiol.,* 141, 391-396.

Asada, K., Endo, T., Mano, J., Miyake, C., (1998). Molecular mechanism for relaxation of and protection from light stress. In: Saton K, Murata N, editors. Stress responses of photosynthetic organisms. Amsterdam: Elsevier. 37-52.

Ashraf, M. Y., Akhtar, K., Sarwar, G., Ashraf, M., (2002). Evaluation of arid and semi-arid ecotypes of guar (*Cyamopsis tetragonoloba* L.) for salinity (NaCl) tolerance. *J. Arid Environ.,* 52, 437-482.

Ashraf, M. Y., Sarwar, G., (2002). Salt tolerance potential in members of Brassicaceae. Physiological studies on water relations and mineral contents. In: Prospects for saline Agriculture. (Eds.): Ahmad, R., Malik, K. A., Kluwer Academic Publishers, Netherlands, 237-245.

Bandurska, H., Stroinski, A., (2005). The effect of salicylic acid on barley response to water deficit. *Acta Physiol. Plant,* 27, 379-386.

Bandurska, H., Stroinski, A., Kubı´s, J., (2003). The effect of jasmonic acid on the accumulation of ABA proline and spermidine and its influence on membrane injury under water deficit in two barley genotypes. *Acta Physiol. Plant,* 25, 279-285.

Bano, A., Yasmeen, S., (2010). Role of phytohormones under induced drought stress in wheat. *Pak. J. Bot.,* 42, 2579-2587.

Bernacchia, G., Salamini, F., Bartels, D., (1996). Molecular characterization of the rehydration process in the resurrection plant *Craterostigma plantagineum. Plant Physiol.,* 111, 1043-1050.

Berry, J. A., Björkman, O., (1980). Photosynthetic response and adaptation to temperature in higher plants. *Annu. Rev. Plant Physiol.,* 31, 491-543.

Biddington, N. L., Robinson, H. T., (1993). High temperature enhances ethylene promotion of anther filament growth in Brussels sprouts (*Brassica oleracea* var. Gemmifera). *Plant Growth Regul.,* 12, 29-35.

Bideshki, A., Arvin, M. J., Darini, M., (2013). Interactive effects of Indole- 3-butyric acid (IBA) and salicylic acid (SA) on growth parameters, bulb yield and allicin contents of garlic (*Allium sativum*) under drought stress in field. *Int. J. Agro. Plant Prod.,* 4, 271-279.

Bode, J., Wild, A., (1984). The influence of 2-chloroethyl trimethyl ammonium chloride (CCC) on growth and photosynthetic metabolism of young wheat plants (*Triticum aestivum* L.). *J. Plant Physiol.,* 116, 435-446.

Bukhov, N. G., Wiese, C., Neimanis, S., Heber, U., (1999). Heat sensitivity of chloroplasts and leaves: leakage of protons from thylakoids and reversible activation of cyclic electron transport. *Photosyn Res.,* 59, 81-93.

Camejo, D., Rodíguez, P., Morales, M. A., Dell'amico, J. M., Torrecillas, A., Alarc´on, J. J., (2005). High temperature effects on photosynthetic activity of two tomato cultivars with different heat susceptibility. *J. Plant Physiol.,* 162, 281-289.

Cao, W. H., Liu, J., He, X. J., Mu, R. L., Zhou, H. L., Chen, S. Y., Zhang, J. S., (2007). Modulation of ethylene responses affects plant salt-stress responses. *Plant Physiol.,* 143, 707-719.

Capiati, D. A., País, S. M., Téllez-Iñón, M. T., (2006). Wounding increases salt tolerance in tomato plants: evidence on the participation of calmodulin-like activities in cross-tolerance signaling. *J. Exp Bot.,* 57, 2391-2400.

Chaves, M. M., Flexas, J., Pinheiro, C., (2009). Photosynthesis under drought and salt stress: regulation mechanisms from whole plant to cell. *Ann. Bot.,* 103, 551-560.

Chaves, M. M., Maroco, J. P., Pereira, J. S., (2003). Understanding plant responses to drought from genes to the whole plant. *Funct. Plant Biol.,* 30, 239-264.

Chernyad'ev, I. I., Monakhova, O. F., (1998). The activity and content of ribulose-1,5-bisphosphate carboxylase/oxygenase in wheat plants as affected by water stress and kartolin-4. *Photosynthetica,* 35, 603-610.

Choi, J., Hwang, I., (2007). Cytokinin: perception, signal transduction, and role in plant growth and development. *J. Plant Biol.,* 50, 98-108.

Choudhary, S. P., Oral, H. V., Bhardwaj, R., Yu, J. Q., Tran, L. S. P., (2012). Interaction of brassinosteroids and polyamines enhances copper stress tolerance in *Raphanus sativus. J. Exp. Bot.,* 63, 5659-5675.

Chu, C., Dai, Z., Ku, M. S., Edwards, G. E., (1990). Induction of Crassulacean acid metabolism in the facultative halophyte *Mesembryanthemum crystallinum* by abscisic acid. *Plant Physiol.,* 93, 1253-1260.

Clarke, S. M., Cristescu, S. M., Miersch, O., Harren, F. J. M., Wasternack, C., Mur, L. A. J., (2009). Jasmonates act with salicylic acid to confer basal thermotolerance in *Arabidopsis thaliana. New Phytol.,* 182, 175-187.

Clarke, S. M., Mur, L. A., Wood, J. E., Scott, I. M., (2004). Salicylic acid dependent signaling promotes basal thermotolerance but is not essential for acquired thermotolerance in *Arabidopsis thaliana*. *Plant J.,* 38, 432-447.

Crafts-Brandner, S. J., Salvucci, M. E., (2000). Rubisco activase constrains the photosynthetic potential of leaves at high temperature and CO_2. *Proc. Natl. Acad. Sci. USA,* 97, 13430-13435.

Creelman, R. A., Mullet, J. E. (1995). Jasmonic acid distribution and action in plants: regulation during development and response to biotic and abiotic stress. *Pro. Nat. Acad. Sci. USA,* 92, 4114-4119.

Creelman, R. A., Mullet, J. E., (1997). Biosynthesis and action of jasmonates in plants. *Annu. Rev. Plant Physiol. Plant Mol. Biol.,* 48, 355-381.

Cvikrová, M., Gemperlová, L., Dobr, J., Martincová, O., Prásil, I. T., (2012). Effect of heat stress on polyamine metabolism in proline-over-producing tobacco plants. *Plant Sci.,* 182, 49-58.

Dalio, R. J. D., Pinheiro, H. P., Sodek, L., Haddad, C. R. B., (2011). The effect of 24-epibrassinolide and clotrimazole on the adaptation of *Cajanus Cajan* (L.) Millsp. tosalinity. *Acta Physiol. Plant,* 33, 1887-1896.

Dash, M., Panda, S. K., (2001). Salt stress induced changes in growth and enzyme activities in germinating *Phaseolus muingo* seeds. *Biol. Plant,* 44, 587-589.

Dat, J. F., Christine, H. Foyer., Ian, M. Scott., (1998). Changes in salicylic acid and antioxidants during induced thermotolerance in mustard seedlings. *Plant Physiol.,* 118, 1455-1461.

De Ronde, J. A. D., Cress, W. A., Kruger, G. H. J., Strasser, R. J., Staden, J. V., (2004). Photosynthetic response of transgenic soybean plants containing an Arabidopsis P5CR gene, during heat and drought stress. *J. Plant Physiol.,* 61, 1211-1244.

Desikan, R., Cheung, M. K., Bright, J., Henson, D., Hancock, J, T., Neill, S. J., (2004). ABA, hydrogen peroxide, and nitric oxide signalling in stomatal guard cells. *J. Exp. Bot.,* 55, 205-212.

Dias, M. C., Monteiro, C., Moutinho-Pereira, J., Correia, C., Gonçalves, B., Santos, C., (2013). Cadmium toxicity affects photosynthesis and plant growth at different levels. *Acta Physiol. Plant,* 35, 1281-1289.

Dijkstra, P., Reegan, H., Kuiper, P. J. C., (1990). Relation between relative growth rate, endogenous gibberellins and the response to applied gibberellic acid on *Plantago ovata. Physiol. Plant,* 79, 629-634.

Distefano, A. M., Garcia-Mata, C., Lamattina, L., Laxalt, A. M., (2008). Nitric oxide-induced phosphatidic acid accumulation: a role for phospholipases C and D in stomatal closure. *Plant Cell Environ.,* 31, 187-194.

Divi, U. K., Krishna, P., (2009). Brassinosteroid: a biotechnological target for enhancing crop yield and stress tolerance. *New Biotech.,* 26, 131-136.

Djanaguiraman, M., Prasad, P. V. V., Boyle, D. L., Schapaugh, W. T., (2013). Soybean pollen anatomy, viability and pod set under high temperature stress. *J. Agro. Crop Sci.,* 199, 171-177.

Dobra, J., Motyka, V., Dobrev, P., Malbeck, J., Prasil, I. T., Haisel D., Gaudinova, A., Havlova, M., Gubis, J., Vankova, R., (2010). Comparison of hormonal responses to heat, drought and combined stress in tobacco plants with elevated proline content. *J. Plant Physiol.* 167, 1360-1370.

Dombrowski, J.E., (2003). Salt stress activation of woundrelated genes in tomato plants. *Plant Physiol.* 132, 2098-2107.

dos Santos, R., Vergauwen R., Pacolet P., Lescrinier E., Van den Ende W., (2013). Manninotriose is a major carbohydrate in red deadnettle (*Lamium purpureum*, Lamiaceae). *Ann. Bot.* 111, 385-393.

Eckhardt, N.A., Portis, A.R. Jr., (1997). Heat denaturation profiles of ribulose-1,5-bisphosphate carboxylase/oxygenase (Rubisco) and Rubisco activase and the inability of Rubisco activase to restore activity of heat-denatured Rubisco. *Plant Physiol.* 113, 243-248.

Eckhardt, N.A., Snyder, G.W., Portis, A.R. Jr., Ogren, W.L., (1997). Growth and photosynthesis under high and low irradiance of *Arabidopsis thaliana* antisense mutants with reduced ribulose-1, 5-bisphosphate carboxylase/oxygenase activase content. *Plant Physiol.*, 113, 575-586.

Enami, I., Kitamura, M., Tomo, T., Isokawa, Y., Ohta, H., Katoh, S., (1994). Is the primary cause of thermal inactivation of oxygen evolution in spinach PSII membranes release of the extrinsic 33 kDa protein or of Mn. *Biochim. Biophys. Acta.*, 1186, 52-58.

Fahad, S., Bano, A., (2012). Effect of salicylic acid on physiological and biochemical characterization of maize grown in saline area. *Pak. J. Bot.*, 44, 1433-1438.

Falkowska, M., Pietryczuk, A., Piotrowska, A., Bajguz, A., Grygoruk, A., Czerpak, R., (2011). The effect of gibberellic acid (GA_3) on growth, metal biosorption and metabolism of the green algae *Chlorella vulgaris* (chlorophyceae) Beijerinck exposed to cadmium and lead stress. *Polish J. Environ. Stud.*, 20, 53-59.

Farooq, M., Wahid, A., Basra, S. M. A., Din, I., (2009). Improving water relations and gas exchange with Brassinosteroids in rice under drought stress. *J. Agron. Crop Sci.*, 195, 262-269.

Fayez, K. A., Bazaid, S. A., (2013). Improving drought and salinity tolerance in barley by application of salicylic acid and potassium nitrate. *J. Saudi Soc. Agri. Sci.*, 13, 45-55.

Feller, U., Anders, I., Mae, T., (2008). Rubiscolytics: fate of Rubisco after its enzymatic function in a cell is terminated. *J. Exp. Bot.*, 59, 1615-1624.

Feng, J., Barker, A. V., (1992). Ethylene evolution and ammonium accumulation by tomato plants under water and salinity stress. *J. Plant Nut.*, 15, 2471-2490.

Filová, A., Sytar, O., Krivosudská, E., (2013). Effects of brassinosteroid on the induction of physiological changes in *Helianthus annuus* l. under copper stress. *Acta Univ. Agric. Silvic. Men. Brun.*, 61, 623-629.

Forde, B. G., (2002). Local and long-range signaling pathways regulating plant responses to nitrate. *Ann. Rev. Plant Biol.*, 53, 203-224.

Fricke, W., Akhiyarova, G., Veselov, D., Kudoyarova, G., (2004). Rapid and tissue-specific changes in ABA and in growth rate in response to salinity in barley leaves. *J. Exp. Bot.*, 55, 1115-1123.

Fukaki, H., Tasaka, M., (2009). Hormone interactions during lateral root formation. *Plant Mol. Biol.*, 69, 437-449.

Gadallah, M. A. A., El-Enany, A. E., (1999). Role of kinetin in alleviation of copper and zinc toxicity in *Lupinus termis* plants. *Plant Growth Regul.*, 29, 151-160.

Galli, M. G., Levi, M., (1982). Increased drought resistance induced by pretreatment with abscisic acid in germinating embryos of *Haplopappus gracilis*. *Physiol. Plant*, 54, 425-430.

Garcia-Mata, C., Lamattina, L., (2002). Nitric oxide and abscisic acid cross talk in guard cells. *Plant Physiol.*, 128, 790-792.

Gemrotová, M., Kulkarni, M. G., Stirk, W. A., Strnad, M., Van Staden, J., Spíchal, L., (2013). Seedlings of medicinal plants treated with either a cytokinin antagonist (PI-55) or an inhibitor of cytokinin degradation (INCYDE) are protected against the negative effects of cadmium. *Plant Growth Regul.,* 71, 137-145.

Georgieva, K., Szigeti, Z., Sarvari, E., Gaspar, L., Maslenkova, L., Peva, V., Peli, E., Tuba, Z., (2007). Photosynthetic activity of homoiochlorophyllous desiccation tolerant plant *Haberlea rhodopensis* during desiccation and rehydration. *Planta,* 225, 955-996.

Gesch, R. W., Kang, I. H., Gallo-Meagher, M., Vu, J. C. V., Boote, K. J., Allen, L. H., Bowes, J. R. G., (2003). Rubisco expression in rice leaves is related to genotypic variation of photosynthesis under elevated growth CO_2 and temperature. *Plant Cell Environ.,* 26, 1941-1950.

Grant, M. R., Jones, J. D., (2009). Hormone (dis) harmony moulds plant health and disease. *Science,* 324, 750-52.

Grossmann, K., Heusen, H., (2001). Ethylene triggered abscisic acid. A principal in plant growth regulation? *Physiol. Plant,* 113, 9-14.

Gurmani, A. R., Bano, A., Najeeb, U., Zhang, J., Khan, S. U., Flowers, T. J., (2013). Exogenously applied silicate and abscisic acid ameliorates the growth of salinity stressed wheat (*Triticum aestivum* L) seedlings through Na^+ exclusion. *Aus. J. Crop Sci.,* 7, 1123-1130.

Gurmani, A. R., Bano, A., Din, J., Khan, S. U., Hussain, I., (2011). Effect of phytohormones on growth and ion accumulation of wheat under salinity stress. *Afr. J. Biotech.,* 8, 1887-1894.

Haisel, D., Pospisilova, J., Synkova, H., Schnablova, R., Batkova, P., (2006). Effects of abscisic acid or benzyladenine on pigment contents, chlorophyll fluorescence, and chloroplast ultrastructure during water stress and after rehydration. *Photosynthetica,* 44, 606-614.

Hajiboland, R., Ebrahimi, N., Poschenrieder, C., (2012). Bound Putrescine, a distinctive player under salt stress in the natrophilic sugar beet in contrast to glycophyte tobacco. *J. Sci. Isl. Rep. Iran.,* 23, 105-114.

Haldimann, P., Feller, U., (2005). Growth at moderately elevated temperature alters the physiological response of the photosynthetic apparatus to heat stress in pea (*Pisum sativum*) leaves. *Plant Cell Environ.,* 28, 302-317.

Hamayun, M., Khan, S. A., Khan, A. L., Shin, J. H., Ahmad, B., Shin, D. H., Lee, I. J., (2010). Exogenous gibberellic acid reprograms soybean to higher growth and salt stress tolerance. *J. Agri. Food Chem.,* 58, 7226-7232.

Hao, L., Zhao, Y., Jin, D., Zhang, L., Bi, X., Chen, H., Xu, Q., Ma, C., Li, G., (2012). Salicylic acid altering Arabidopsis mutants response to salt stress. *Plant Soil,* 354, 81-95.

Heide, O. M., Bush, M. G., Evans, L. T., (1985). Interaction of photoperiod and gibberellin on growth and photosynthesis of high latitude *Poa pratensis*. *Physiol. Plant,* 65, 135-145.

Helmy, Y. H., El Abd, S. O., Abou Hadid, A. F., El Beltag, U., El Betagy, A. S., (1994). Ethylene production from tomato and cucumber plants under saline conditions. *Egypt. J. Hort.,* 21, 153-160.

Hirose, N., Takei, K., Kuroha, T., Kamada-Nobusda, T., Hayashi, H., Sakakibara, H., (2008). Regulation of cytokinin biosynthesis, compartmentalization and translocation. *J. Exp. Bot.,* 59, 75-83.

Houghton, J. T., Collander, B. A., Ephraums, J. J., (1990). Climate Change – The IPCC Scientific Assessment. Cambridge University Press, Cambridge. 135.

Hsu, Y. T., Kao, C. H., (2008). Accumulation of ammonium ion in cadmium tolerant and sensitive cultivars of *Oryza sativa. Plant Growth Regul.,* 39, 271-276.

Hu, Y. F., Zhou, G., Na, X. F., Yang, L., Nan, W. B., Liu, X., Zhang, Y. Q., Li, L. J., Bi, Y. R., (2013). Cadmium interferes with maintenance of auxin homeostasis in Arabidopsis seedlings. *J. Plant Physiol.,* 170, 965-975.

Iqbal, M., Ashraf, M., (2005). Presowing seed treatment with cytokinins and its effect on growth, photosynthetic rate, ionic levels and yield of two wheat cultivars differing in salt tolerance. *J. Integ. Plant Biol.,* 47, 1315-1325.

Iqbal, M., Ashraf, M., (2013a). Salt tolerance and regulation of gas exchange and hormonal homeostasis by auxin-priming in wheat. *Pesq. Agro. Bras. Brasília,* 48, 1210-1219.

Iqbal, M., Ashraf, M., (2013b). Gibberellic acid mediated induction of salt tolerance in wheat plants: Growth, ionic partitioning, photosynthesis, yield and hormonal homeostasis. *Environ. Exp. Bot.,* 86, 76-85.

Iqbal, M., Ashraf, M., Jamil, A., (2006). Seed enhancement with cytokinins: changes in growth and grain yield in salt stressed wheat plants. *Plant Growth Regul.,* 50, 29-39.

Iqbal, N., Khan, N. A., Nazar, R., Silva, J. A., (2012). Ethylene-stimulated photosynthesis results from increased nitrogen and sulfur assimilation in mustard types that differ in photosynthetic capacity. *Environ. Exp. Bot.,* 78, 84-90.

Iqbal, N., Nazar, R., Syeed, S., Masood, A., Khan, N. A., (2011). Exogenously-sourced ethylene increases stomatal conductance, photosynthesis, and growth under optimal and deficient nitrogen fertilization in mustard. *J. Exp. Bot.,* 62, 4955-4963.

Iqbal, N., Trivellini, A., Masood, A., Ferrante, A., Khan, N. A., (2013). Current understanding on ethylene signaling in plants: The influence of nutrient availability. *Plant Physiol. Biochem.,* 73, 128-138.

Jae-Ung, H., Youngsook, L., (2001). Abscisic acid-induced actin reorganization in guard cells of day flower is mediated by cytosolic calcium levels and by protein kinase and protein phosphatise activities. *Plant Physiol.,* 125, 2120-2128.

Jhanji, S., Setia, R. C., Kaur, N., Kaur, P., Setia, N., (2012). Role of nitric oxide in cadmium-induced stress on growth, photosynthetic components and yield of *Brassica napus* L. *J. Environ. Biol.,* 33, 1027-1032.

Jordan, D. B., Ogren, W. L., (1984). The CO_2/O_2 specificity of ribulose-1,5-bisphosphate carboxylase oxygenase dependence on ribulose bisphosphate concentration, pH and temperature. *Planta,* 161, 308-313.

Jung, J. Y., Shin, R., Schachtman, D. P., (2009). Ethylene mediates response and tolerance to potassium deprivation in Arabidopsis. *Plant Cell,* 21, 607-621.

Kabar, K., (1987). Alleviation of salinity stress by plant growth regulators on seed germination. *J. Plant Physiol.,* 128, 179-183.

Kagale, S., Divi, U. K., Krochko, J. E., Keller, W. A., Krishna, P., (2007). Brassinosteroids confers tolerance in *Arabidopsis thaliana* and *Brassica napus* to a range of abiotic stresses. *Planta,* 225, 353-364.

Karagezov, T. G., (2002). Ethylene as a factor in the adaptability of wheat cells to culture. *Pro. Azerb. Nat. Acad. Sci. Biol. Sci. Ser.,* 1, 344-354.

Karlidag, H., Yildirim, E., Turan, M., (2009). Salicylic acid ameliorates the adverse effect of salt stress on strawberry. *Sci. Agric. (Piracicaba, Braz.),* 66, 180-187.

Kaya, C., Tuna, A. L., Yokas, I., (2009). The role of plant hormones in plants under salinity stress. *Book Salinity Water Stress*, 44, 45-50.

Kaya, M. D., Okçu, G., Atak, M., Çıkılı, Y., Kolsarıcı, Ö., (2006). Seed treatments to overcome salt and drought stress during germination in sunflower (*Helianthus annuus* L.). *Eur. J. Agron.,* 24, 291-295.

Kende, H., Zeevaart, J. A. D., (1997). The five classical plant hormones. *Plant Cell,* 9, 1197-1210.

Keramat, B., Kalantari, K. M., Arvin, M. J., (2009). Effects of methyl jasmonate in regulating cadmium induced oxidative stress in soybean plant (*Glycine max* L.). *Afr. J. Biotech.,* 3, 240-244.

Khadri, M., Tejera, N. A., Lluch, C., (2006). Alleviation of salt stress in common bean (*Phaseolus vulgaris*) by exogenous abscisic acid supply. *J. Plant Growth Regul.,* 25, 110-119.

Khalid, S., Parvaiz, M., Nawaz, K., Hussain, K., Arshad, A., Shawakat, S., Sarfaraz, Z. N., Waheed, T., (2013). Effect of indole acetic acid (IAA) on morphological, biochemical and chemical attributes of two varieties of maize (*Zea mays* L.) under salt stress. *World App. Sci.* J., 26, 1150-1159.

Khan, M. I. R., Asgher, M., Khan, N. A., (2014) Alleviation of salt-induced photosynthesis and growth inhibition by salicylic acid involves glycinebetaine and ethylene in mungbean (*Vigna radiate* L.), *Plant Physiol. Biochem.,* 80, 67-74.

Khan, M. I. R., Khan, N. A., (2014). Ethylene reverses photosynthetic inhibition by nickel and zinc in mustard through changes in PS II activity, photosynthetic nitrogen use efficiency, and antioxidant metabolism. *Protoplasma,* DOI 10.1007/s00709-014-0610-7.

Khan, M. I. R., Khan, N. A., (2013). Salicylic Acid and Jasmonates: Approaches in Abiotic Stress Tolerance. *J. Plant Biochem Physiol.,* 1:4.

Khan, M. I. R., Iqbal, N., Masood, A., Per, T. S., Khan, N. A., (2013a). Salicylic acid alleviates adverse effects of heat stress on photosynthesis through changes in proline production and ethylene formation. *Plant Signal. Behav.,* 8, 11, e26374.

Khan, M. I. R., Asgher, M., Khan, N. A., (2013b). Rising temperature in the changing environment: A serious threat to plants. *Clim. Chan. Environ. Sust.,* 1, 25-36.

Khan, M. I. R., Iqbal, N., Masood, A., Khan, N. A., (2012a). Variation in Salt Tolerance of Wheat Cultivars: Role of Glycinebetaine and Ethylene. *Pedosphere,* 22, 746-754.

Khan, M. I. R., Syeed, S., Nazar, R., Naser, A. A., (2012b). An Insight into the Role of Salicylic Acid and Jasmonic Acid in Salt Stress Tolerance. Phytohormones and abiotic stress tolerance in plants. (eds): Khan, N. A., Nazar, R., Iqbal, N., Naser, A. A., 277-300.

Khan, N. A., (1996). Effect of gibberellic acid on carbonic anhydrase, photosynthesis, growth and yield of mustard. *Biol. Plant,* 38, 145–147.

Khan, N. A., (2003). NaCl-inhibited chlorophyll synthesis and associated changes in ethylene evolution and antioxidative enzyme activities in wheat. *Biol Plant,* 47, 437–440.

Khan, N. A., (2004). Activity of 1-aminocyclopropane carboxylic acid synthase in two mustard (*Brassica juncea* L.) cultivars differing in photosynthetic capacity. *Photosynthetica,* 42, 477-480.

Khan, N. A., Mobin, M., Samiullah, (2005). The influence of gibberellic acid and sulfur fertilization rate on growth and S-use efficiency of mustard (*Brassica juncea*). *Plant Soil,* 270, 269-274.

Khan, N. A., Samiullah, (2003). Comparative effect of modes of gibberellic acid application on photosynthetic biomass distribution and productivity of rapeseed-mustard. *Physiol. Mol. Biol. Plants,* 9, 141-145.

Khan, N. A., Syeed, S., Masood, A., Nazar, R., Iqbal. N., (2010). Application of salicylic acid increases contents of nutrients and antioxidative metabolism in mungbean and alleviates adverse effects of salinity stress. *Int. J Plant Biol.,* 1:e1

Kholodova, V. P., Meshcheryakov, A. B., Rakitin, Yu, V., Karyagin, V. V., Kuznetsov, Vl. V., (2006). Hydraulic signal as a "primary messenger of water deficit" under salt stress in plants. *Dokl. Biol. Sci.,* 407, 155-157.

Khripach, V. A., Zhabinskii, V. N., de Groot, A., (1999). Brassinosteroids Đa new class of plant hormones. San Diego: Academic Press.

Kim, K., Portis, Jr. A. R., (2005). Temperature dependence of photosynthesis in Arabidopsis plants with modifications in Rubisco activase and membrane fluidity. *Plant Cell Physiol.,* 46, 522–530.

Kobza, J., Edwards, G. E., (1987). Influences of leaf temperature on photosynthetic carbon metabolism in wheat. *Plant Physiol.,* 83, 69-74.

Krishna, P., (2003). Brassinosteroid-mediated stress responses. *J. Plant Growth Regul.,* 22, 289-297.

Kulaeva, O. N., Kuznetsov, V. V., (2002). Recent advances and horizons of the cytokinin studying. *Russ. J. Plant Physiol.,* 49, 561-574.

Kumar, B., Pandey, D. M., Goswami, C. L., Jain, S., (2001). Effect of growth regulators on photosynthesis, transpiration, and related parameters in water stressed cotton. *Biol. Plant,* 44, 475-478.

Kumari, A., Sheokand, S., Swaraj, K., (2010). Nitric oxide induced alleviation of toxic effects of short term and long term Cd stress on growth, oxidative metabolism and Cd accumulation in Chickpea. *Braz. J. Plant Physiol.,* 22, 271-284.

Larkindale, J., Knight, M. R., (2002). Protection against heat stress-induced oxidative damage in Arabidopsis involves calcium, abscisic acid, ethylene, and salicylic acid. *Plant Physiol.,* 128, 682-695.

Lawlor, D., Tezara, W., (2009). Causes of decreased photosynthetic rate and metabolic capacity in water-deficient leaf cells: a critical evaluation of mechanisms and integration of processes. *Ann. Bot.,* 103, 561-579.

Lefebvre, S., Lawson, T., Zakhleniuk, O. V., Lloyd, J. C.,Raines, C. A., (2005). Increased sedoheptulose-1,7-bisphosphatase activity in transgenic tobacco plants stimulates photosynthesis and growth from an early stage in development. *Plant Physiol.,* 138, 451-460.

Leterrier, M., Chaki, M., Airaki, M., Valderrama, R., Palma, J. M., Barroso, J. B., Corpas, F. J., (2011). Function of S-nitrosoglutathione reductase (GSNOR) in plant development and under biotic/abiotic stress. *Plant Signal. Behav.,* 6, 789-793.

Leterrier, M., Valderrama, R., Chaki, M., Airaki, M., Palma, J. M., Barroso, J. B., Corpas, F. J., (2012). Function of nitric oxide under environmental stress conditions. In: Khan N., Nazar R., Iqbal N., Anjum N., editors. Phytohormones and abiotic stress tolerance in plants. Springer Heidelberg Dordrecht: London, New York, 99-113.

Li, X. J., Yang, M. F., Chen, H., Qu, L. Q., Chen, F., Shen, S. H., (2010). Abscisic acid pretreatment enhances salt tolerance of rice seedlings: proteomic evidence. *Bio. chim. Biophys. Acta.,* 1804, 929–940.

Liao, W. B., Huang, G. B., Yu, J. H., Zhang, M. L., Shi, X. L., (2011). Nitric oxide and hydrogen peroxide are involved in indole-3-butyric acid-induced adventitious root development in marigold. *J. Horti. Sci. Biotechnol.,* 86, 159–165.

Lin, Z., Zhong, S., Grison, D., (2009). Recent advances in ethylene research. *J. Exp. Bot.,* 60, 3311-3336.

Lin, Z. H., Chen, L. S., Chen, R. B., Zhang, F. Z., Jiang, H. X., Tang, N., (2009). CO_2 assimilation, ribulose-1, 5-bisphosphate carboxylase/oxygenase, carbohydrates and photosynthetic electron transport probed by the JIP-test, of tea leaves in response to phosphorus supply. *BMC Plant Biol.,* 9, 43.

Liu, F., Guo, F. Q., (2013). Nitric oxide deficiency accelerates chlorophyll breakdown and stability loss of thylakoid membranes during dark-induced leaf senescence in Arabidopsis. *PLoS ONE,* 8, e56345.10.1371.

Loggini, B., Scartazza, A., Brugnoli, E., Navari-Izzo, F., (1999). Antioxidant defense system, pigment composition and photosynthetic efficiency in two wheat cultivars subjected to drought. *Plant Physiol.,* 119, 1091—1099.

Lohse, G., Hedrich, R., (1992). Characterization of the plasma-membrane H+-ATPase from *Vicia faba* guard cells. Modulation by extracellular factors and seasonal changes. *Planta,* 188, 206–214.

Longstreth, D. J., Nobel, P. S., (1980). Nutrient Influences on Leaf Photosynthesis: effects of nitrogen, phosphorus, and potassium for *Gossypium hirsutum* L. *Plant Physiol.,* 65, 541-543.

Lopez-Delgado, H., Dat, J. F., Foyer, C. H., Scott, I. M., (1998) Induction of thermotolerance in potato micro plants by acetylsalicylic acid and H_2O_2. *J. Exp. Bot.,* 49, 713-720.

Marcińska, I., Czyczyło-Mysza, I., Skrzypek, E., Grzesiak, M., Janowiak, F., Filek, M., et al. (2013). Alleviation of osmotic stress effects by exogenous application of salicylic or abscisic acid on wheat seedlings. *Int. J. Mol. Sci.,* 14, 13171-13193.

Maslenkova, L. T., Zanev, Yu., Popova, L. P., (1989). Effect of abscisic acid on the photosynthetic oxygen evolution in barley chloroplasts. *Photosynth. Res.,* 21, 45-50.

Maslenkova, L. T., Zanev, Yu., Popova, L. P., (1990). Oxygen-evolving activity of thylakoids from barley plants cultivated on different concentrations of jasmonic acid. *Plant Physiol.,* 93, 1316-1320.

Mason, H. S., Mullet, J. E., (1990). Expression of two soybean vegetative storage protein genes during development and in response to water deficit, wounding, and jasmonic acid. *Plant Cell Online,* 2, 569-579.

Masood, A., Iqbal, N., Khan, N. A., (2012). Role of ethylene in alleviation of cadmium-induced photosynthetic capacity inhibition by sulphur in mustard. *Plant Cell Environ.,* 35, 524-533.

Mcdonald, G. K., Paulsen, G. M., (1997). High temperature effects on photosynthesis and water relations of grain legumes. *Plant Soil,* 196, 47-58.

Meinzer, F. C., Zhu, J., (1998). Nitrogen stress reduces the efficiency of the C4 CO_2 concentrating system, and therefore quantum yield, in *Saccharum* (sugarcane) species. *J. Exp. Bot.,* 49, 1227-1234.

Metwally, A., Tsonev, T., Zeinalov, Y., (1998). Effect of cytokinins on the photosynthetic apparatus in water-stressed and rehydrated bean plants. *Photosynthetica*, 34, 563-567.

Meyer, S., Genty, B., (1998). Mapping intercellular CO_2 mole fraction (Ci) on *Rosa Rubiginosa* leaves fed with abscisic acid by using chlorophyll fluorescence imaging. *Plant Physiol.,* 116, 947-957.

Mobin, M., Khan, N. A., (2007). Photosynthetic activity, pigment composition and antioxidative response of two mustard (*Brassica juncea*) cultivars differing in photosynthetic capacity subjected to cadmium stress. *J. Plant Physiol.,* 164, 601-610.

Moniuszko, G., Skoneczny, M., Zientara-Rytter, K., Wawrzyńska, A., Głów, D., Simona, M., Cristescu Frans, J. M., Sirko H. A., (2013). Tobacco LSU-like protein couples sulphur-deficiency response with ethylene signalling pathway. *J. Exp. Bot.,* 64, 5173-5182.

Monreal, J. A., Arias-Baldrich, C., Tossi, V., Feria, A. B., Rubio-Casal, A., García-Mata, C., Lamattina, L., García-Mauriño, S., (2013). Nitric oxide regulation of leaf phosphoenolpyruvate carboxylase-kinase activity: implication in sorghum responses to salinity. *Planta,* 238, 859-869.

Monson, R. K., Stidham, M. A., Williams, G. J. III., Edwards, G. E., Uribe, E. G., (1982). Temperature dependence of photosynthesis in *Agropyron smithii* Rybd: factors affecting net CO2uptake in intact leaves and contribution fromribulose-1,5-bisphosphate carboxylase measured in vivo and in vitro. *Plant Physiol.,* 69, 921-928.

Morillon, R., Chrispeels, M. J., (2001). The role of ABA and the transpiration stream in the regulation of the osmotic water permeability of leaf cells. *Proc. Natl. Acad. Sci. USA,* 98, 14138-14143.

Munemasa, S., Oda, K., Wantanabe-Suigimoto, M., Nakamura, Y., Shimoishi, Y., Murata, Y., (2007). The coronatine-insensetive 1 mutation reveals the hormonal signaling interaction between ABA and MeJA in Arabidopsis guard cells. Specific impairment of ion channel activation and second messenger production. *Plant Physiol.,* 143, 1398-1407.

Munné-Bosch, S., Peñuelas, J., Asensio, D., Llusia, J., (2004). Airborne ethylene may alter antioxidant protection and reduce tolerance of holm oak to heat and drought stress. *Plant Physiol.,* 136, 2937-2947.

Mutlu, F., Bozcuk, S., (2000). Effects of exogenous spermine on seed germination and early growth of sunflower Seeds under salinity stress. *Turk. J. Biol.,* 24, 635-643.

Nagel, O. W., Lambers, H., (2002). Changes in acquisition and partitioning of carbon and nitrogen in the gibberellin deficient mutants A70 and W335 of tomato (*Solanum lycopersicumi*). *Plant Cell Environ.,* 25, 883-891.

Nazar, R., Iqbal, N., Syeed, S., Khan, N. A., (2011). Salicylic acid alleviates decreases in photosynthesis under salt stress by enhancing nitrogen and sulfur assimilation and antioxidant metabolism differentially in two mungbean cultivars. *J. Plant Physiol.,* 168, 807-815.

Nazar, R., Khan, M. I. R., Iqbal, N., Masood, A., Khan, N. A., (2014). Involvement of ethylene in reversal of salt-inhibited photosynthesis by sulphur in mustard. *Physiol. Plant,* doi:10.1111/ppl.12173

Neill, S., Barros, R., Bright, J., Desikan, R., Hancock, J., Harrison, J., Morris, P., Ribeiro, D., Wilson, I., (2008). Nitric oxide, stomatal closure, and abiotic stress. *J. Exp. Bot.,* 59, 165-176.

Noreen, S., Ashraf, M., Akram, N. A., (2012). Does exogenous application of salicylic acid improve growth and some key physiological attributes in sunflower plants subjected to salt stress? *J. App. Bot. Food Qual.*, 84, 169.

Noriega, G. D., Santa Cruz, A., Batlle, M., Tomaro, Balestrasse, K., (2012). Heme oxygenase is involved in the protection exerted by jasmonic acid against cadmium stress in soybean roots. *Plant Growth Regul.*, 31, 79-89.

Ntatsi, G., Savvas, D., Huntenburg, K., Druege, U., Hincha, D. K., Zuther, E., Schwarz, D., (2014). A study on ABA involvement in the response of tomato to suboptimal root temperature using reciprocal grafts with notabilis, a null mutant in the ABA-biosynthesis gene LeNCED1. *Environ. Exp. Bot.*, 97, 11-21.

Ogweno, J. O., Song, X. S., Shi, K., Hu, W. H., Mao, W. H., Zhou, Y. H., You, J. Q., Nogues, S., (2008). Brassinosteroids alleviate heat-induced inhibition of photosynthesis by increasing carboxylation efficiency and enhancing antioxidant systems in *Lycopersicon esculentum*. *J. Plant Growth Regul.*, 27, 49-57.

Ouzounidou, G., Ilias, I., (2005). Hormone-induced protection of sunflower photosynthetic apparatus against copper toxicity. *Biol. Plant,* 49, 223-228.

Pancheva, T. V., Popova, L. P., (1998). Effect of salicylic acid on the synthesis of ribulose 1, 5-bisphosphate carboxylase/oxygenase in barley leaves. *J. Plant Physiol.,* 152, 381-386.

Pandey, D. M., Goswami, C. L., Kumar, B., Jain, S., (2000). Hormonal regulation of photosynthetic enzymes in cotton under water stress. *Photosynthetica,* 38, 403-407.

Parmar, P., Kumari, N., Sharma, V., (2013). Structural and functional alterations in photosynthetic apparatus of plants under cadmium stress. *Bot. Stud.*, 54:45.

Paul, M. J., Pellny, T. K., (2003). Carbon metabolite feedback regulation of leaf photosynthesis and development. *J. Exp. Bot.*, 54, 539-547.

Pietrini, F., Iannelli, M. A., Massacci, A., (2002). Anthocyanin accumulation in the illuminated surface of maize leaves enhances protection from photo-inhibitory risks at low temperature, without further limitation to photosynthesis. *Plant Cell Environ.,* 25, 1251-1259.

Piotrowska-Niczyporu, A., Bajguz, A., Zambrzycka, E., Godlewska-Żyłkiewicz, B., (2012). Phytohormones as regulators of heavy metal biosorption and toxicity in green algae *Chlorella vulgaris* (Chlorophyceae). *Plant Physiol. Biochem.,* 52, 52-65.

Poljakoff-Mayber, A., Lerner, H. R., (1994). Plants in saline environments. In: Pessarakli M (ed) Handbook of plant and crop stress. Dekker, New York. 65-96.

Popova, L. P., Lazova, G. H., Miteva, T. S., (1991). Carbonic anhydrase activity in barley leaves after treatment with abscisic acid and jasmonic acid. *Com. Ren ABS.,* 44, 51-54.

Pustovoitova, T. N., Zhdanova, N. E., Zholkevich, V. N., (2004). Changes in the levels of IAA and ABA in cucumber leaves under progressive soil drought. *Plant Physiol. (Russia),* 51, 513-517.

Pustovoitova, T. N., Zhdanova, N. E., Zholkevich, V. N., (2001). Epibrassinolide increases plant water resistance. *Dokl. Biochem. Biophys.,* 376, 36-38.

Qin, F., Sakuma, Y., Tran, L. S., Maruyama, K., Kidokoro, S., Fujita, Y., Fujita, M., Umezawa, T., Sawano, Y., Miyazono, K., Tanokura, M., Shinozaki, K., Yamaguchi-Shinozaki, K., (2008). Arabidopsis DREB2A-interacting proteins function as RING E3 ligases and negatively regulate plant drought stress responsive gene expression. *Plant Cell,* 20, 1693-1707.

Rajasekaran, L. R., Blake, T. J., (1999). New plant growth regulators protect photosynthesis and enhance growth under drought of jack pine seedlings. *J. Plant Growth Regul.,* 18, 175-181.

Raschke, K., Hedrich, R., (1985). Simultaneous and independent effects of abscisic acid on stomata and the photosynthetic apparatus in whole leaves. *Planta,* 163, 105-118.

Reddy, M. P., A. B. Vora., (1986). Changes in pigment composition. Hill reaction activity and saccharides metabolism in bajra (*Penisetum typhoides* S & H) leaves under NaCl salinity. *Photosynthetica,* 20, 50-55.

Rensen, V. J. J., Curwiel, V. B., (2000). Multiple functions of photosystem II. *Indian J. Biochem. Biophys.,* 37, 377-382.

Rubio, V., Bustos, R., Irigoyen, M. L., Cardona-Lopez, X., Rojas-Triana, M., Paz-Ares, J., (2009). Plant hormones and nutrient signaling. *Plant Mol. Biol.,* 69, 361-373.

Saeidi-Sar, S., Khavari-Nejad, R., Fahimi, H., Ghorbanli, M., Majd, A., (2007). Interactive effects of gibberellin A3 and ascorbic acid on lipid peroxidation and antioxidant enzyme activities in Glycine max seedlings under nickel stress. *Russ. J. Plant Physiol.,* 54, 74-79.

Sage, R. F., Sharkey, T. D., (1987). The effect of temperature on the occurrence of O_2 and CO_2 insensitive photosynthesis in field grown plants. *Plant Physiol.,* 84, 658-664.

Saibo, N. J. M., Vriezen, W. H., Beemster, G. T. S., Van Der Straeten, D., (2003). Growth and stomata development of Arabidopsis hypocotyls are controlled by gibberellins and modulated by ethylene and auxins. *Plant J.,* 33, 989-1000.

Saidia, I., Ayounia, M., Dhiebb, A., Chtouroub, Y., Chaïbia, W., Djebali, W., (2013). Oxidative damages induced by short-term exposure to cadmium in bean plants: Protective role of salicylic acid. *South Afr. J. Bot.,* 85, 32-38.

Sairam, R. K., Shukla, D. S., Deshmuk, P. S., (1996). Effect of homobrassinolide seed treatment on germination, α- amylase activity and yield of wheat under moisture stress condition. *Indian J. Plant Physiol.,* 1, 141-144.

Sairam, R. K., Deshmukh, P. S., Shukla, D. S., (1991). Influence of chloromequate chloride on photosynthesis and nitrate assimilation in wheat genotypes under water stress. *Indian J. Plant Physiol.,* 34, 222-227.

Salvucci, M. E., Osteryoung, K. W., Crafts-Brandner, S. J., (2001). Thermal denaturation in vitro and in vivo. *Plant Physiol.,* 127, 1053-1064.

Santarius, K. A., Müller, M., (1979). Investigations on heat resistance of spinach leaves. *Planta,* 146, 529-538.

Sayed, S. A., (1999). Effect of lead kinetin on the growth and some physiological components of safflower. *Plant Growth Regul.,* 29, 167-174.

Sayed, S. A., Gadallah, M. A. A., (2013). Gibberellic acid ameliorates the adverse effects of acid mist and improved antioxidant defense, water status and growth of acid misted sunflower plants. *J. Biol. Earth Sci.,* 3, B275-B285.

Sazanova, K. A., Bashmakov D. I., Brazaitytė A., Bobinas Č, Duchovskis P., Lukatkin A. S., (2012). The effect of heavy metals and thidiazuron on winter wheat (*Triticum aestivum* L.) seedlings. *Žemd. Agri.,* 99, 273-278.

Schilling, G., Schiller, C., Otto, S., (1991). Influence of brassinosteroids on organ relation and enzyme activities of sugar beet plants. In Brassinosteroids–Chemistry, Bioactivity and Application ACS Symposium Ser. HG Cutler, T. Yokota G. Adam, eds. American Chemistry Society, Washington DC. 208-219.

Schroeder, J. I., Kwak, J. M., Allen, G. J., (2001). Guard cell abscisic acid signalling and engineering drought hardiness in plants. *Nature,* 410, 327-330.

Seckin, B., Sekmen, A. H., Turkan, I., (2009). An enhancing effect of exogenous mannitol on the antioxidant enzyme activities in roots of wheat under salt stress. *J. Plant Growth Regul.,* 28, 12-20.

Senaratna, T., Touchell, D., Bunn, E., Dixon, K., (2000). Acetyl salicylic acid (Aspirin) and salicylic acid induce multiple stress tolerance in bean and tomato plants. *Plant Growth Regul.,* 30, 157-161.

Seneweera, S., Aben, S. K., Basra, A. S., Jones, B., Conroy, J. P., (2003). Involvement of ethylene in the morphological and developmental response of rice to elevated atmospheric CO_2 concentrations. *Plant Growth Regul.,* 39, 143-153.

Shah., (2007). Effects of salt stress on mustard as affected by gibberellic acid application. *Gen. Appl. Plant Physiol.,* 33, 97-106.

Sharkey, T. D., (2005). Effects of moderate heat stress on photosynthesis: importance of thylakoid reactions, Rubisco deactivation, reactive oxygen species, and thermotolerance provided by isoprene. *Plant Cell Environ.,* 28, 269-277.

Sharma, P., Kaur, H., Geetika, S., (2013). Effect of Jasmonic Acid on Photosynthetic Pigments and Stress Markers in *Cajanus cajan* (L.) Millsp. Seedlings under Copper Stress. *Amr. J Plant Sci.,* 4, 817-823.

Shevyakova, N. I., Musatenko, L. I., Stetsenko, L. A., Vedenicheva, N. P., Voitenko, L. V., Sytnik, K. M., Kuznetsov, Vl. V., (2013). Effects of abscisic acid on the con tents of polyamines and proline in common bean plants under salt stress. *Russ. J. Plant Physiol.,* 60, 200–211.

Shu, S., Yuan, L. Y., Guo, S. R., Liu, J. S. C., (2012). Effects of exogenous spermidine on photosynthesis, xanthophyll cycle and endogenous polyamines in cucumber seedlings exposed to salinity. *Afr. J. Biotech.,* 11, 6064-6074.

Shukla, N., Awasthi, R. P., Rawat, L., Kumar, J., (2012). Biochemical and physiological responses of rice (*Oryza sativa* L.) as influenced by *Trichoderma harzianum* under drought stress. *Plant Physiol. Biochem.,* 54, 78-88.

Siddiqui, M. H., Al-Whaibi, M. H., Basalah M. O., (2011). Role of nitric oxide in tolerance of plants to abiotic stress. *Protoplasma,* 248, 447-455.

Siege, S. M., Kristensen, B. K., Robson, C. A., Amirsadeghi, S., Eng, E. W., Abdel-Mesih, A., Møller, M. I., Vanlerberghe, G. C., (2005). The role of alternative oxidase in modulating carbon use efficiency and growth during macronutrient stress in tobacco cells. *J. Exp. Bot.,* 56, 1499-1515.

Sieger, S. M., Kristensen, B. K., Robson, C. A. et al., (2005). The role of alternative oxidase in modulating carbon use efficiency and growth during macronutrient stress in tobacco cells. *J. Exp. Bot.,* 56, 1499-1515.

Singh, B., Usha, K., (2003). Salicylic acid induced physiological and biochemical changes in wheat seedlings under water stress. *Plant Growth Regul.,* 39, 137-141.

Singh, D. V., Srivastava, G. C., Abdin, M. Z., (2001). Amelioration of negative effect of water stress in *Cassia angustifolia* by benzyladenine and/or ascorbic acid. *Biol. Plant,* 44, 141- 143.

Sokolovski, S., Hills, A., Gay, R., Garcia-Mata, C., Lamattina, L., Blatt, M. R., (2005). Protein phosphorylation is a prerequisite for intracellular Ca^{2+} release and ion channel control by nitric oxide and abscisic acid in guard cells. *Plant J.,* 43, 520–529.

Soussi, M., Ocana, A., Lluch, C., (1998). Effects of salt stress on growth, photosynthesis and nitrogen fixation in chickpea (*Cicer arietinum* L.). *J. Exp. Bot.,* 49, 1329-1337.

Stoeva, N., Berova, M., Vassilev, A., Zlatev, Z., (2006). Effect of exogenous polyamine diethylenetriamine on oxidative changes and photosynthesis in As-treated maize plants (*Zea mays* L.). *J. Cent. European Agri.,* 6, 3.

Stoeva, N., Berova, M., Zlatev, Z., (2005). Effect of arsenic on some physiological parameters in bean plants. *Biol Plant,* 49, 293-296.

Subrahmanyam, D., Rathore, V. S., (1992). Influence of ethylene on carbon-14 labeled carbondioxide assimilation and partitioning in mustard. *Plant Physiol. Biochem.,* 30, 81-86.

Sung, C. H., Hong, J. K., (2010). Sodium nitroprusside mediates seedling development and attenuation of oxidative stresses in Chinese cabbage. *Plant Biotech.,* 4, 243-251.

Swamy, K. N., Rao, S. S. R., (2006). Effect of 24-epibrassinolide on growth, photosynthesis, and essential oil content of *Pelargonium graveolens* (L.) Herit Russ. *J. Plant Physiol.,* 56, 616–620.

Tabur, S., Demir, K., (2010). Role of some growth regulators on cytogenetic activity of barley under salt stress. *Plant Growth Regul.,* 60, 99-104.

Takei, K., Sakakibara, H., Taniguchi, M., Sugiyama, T., (2001). Nitrogen-dependent accumulation of cytokinins in root and the translocation to leaf: implication of cytokinin species that induces gene expression of maize response regulator. *Plant Cell Physiol.,* 42, 85-93.

Takei, K., Takahashi, T., Sugiyama, T., Yamaya, T., Sakakibara, H., (2002). Multiple routes communicating nitrogen availability from roots to shoots: a signal transduction pathway mediated by cytokinin. *J. Exp. Bot.,* 53, 971-977.

Tanaka, Y., Sano, T., Tamaoki, M., Nakajima, N., Kondo, N., Hasezawa, S., (2006). Cytokinin and auxin inhibit abscisic acid-induced stomatal closure by enhancing ethylene production in Arabidopsis. *J. Exp. Bot.,* 57, 2259-2266.

Tanaka, Y., Sano, T., Tamaoki, M., Nakajima, N., Kondo, N., Hasezawa, S., (2005). Ethylene inhibits ABA-induced stomatal closure in arabidopsis. *Plant Physiol.,* 138, 2337-2343.

Tezara, W., Martínez, D., Rengifo, E., Herrera, A., (2003). Photosynthetic responses of the tropical spiny shrub *Lycium nodosum* (Solanaceae) to drought, soil salinity and saline spray. *Ann. Bot.,* 92, 757-765.

Thetford, M., Warren, S. L., Blazick, F. A., Thomas, J. F., (1995). Response of Forsynthia-x intermedia Spectabilisto unicongzole. 2. Leaf and stem anatomy, chlorophyll and photosynthesis. *J. Am .Soc. Hort. Sci.,* 120, 983-988.

Thomas, C. D., Cameron, A., Green, R. E., Bakkenes, M., Beaumont, L. J., et al., (2004). Extinction risk from climate change. *Nature,* 427, 145-148.

Tsonev, T. D., Lazova G. N., Stoinova, Z. G., Popova, L. P., (1998). A possible role for jasmonic acid in adaptation of barley seedlings to salinity stress. *J. Plant Growth Regul.,* 17, 153-159.

Upreti, K. K., Murti, G. S. R., (2004). Effects of brassinosteroids on growth, nodulation, phytohormones content, nitrogenise activity in French bean under water stress. *Biol. Plant,* 48, 407-411.

Vardhini, B. V., Rao S. S. R., (2003). Amelioration of water stress by brassinosteroids on seed germination and seedling growth of three varieties of sorghum. *Plant Growth Regul.,* 41, 25-31.

Veerasamy, M., He, Y., Huang, B., (2007). Leaf senescence and protein metabolism in creeping bentgrass exposed to heat stress and treated with cytokinins. *J. Amer. Soc. Horti. Sci.*, 132, 467-472.

Vettakkorumakankav, N. N., Falk, D., Saxena, P., Fletcher, R. A., (1999). A crucial roles of gibberellins in stress protection of plants. *Plant Cell Physiol.*, 40, 1542-1548.

Vu, J. C. V., Gesch, R. W., Pennanen, A. H., Allen, L. H. Jr., Boote, K. J., Bowes, G., (2001). Soybean photosynthesis, Rubisco, and carbohydrate enzymes function at supra optimal temperatures in elevated CO_2. *J. Plant Physiol.*, 158, 295-307.

Walia, H., Wilson, C., Condamine, P., Liu, X., Ismail, A. M., Close, T. J., (2007). Large-scale expression profiling and physiological characterization of jasmonic acid-mediated adaptation of barley to salinity stress. *Plant Cell Environ.*, 13, 410-421.

Wang, L. J., Fan, L., Loescher, W., Duan, W., Liu, G. J., Cheng, J. S., Luo, H. B., Li, S. H., (2010) Salicylic acid alleviates decreases in photosynthesis under heat stress and accelerates recovery in grapevine leaves. *BMC Plant Biol.*, 10:34.

Wang, L. J., Li, S. H., (2006). Thermotolerance and related antioxidant enzyme activities induced by heat acclimation and salicylic acid in grape (*Vitis vinifera* L.) leaves. *Plant Growth Regul.*, 48, 137-144.

Wang, X. W., Vinocur, B., Altman, A., (2003). Plant response to drought, salinity and extreme temperatures: towards genetic engineering for stress tolerance. *Planta,* 218, 1-14.

Wang, Y., Zhang, W., Li, K., Sun, F., Han, C., Wang, Y., Li, X., (2008). Salt induced plasticity of root hair development is caused by ion disequilibrium in *Arabidopsis thaliana*. *J. Plant Res.*, 121, 87-96.

Weis, E., (1981). Temperature regulation of photosynthesis: a heatinactivation of the ribulose-1,5-biphosphat carboxylase of spinach. *Beri der Deuts. Bota. Gesel.*, 94, 135-142.

Wise, R. R., Olson, A. J., Schrader, S. M., Sharkey, T. D., (2004). Electron transport is the functional limitation of photosynthesis in field grown Pima cotton plants at high temperature. *Plant Cell Environ.*, 27, 717-724.

Wodala, B., Deak, Z., Vass, I., Erdei, L., Altorjay, I., Horvath, F., (2008). In vivo target sites of nitric oxide in photosynthetic electron transport as studied by chlorophyll fluorescence in pea leaves. *Plant Physiol.*, 146, 1920-1927.

Wodala, B., Deák, Z., Vass, I., Erdei, L., Horváth, F., (2005). Nitric oxide modifies photosynthetic electron transport in pea leaves. *Acta Biol. Szegediensis,* 49, 7-8.

Wu, H., Wu, X., Li, Z., Duan, L., Zhang, M., (2012). Physiological evaluation of drought stress tolerance and recovery in cauliflower (*Brassica oleracea* L.) seedling treated with methyl jasmonate and Coronatine. *J. Plant Growth Regul.*, 31, 113-123.

Wu, X. X., Zhu, X. X., Chen, J. L., Yang, S. J., Ding, H. D., Zha, D. S., (2013). Nitric oxide alleviates adverse salt-induced effects by improving the photosynthetic performance and increasing the anti-oxidant capacity of eggplant (*Solanum melongena* L.). *J. Horti. Sci. Biotech.*, 88, 352-360.

Wu, X., Zhu, Z., Li, X., Zha, D., (2012). Effects of cytokinin on photosynthetic gas exchange, chlorophyll fluorescence parameters and antioxidative system in seedlings of eggplant (*Solanum melongena* L.) under salinity stress. *Acta. Physiol. Plant,* 34, 2105-2114.

Wu, X. X., Hai-Dong, Ding, H. D., Chen, J. L., Zhang, H. J., Zhu, W. M., (2010). Attenuation of salt-induced changes in photosynthesis by exogenous nitric oxide in tomato (*Lycopersicon esculentum* Mill. L.) seedlings. *Afr. J. Biotech.*, 9, 7837-7846.

Xiang, C., Oliver, D. J., (1998). Glutathione metabolic genes coordinately respond to heavy metals and jasmonic acid in Arabidopsis. *Plant Cell,* 10, 1539-1550.

Xie, Z. J., Jiang, D., Cao, W. X., Dai, T. B., Jing, Q., (2003). Relationships of endogenous plant hormones to accumulation of grain protein and starch in winter wheat under different post-anthesis soil water statuses. *Plant Growth Regul.,* 41, 117-127.

Xu, J., Wang, W., Yin, H., Liu, X., Sun, H., Mi, Q., (2010). Exogenous nitric oxide improves antioxidative capacity and reduces auxin degradation in roots of *Medicago truncatula* seedlings under cadmium stress. *Plant Soil,* 326, 321–330.

Yamori, W., Suzuki, K., Noguchi, K., Nakai, M., Terashima, I., (2006). Effects of Rubisco kinetics and Rubisco activation state on the temperature dependence of the photosynthetic rate in spinach leaves from contrasting growth temperatures. *Plant Cell Environ.,* 29, 1659-1670.

Yang, Y., Qin, Y., Xie, C., Zhao, F., Zhao, J., Liu, D., Chen, S., Fuglsang, A. T., Palmgren, M. G., Schumaker, K. S., Deng, X. W., Guo, Y. (2010). The Arabidopsis chaperone J3 regulates the plasma membrane H^+-ATPase through interaction with the PKS5 kinase. *Plant Cell,* 22, 1313-1332.

Yi Ting, H. S. U., Ching Huei, K. A. O., (2008). Distinct roles of abscisic acid in rice seedlings during cadmium stress at high temperature. *Bot. Stud.,* 49, 335-342.

Ying, Y., Yue, Y., Huang, X., Wang, H., Mei, L., Yu, W., Zheng, B., Wu, J., (2013). Salicylic acid induces physiological and biochemical changes in three Red bayberry (*Myric rubra*) genotypes under water stress. *Plant Growth. Regul.,* 71, 181-189.

Yoon, J. Y., Hamayun, M., Lee, S. K., Lee, I. J., (2009). Methyl jasmonate alleviated salinity stress in soybean. *J. Crop Sci. Biotech.,* 12, 63-68.

Yu, C. C., Hung, K. T., Kao, C. H., (2005). Nitric oxide reduces Cu toxicity and Cu-induced NH_{4+} accumulation in rice leaves. *J. Plant Physiol.,* 162, 1319-1330.

Yu, J. Q., Huang, L. F., Hu, W. H., Zhou, Y. H., Mao, W. H., Ye, S. F., Nogues, S., (2004). A role for brassinosteroids in the regulation of photosynthesis in *Cucumis sativus. J. Exp. Bot.,* 55, 1135-1143.

Yuan, G. F., Jia, C. G., Li, Z., Sun, B., Zhang, L. P., Liu, N., Wang, Q. M., (2010). Effect of brassinosteroids on drought resistance and abscisic acid concentration in tomato under water stress. *Sci. Horti.,* 126, 103-108.

Yuan, L., Xu, D. Q., (2001). Stimulation effect of gibberellic acid short-term treatment on leaf photosynthesis related to the increase in Rubisco content in broad bean and soyabean. *Photosynth. Res.,* 68, 39-47.

Zhang, M., Zhai, Z., Tian, X., Duan, L., Li, Z., (2008). Brassinolide alleviated the adverse effect of water deficit on photosynthesis and the antioxidant of soybean (*Glycine max* L.). *Plant Growth Regul.,* 56, 257-264.

Zheng, J., Fu, J., Gou, M., Huai, J., Liu, Y., Jian, M., Huang, Q., Guo, X., Dong, Z., Wang, H., Wang, G., (2010). Genome wide transcriptome analysis of two maize inbred lines under drought stress. *Plant Mol. Biol.,* 72, 407-421.

Zimmermann, P., Hirsch-Hoffmann, M., Hennig, L., Gruissem, W. M., (2004). GENEVESTIGATOR. Arabidopsis microarray database and analysis toolbox. *Plant Physiol.,* 136, 2621–2632.

In: Photosynthesis
Editor: Nafees Khan

ISBN: 978-1-63463-304-8

Chapter 8

PHOTOSYNTHETIC INHIBITION UNDER SALINITY CHALLENGED ENVIRONMENT: AN INSIGHT INTO REGULATION OF RUBISCO

Noushina Iqbal[1,•], Nafees A. Khan[2] and Shahid Umar[1]
[1]Department of Botany, Jamia Humdard University, New Delhi, India
[2]Aligarh Muslim University, Aligarh, Uttar Pradesh, India

ABSTRACT

Photosynthesis is undoubtedly a vital biological process which contributes substantially towards plant's growth and development. The enzyme ribulose 1, 5-bisphosphate carboxylase (Rubisco) is the most abundant enzyme on planet and plays an important role in autotrophic carbon fixation. It regulates the pool size of important photosynthetic intermediates. It is the key enzyme of the Calvin cycle which catalyzes, the carboxylation and oxygenation of ribulose 1, 5-bisphosphate (RuBP) and is the main enzyme fixing the vast majority of atmospheric carbon, acting as the natural sink for CO_2. Abiotic stress hampers the process of photosynthesis in most of the plants by modifying the ultrastructure of the organelles, concentration of various pigments and metabolites including enzymes involved in this process. It also influences stomatal regulation. Salinity is currently the most important abiotic stress that unfavorably affects the plant growth process and production potential through its impact on various physiological and biochemical process, including the imperative process of photosynthesis. One of the most prominent effects of stresses is the stomata closure. It results in lowering the concentration of intercellular CO_2, which in turn causes deactivation of Rubisco. Understanding the role of Rubisco in the biosphere is very important in today's context, with increased rate of fossil carbon liberated by human activities and flaring the problem of global warming. Further, it is important to enhance our understanding on the expression patterns of Rubisco genes under salt stress, involved in photosynthesis, in order to develop transgenic lines of different crops with enhanced photosynthetic capacity under stressful conditions. Understanding the details of the adverse effect of salinity stress on photosynthesis, focusing on Rubisco, can lead to the design of new salt-

• E-mail: naushina.iqbal@gmail.com.

tolerant genotypes with better abitity to protect photosynthetic potential under salinity stress. The present study focuses on structural and functional aspects of the Rubisco in photosynthesis under optimal and salt stress conditions.

Keywords: Photosynthesis, Rubisco, Salinity

INTRODUCTION

The expanding trend of land degradation due to increasing soil salinity is decreasing the plant's production potential at a disturbing rate. The world population is expected to increase by 34% in contrast with today's population, which demands an increase of food production by 70% to fulfil the food requirements (FAO 2009). However, realizing the full potential of crop plants is hindered due to increasing the area under soil salinity. According to an estimate salt affected area is about 6% of the arable land, covering over 400 Mha of lands area (FAO 2008). Photosynthesis is the key phenomenon that significantly influences plant growth and development. The adverse effects of salt stress inhibits photosynthetic potential of plants and, thus growth and productivity. Salinity alters the ultrastructure of chloroplast and plays a premier role in augmentation of photosynthesis (Biswal et al. 2008; Saravanavel et al. 2011). It reduces the concentration of photosynthetic pigments and metabolites, and also stomatal regulation and enzymes involved in the process of photosynthesis.

The inhibitory effect of stress on photosynthetic rate is due to stomatal or non-stomatal limitations (Saibo et al. 2009; Rahnama et al. 2010). Abiotic stress suppresses the mesophyll conductance and closes stomata (Flexas et al. 2004; Chaves et al. 2009). The closure of stomata in response to drought and salinity stress is generally due to the decreased leaf turgor and atmospheric vapor pressure, along with root-generated chemical signals (Chaves et al. 2009).The effects of salinity and drought on photosynthesis are attributed directly to the stomatal limitations for the diffusion of gases, which ultimately alters photosynthesis and the mesophyll metabolism (Parida et al. 2005; Chaves et al. 2009). Under salt stress, a shift in the mode of CO_2 fixation from C_3 to CAM (crassulacean acid metabolism) has been observed in ice plant *(Mesembryanthemum crystallinum)* due to induction of some of the enzymatic machinery for CAM metabolism, e.g., phosphoenolpyruvate (PEP) carboxylase within few hours of salt stress. This change under salt stress was a strategy to increase water use efficiency (WUE) by reducing evaporative water loss owing to closure of stomata during day time (Jouyban 2012). Salinity reduces net photosynthetic rate, stomatal conductance and intercellular CO_2 concentration (Nazar et al. 2011; Khan et al. 2014) and impairs ribulose 1,5-bisphosphate carboxylase/oxygenase (Rubisco) enzyme activity. Meyer and Genty (1998) reported the de-activation of the carboxylating enzyme Rubisco by the low intercellular CO_2 concentration. Chaves et al. (2009) have reported that salt stress decreases stomatal conductance and restricts the supply of CO_2 to Rubisco, which predisposes the photosynthetic apparatus to increased energy dissipation and results in down-regulation of photosynthesis. Rubisco is the main enzymes responsible for CO_2 fixation and its decreased activity decreases photosynthesis. Delfine et al. (1999) suggested that the damage to the photosynthetic machinery could occur due to non-stomatal limitations, i.e., decrease in Rubisco activity. However, He et al. (2014) have shown that salinity stress decreases net photosynthetic rate partially through Rubisco degradation.

The present review revisits the adverse effects of salt stress on photosynthesis with special emphasis on Rubisco activity. The review will improve our understanding on the modulation of Rubisco activity in view of maintaining/enhancing photosynthesis under salt stress.

PHOTOSYNTHETIC RESPONSES OF PLANTS TO SALINITY STRESS

Photosynthesis is the most essential process for plant growth and dry mass production, and is adversely influenced by salinity stress. As the process determines the rate of carbon fixation and overall size of the organic pool, it is the most obvious target for genetic intervention in terms of yield enhancement under optimal and stressful environments (Primrose and Twyman 2013).

Salinity stress reduces the photosynthetic rate (Hui et al. 2004; Munns 2002), stomatal conductance, intercellular CO_2 concentration and transpiration rate (Nazar et al. 2011; Wen-Yuan et al. 2012). Several reasons have been attributed to the decrease in net photosynthesis, such as reduced stomatal conductance, decreased intercellular partial pressure of CO_2 in leaves, reduced chlorophyll content (Downtown et al. 1985), changed ultrastructure of chloroplasts (Suleyman et al. 2002; Salama et al. 1994), decreased photochemical and carboxylation reactions (Miteva et al. 1992), and increase in the level of soluble sugars in the tissues (Dubey and Singh 1999; Flowers et al. 1977). Under salt stress, the combined effects of osmotic stress and Na^+ toxicity reduces stomatal conductance and thereby net photosynthesis (Plaut 1995; Downtown et al. 1985). The decrease in photosynthetic pigments following salinity treatments is also a major contributing factor for the decline in photosynthesis. High concentration of Na^+ has been accounted for the breakdown of chlorophyll (Chl), an important photosynthetic pigment, thereby decreasing photosynthesis (Pinheiro et al. 2008; Li et al. 2010; Yang et al. 2011). This decrease in Chl could be attributed to either Chl degradation or to its decreased synthesis. Experiments with sunflower callus and plants (Santos and Caldeira 1999; Santos et al. 2001; Santos 2004; Akram and Ashraf 2011) have shown that the important precursors of Chl, i.e., glutamate and 5-aminolaevulinic acid (ALA), decreases in salt-stressed calli and leaves, which indicates that salt stress affects more markedly Chl biosynthesis than Chl breakdown. Photosynthetic pigments present in the photosystems are believed to be damaged by the stress factors resulting in reduced light-absorbing efficiency of both photosystems (PSI and PSII) and hence a reduced photosynthetic capacity (Geissler et al. 2009; Zhang et al. 2011). Zeaxanthins are also reported to be widely used in alleviating oxidative damage to the membranes (Davison et al. 2002; Verma and Mishra 2005; Isaksson and Andersson, 2008). Chlorophyll level in salinity stressed plants is also partly attributed to the increased activity of Chl degrading enzyme chlorophyllase (Bertrand and Schoefs 1999). It has also been reported that salinity stress led reduction in leaf Chl content in mature trees of *Prunus salicina* was apparently due to the increased leaf chloride content and decreased CO_2 assimilation capacity (Ziska et al. 1990).

Salinity stress has been accounted for hindering the transcription and translation of quinone-binding protein (D1) genes,thus inhibits the repair of photodamaged PSII (Suleyman et al. 2002). Mishra et al. (1991) reported that the ratio of variable fluorescence to maximal fluorescence (Fv/Fm), steady state levels of photochemical quenching coefficient (qP) and non-photochemical quenching coefficient (qn) remained unaffected, whereas, the peak of the induction transient, Fp (expressed as fraction of Fm) was 20% higher than control in wheat plants (*Triticum aestivum* L. var. HD 2329) exposed to salt stress. Singh and Dubey (1995) studied photosynthesis in rice cultivars Ratna under salinity stress and observed that electron transport activity decreased by 53% and PSII activity by 71% in chloroplasts compared to non-stressed plants. In sugar cane, Gomathi and Rakkiyapan (2011) found that on application of 7-8 dS m^{-1} of salt stress at various plant growth stages, there occurred a marked reduction in Chl and carotenoid contents in salt-sensitive varieties but salt-tolerant varieties exhibited higher membrane stability and pigment contents.

Salinity causes both stomatal and non-stomatal limitations. It may damage chloroplast or decrease photosynthesis enzyme activity. Chaves et al. (2009) studied the effects of salinity and drought on genes associated with photosynthesis (AraCyc Pathways, www.arabidopsis.org/tools/aracyc) in *Arabidopsis* seedlings using the expression profiles available (generated with the Affymetrix ATH1 chip; Kilian et al. 2007). They evaluated the effects of stress on genes after 24 hours and observed that out of 139 photosynthesis-related genes (AraCyc Pathways), the AtGenExpress salt and drought microarray gave information for 102 genes. They observed that among the two stresses, the effect of salt stress was more intense reflecting that salinity caused both dehydration and osmotic stress. Both stresses led to gene down-regulation. Thus, salinity adversely affects photosynthesis by stomatal or non-stomatal limitations. It may reduce the light reaction or the dark reaction of photosynthesis. It may close stomata to reduce gaseous exchange or reduce the activity of enzymes involved in the process of photosynthesis.

The next section deals with the photosynthetic enzyme Rubisco and its involvement in photosynthesis under normal and salt stress conditions.

RUBISCO: STRUCTURE AND FUNCTION

Rubisco comprises of a number of identical small (S) and identical large (L) subunits. Although there are several variant forms of Rubisco, with differing numbers of subunits, the most abundant form in nature, present in cyanobacteria, green algae and plants is called form I, and can be described as L8S8. Form I of Rubisco comprises of eight L catalytic subunits (about 55 kDa each) and eight S subunits (about 14 kDa each), giving molecular mass of about 560 kDa for the complete protein (L8S8). The 8 L subunits of form I are arranged as an octameric core surrounded by two layers of four S subunits, with each layer located on the opposite sides of the molecule (Taiz and Zeiger 2010). In plants and algal cells, the S subunits are encoded by nuclear genes, while the L subunits are encoded by genes located in the chloroplast genome (Roy and Andrews 2000). The L subunit contains the catalytic site but the S subunits are essential probably for catalytic viability (Spreitzer 2003). In order to engineer improvements in Rubisco, complementary changes to both subunits are required because the amino acid substitutions in the S subunits can augment changes made in the L subunit (Read and Tabita 1992; Spreitzer et al. 2005). The Rubisco large subunit (rbcL) and Rubisco small subunit (rbcs) are both, therefore, interdependent to bring about a change in enzyme activity under stress.

Whitney and Sharwood (2007) linked both the Rubisco subunits together as a single fusion peptide. The rbcL has two major structural domains, a N-terminal domain and a larger C-terminal domain which is an alpha/beta barrel. Most of the active site residues (that interact with substrate and/or substrate analogues) are contributed by loops at the mouth of the alpha/beta barrel with the remaining residues being supplied by two loop regions in the N-terminal domain of the second large subunit within a dimer. The availability of high resolution 3-D structures has provided detailed insight into the catalytic mechanisms of the enzyme and enabled properties to be related to sequences (Parry et al. 1987; Andersson 1996; 2008; Cleland et al. 1998; Spreitzer and Salvucci 2002; Portis and Parry 2007).

A portion of the activation reaction is catalyzed by another enzyme Rubisco activase (Portis 1995), and in most of the leaves the amount of the activated enzyme in the steady-state is regulated such that it closely parallels the photon flux density-driven rate of electron transport and therefore, the prevailing photosynthetic rate (Portis 1992; Woodrow and Berry 1988). The function of Rubisco activase is to promote and maintain the catalytic activity of Rubisco (Portis 2003). It mediates the release of inhibitory sugar phosphates, such as ribulose 1, 5-bisphosphate (RuBP) from the active sites of Rubisco to enable CO_2 to activate the enzyme by carbamylation (Portis 2003).

The reaction of Rubisco starts with the activity of regulatory enzyme Rubisco activase. It is followed by the addition of CO_2 to the critical amino acid residue in the active site and then by Mg^{2+} to the active site activating the enzyme. Once the enzyme is active, the RuBP substrate forms a coordinate bond with the Mg^{2+}. Binding of RuBP leads to a strain in the bond as it aligns with Mg^{2+} causing it to deprotonate and form a reactive enediolate intermediate structure. This intermediate form of RuBP is receptive towards CO_2 and forms a covalent bond with it, initiating the carboxylation reaction (Monson and Baldocchi 2014).

Rubisco plays an important role in the process of photosynthesis. It is located in the chloroplast of stroma where the dark reaction of photosynthesis occurs. Rubisco catalyzes the carboxylation of RuBP in photosynthetic carbon assimilation (Ellis 1979). However, the catalytic limitations of Rubisco compromise the efficiency of photosynthesis (Parry et al. 2007). Compared to the other enzymes of the Calvin cycle, Rubisco has a low turnover number; meaning, relatively large amount of the enzyme must be present to sustain sufficient rates of photosynthesis. Besides its carboxylation reaction Rubisco also reacts with O_2, leading to photorespiration which is apparently a wasteful reaction that lowers the efficiency of carbon incorporation to the organic compounds. Since both CO_2 and O_2 compete for Rubisco, the carboxylation of the 2,3-enediol intermediate of RuBP depends on the kinetic properties of the Rubisco and the ratio of CO_2/O_2 at the catalytic site of the enzyme and on an intrinsic feature of the enzyme (i.e., the specificity factor φ (sigma)). A high value of φ indicates greater specificity for CO_2 (Liang et al. 1974; Taiz and Zeiger 2010).

$$V_c/V_o = (V_cK_o/V_oK_c)([CO_2]/[O_2])$$

Where, V_c and V_o are the velocities of carboxylation and oxygenation respectively, V_c and V_o are the maximal velocities of the two reactions and K_c and K_o the Michaelis constants for CO_2 and O_2. The substrate specificity factor, V_cK_o/V_oK_c, φ, determines the relative rates of the two reactions at any given CO_2 and O_2 concentrations.

Growth studies on transgenic plants with decreased amounts of Rubisco has confirmed that photosynthetic rate is highly correlated with the amount of Rubisco, under field conditions with intense or variable irradiance (Hudson et al. 1992). This relationship is important in improving resource use , particularly that of nitrogen and water (Lea and Azevedo 2006; 2007; Parry et al. 2005; 2007; Parry and Reynolds 2007; Tambussi et al. 2007; Tuberosa et al. 2007) by increasing Rubisco activity under stress in order to sustain plant growth.

Studies have shown that chloroplast-encoded rbcL has been the nain focus for manipulating the increase in net CO_2 fixation and increasing photosynthesis under optimal and stressful environments with the aim to increase the production of food and fibre, although both small and large subunit of Rubisco are important. The large subunit contains the active site for binding of the substrate and the small subunit influences carboxylation catalytic efficiency and the CO_2/O_2 specificity of the enzymes. It has been reported that the small subunits contains the structural elements responsible for targeting Rubisco to the algal pyrenoid, which is the site where CO_2 is concentrated for optimal photosynthesis (Genkov et al. 2010).

The significant benefits of replacing the Rubisco in C_3 crops with additional proficient variants from some C_4 plants and the more CO_2-specific forms naturally found in non-green algae have been recognized (Whitney et al. 2001; Andrews and Whitney 2003; Zhu et al. 2004; Zhu et al. 2007). The key point of interest with such modulation is that it takes a stab at no extra energy or resource cost. This challenge has provoked significant improvements in CO_2 fixing capacity of crop plants.

INFLUENCE OF SALINITY STRESS ON RUBISCO

Salinity adversely affects plants photosynthetic potential through its adverse effect on Rubisco activity and function. It has been reported that salinity stress negatively affects Rubisco level as well as carboxylating capacity of the enzyme (Flowers et al. 1977; Mitiva et al. 1992; Seeman and Sharkey 1986). Decreased Rubisco activity results in reduced photosynthetic efficiency (Seemann and Sharkey 1986; Gutteridge 1990; Sharkey 1989; Woodrow and Berry 1988). In winged bean (*Phosphocarpus tetragonolobus*) plants, salinity decreased the activities of Rubisco, as well as PEPCase and the rate of CO_2 fixation (Rajmane and Karadge 1986). Miteva et al. (1992) reported that salinity reduced the Rubisco level by approximately 80% in barley seedlings. Such a kind of salinity induced reduction in the amount of Rubisco protein might imply an effect of salt at the level of transcription, translation or gene regulation (Ziska et al. 1990).

The accumulation of compatible osmolytes, such as proline and its analogs in cytoplasm or chloroplasts under salt stress helps in protection of Rubisco against salinity (Solomon et al. 1994). Under in vitro conditions, the accumulation of proline related analogs, N-methyl-L-proline and N-methyl-trans-4-hydroxy-L-proline have been shown to ameliorate the inhibition of the activity of Rubisco by NaCl (Solomon et al. 1994). They have reported that the NaCl adapted plants of *Tamarix jordanis* showed production of higher level of compatible solutes accompanied with higher level of Rubisco.

Seeman and Critchley (1985) have shown that the inhibition of Rubisco activity under salt stress was associated with the sensitivity of this enzyme to chloride ions. Rodermel et al. (1988) produced a series of transformants with decreased expression of Rubisco. They transformed tobacco plants with antisense sequences to the rbcs. It was observed that photosynthetic proteins or Chl remained unchanged until most of the Rubisco was removed (Quick et al. 1991). It was found that Rubisco was reduced to 60% of the average found in wild-type plants, without a strong inhibition of photosynthesis. However, further reduction in Rubisco correlated with the inhibition of photosynthesis (Quick et al. 1991; Stitt et al. 1991).

Manaa et al. (2013) reported that there was up-regulation of Rubisco activases and rbcL in the salt tolerant genotype under salt stress. This increase was correlated with an increase in protein levels that were involved in energy metabolism i.e., pyruvate dehydrogenase, glucose-6-phosphate dehydrogenase, malate dehydrogenase, especially in salt-tolerant genotype. Proteomic studies have shown that abiotic stress degrades some proteins. Yan et al. (2006) reported that rbcL was detected as 19 different fragments plus the intact protein in salinity-treated rice roots. Salinity stress decreased Rubisco activity, Rubisco activation state, transcriptional level of rbcL gene and expression level of rbcL in perennial ryegrass. However, these were comparatively higher in salt-tolerant genotype after 8 days of salt treatment (Hu et al. 2013). Further, they reported that salinity-induced inhibition of photosynthesis in perennial ryegrass was mainly the result of stomatal limitation during early salt stress and metabolic limitation associated with the inhibition of relative water content, activity of Rubisco, expression level of rbcL gene and LSU under a prolonged period of severe salinity (Hu et al. 2013).

Ziaf et al. (2009) found that Rubisco activase exhibited a general increase in intensity under salt stress which was higher in salt-tolerant genotype. The up-regulation of Rubisco activase under salinity stress has been reported previously (Kim et al. 2005; Bandehagh et al. 2011). Parker et al. (2006) reported that up-regulation of Rubisco activase activity in rice leaf lamina was required in order to tolerate salt stress, owing to a direct reduction in stomatal conductance and subsequent low CO_2 level. Rubisco activase functions as a chaperon during stress (Rokka et al. 2001). It functions to reactivate Rubisco for fixing CO_2 or in protection of energy dissipation under stress. The reactivation of Rubisco by Rubisco activase is particularly important when stress-induced stomatal closure results in decreased leaf internal CO_2 concentration. The reduction in Rubisco protein under salinity stress is also associated with stress-generated ROS, which contribute to the degradation of proteins under stress conditions (Kingston-Smith and Foyer 2000; Feller et al. 2008). Hajduch et al. (2001) reported that the number of rbcL breakdown products increased in metal-stressed rice leaves, while Caruso et al. (2008) have shown that rbcL was up-regulated in salt-stressed leaves of wheat (*Triticum durum* L.).

Ranieri et al. (2001) reported that there occurred decrease in Rubisco SSU in response to salinity stress and this reduction was accompanied with limited CO_2 fixation and a simultaneous decrease in enzymes associated with the Calvin cycle. Hence, the down-regulation of the Calvin cycle leads to the accumulation of ATP and NADPH in the chloroplasts and protects plants from photo-oxidative damage.

In salinity stressed rice, the activity of Rubisco-degrading endoproteinases increases causing a paramount decrease in Rubisco (Kang and Titus 1989). Similar decrease was reported in the Rubisco content of barley, spinach, and Amaranthus tricolor (Miteva et al. 1992; Delfine et al. 1998; Wang and Nii 2000) and in Rubisco activity of Spinach, sunflower and tomato (Martino et al. 1999; Rivelli et al. 2002; Taleisnik 1987). However, in tomato, salinity did not affect Rubisco activity, but reduced the potential of enzyme activation and salt might directly inhibit catalysis by acting as a competitive inhibitor of RuBP (Heuer 2005). It was also reported that the specific activity of Rubisco in stressed and non-stressed plants was equal.

Recent study of He et al. (2014) has shown that salinity stress induced chloroplast protrusions (CPs) and production of many more Rubisco containing bodies (RCBs) responsible for degradation of Rubisco under salt stress. Chloroplast protrusions are the pathway of Rubisco exclusion from chloroplasts. Delfine et al. (1999) reported that spinach leaves exhibited substantial reduction in photosynthesis through reduced conductance to CO_2 diffusion at the stomata and in the mesophyll of salt-stressed leaves. They also reported that lower photosynthetic rates indicated a reduction of Rubisco characteristics in salt-stressed leaves. The decrease in Rubisco content induces the operation of the cyclic electron flow around PS I, which enables to prevent photoinhibitionIn order to manipulate the specificity factor of Rubisco attempts are focused on the C-terminal loop 6 region of the LSU, which is critical to specificity for the gaseous substrates (Chen and Spreitzer 1989). Mutation of valine 331 of the LSU to alanine diminished the specificity factor by nearly 40%. Loop 6, at the mouth of the alpha/beta barrel, is an integral part of the catalytic site. The investigation of various 3-D structures prescribes that the loop is flexible in the initial stages of carboxylation. This closed conformation of loop 6 is maintained by the residues of the C-terminal tail (Knight et al. 1990) and the N-terminal loops (Newman and Gutteridge 1993) of the LSU. Valine 331 at the N terminal end of loop 6 is the 'spring', responsible for the movement of the loop (Newman and Gutteridge 1993). Characterization of a loop 6 deletion mutant has demonstrated that loop 6 is vital to normal processing of the enediolate intermediate (Larson et al. 1995). Many of the loop 6 residues are conserved, but only lysine 334 at the apex of the loop has ionic interactions with the transition state analogue 2-carboxyarabinitol 1,5-bisphosphate (CABP). During carboxylation, lysine 334 polarizes both oxygen of CO_2, thereby enhancing the electrophilic status of the carbon atom of CO_2, promoting electrophilic attack by CO_2 on C_2 of the enediol to form the 2-carboxy, 3-keto intermediate (Cleland et al. 1998). Site-directed mutants of lysine 334 catalyzed enediolate formation, however, they were unable to catalyze the reaction of the enediolate with CO_2 or form a stable complex with CABP (Soper et al. 1988; Hartman and Lee 1989; Gutteridge et al. 1993). Thus, lysine 334 is thought to play a specific role in stabilizing the transition state intermediates of both the carboxylation and oxygenation reactions, thereby facilitating the reaction between the gaseous substrate and the enediolate.

Agrobacterium-mediated transformation of tobacco lacking the chloroplast rbcL with the rbcL coding region preceded by a plastid targeting sequence was able to supply the defective plastids with fully functional Rubisco (Kanevski and Maliga 1994). Conversely, rbcS relocated to the tobacco plastid genome folded correctly and assembled into active holoenzyme and contributed less than 1% of the total small subunits in the holoenzyme. The scarcity of the transplastomic small subunits may result from inefficient translation or assembly, although the assembled small subunits were as stable as the native counterparts (Whitney and Andrews 2001).

In addition to the stomatal and non-stomatal limitations to photosynthesis under salt stress, the decrease in photosynthesis is also attributed to the decreased nitrogen (N) content of the Rubisco. Salinity stress decreases leaf N content (Nazar et al. 2011). Under salinity stress, accumulation of excess Na^+ and Cl^- causes negative impacts on the acquisition and homeostasis of essential nutrients (Greenway and Munns 1980; Fageria et al. 2008) making it more difficult for water and nutrients to move through the root membrane (Volkmar et al. 1998) decreasing their concentration in leaf (Ashraf and Foolad 2007; Khorshidi et al. 2009; Khan et al. 2010). Makino et al. (1984) reported increased Rubisco content with increased leaf-N content. Makino (2003) reported that increased leaf N content increased photosynthetic rate (von Caemmerer and Farquhar 1981 Makino et al. 1994). Suzuki et al. (2001) reported that variation in the amount of rbcS and rbcL mRNAs was correlated with changes in leaf N influx to leaves. Table 1 shows the impact of salt stress on Rubisco content or activity in different plants.

Transgenic rice with theoretically optimal Rubisco contents at elevated (CO_2) by antisense rbcS shows higher rates of photosynthesis only under elevated (CO_2) conditions (Makino 2003). Increased CO_2 concentration diminished the inhibitory effect of oxygen on photosynthesis and Rubisco since the carboxylation and oxygenation reactions are catalyzed at the same active site on the enzyme and CO_2 and O_2 are competitive substrates (Andrews and Lorimer 1978). Thus, salinity stress adversely affects Rubisco and photosynthesis. An insight into the process of regulation of Rubisco by salinity could provide efficient strategy for reducing the salinity induced growth reduction through manipulating Rubisco content and activity, and thus photosynthesis.

Table 1. Response of salt stress on Rubisco

Regulation of Rubisco	Plant	Response	Reference
Rubisco content Rubisco containing bodies	Glycine max Glycine max	Decreased Increased	He et al. 2014 He et al. 2014
Rubisco content	Vigna unguiculata Vita 3 Vita 5	 Decreased Increased	de Aragão et al. 2005
Rubisco LSU	Solanum lycopersicum Patio F144	 Decreased Increased	Chen et al. 2009
Rubisco LSU Rubisco activase Rubisco SSU	Solanum lycopersicum Solanum lycopersicum Lycopersicon pennellii	Increased Increased Decreased	Manaa et al. 2013 Manaa et al. 2013 Manaa et al. 2013
Rubisco carboxylation efficiency	Hordeum vulgare	Decreased	Tsonev et al. 1998
Rubisco activity	Phosphocarpus tetragonolobus	Decreased	Rajmane and Karadge 1986
Rubisco synthesis	Hordeum vulgare	Decreased	Miteva et al. 1992
Rubisco activity	Phaseolus vulgaris	Decreased	Seemann and Sharkey 1986

Conclusion and Future Prospects

Photosynthesis is the most important phenomenon which governs plant growth and production potential. Emphasis to increase food production in order to support increasing population under stress conditions demands increased knowledge on developing transgenic with increased photosynthetic capacity. Rubisco, an important photosynthetic enzyme decreases under salt stress and could be a target for manipulating and increasing plants photosynthesis under stress. In order to increase Rubisco content and activity, altering the specificity factor is required. Altering specificity is a difficult goal as gas does not bind directly to the active site. Both CO_2 and O_2 can react directly with the enediol of RuBP in the active Rubisco. Thus, both carboxylation and oxygenation reactions follow the same step. Several residues have been shown to influence specificity, but in many cases the effect must be indirect since many of these are not close enough to interact directly with the substrate or reaction intermediates. Manipulating the rbcS or rbcL through formation of transformants is another important focus. In addition, increasing N fertilization could also enhance Rubisco content and photosynthesis. More importantly targeting for increased biosynthesis and accumulation of osmolytes under salt stress may also provide a better option for Rubisco protection and optimization of photosynthesis.

Acknowledgment

First author is thankful to the University Grants Commission, New Delhi for DS Kothari Post Doctoral Fellowship (File No.-No.F.4-2/2006 (BSR)/13-848/2013(BSR))and to the Head, Department of Botany, Jamia Hamdard University, New Delhi for providing research facilities.

References

Akram, M. S., Ashraf, M., 2011. Exogenous application of potassium dihydrogen phosphate can alleviate the adverse effects of salt stress on sunflower (*Helianthus annuus* L.). *J. Plant Nutr.*, 34, 1041-1057.

Andersson, I., 1996. Large structures at high resolution: the 1.6 angstrom crystal structure of spinach ribulose-1,5-bisphosphate carboxylase/oxygenase complexed with 2-carboxyarabinitol bisphosphate. *J. Mol. Biol.*, 259, 160-174.

Andersson, I., 2008. Catalysis and regulation in Rubisco. *J. Exp. Bot.*, 59, 1555-1568.

Andrews, T. J., Lorimer, G. H., 1978. Photorespiration still unavoidable? *FEBS Letters,* 90, 1-9.

Andrews, T. J., Whitney, S. M., 2003. Manipulating ribulose bisphosphate carboxylase/ oxygenase in the chloroplasts of higher plants. *Arch. Biochem. Biophys.*, 414, 159-169.

Ashraf, M., Foolad, M. R., 2007. Improving plant abiotic-stress resistance by exogenous application of osmoprotectants glycinebetaine and proline. *Environ. Exp. Bot.*, 59, 206-216.

Bandehagh, A., Salekdeh, G. H., Toorchi, M., Mohammadi, A., Komatsu, S., 2011. Comparative proteomic analysis of canola leaves under salinity stress. *Proteomics,* 11, 1965-1975.

Bertrand, M., Schoefs, B., 1999. Photosynthetic pigment metabolism in plants during stress. In: Pessarakli M, ed. Handbook of Plant and Crop stress, 2nd ed. New York: Marcel Dekker. 527-543.

Biswal, B., Raval, M. K., Biswal, U. C., Joshi, P., 2008. Response of photosynthetic organelles to abiotic stress: modulation by sulfur metabolism. In: Khan, N. A., Singh, S., Umar, S. (Eds.): Sulfur Assimilation and Abiotic Stress in Plants, Springer-Verlag, Berlin- Heidelberg, 167-191.

Caruso, G., Cavaliere, C., Guarino, C., Gubbiotti, R., Foglia, P., Laganà, A., 2008. Identification of changes in *Triticum durum* L. leaf proteome in response to salt stress by two dimensional electrophoresis and MALDI-TOF mass spectrometry. *Anal. Bioanal. Chem.*, 391, 381-390.

Chaves, M. M., Flexas, J., Pinheiro, C., 2009. Photosynthesis under drought and salt stress: regulation mechanisms from whole plant to cell. *Ann. Bot.*, 103, 551–560.

Chen, S., Gollop, N., Heuer, B., 2009. Proteomic analysis of salt-stressed tomato (*Solanum lycposersicum*) seedlings: effect of genotype and exogenous application of glycinebetaine. *J. Exp. Bot.,* 60, 2005-2019.

Chen, Z., Spreitzer, R. J., 1989. Chloroplast intragenic suppression enhances the low specificity of mutant ribulose-1,5-bisphosphate carboxylase/oxygenase. *J. Biol. Chem.*, 264, 3051-3053.

Cleland, W. W., Andrews, T. J., Gutteridge, S., Hartman, F. C., Lorimer, G. H., 1998. Mechanism of Rubisco: the carbamate as general base. *Chem. Rev.*, 98, 549-561.

Davison, P. A., Hunter, C. N., Horton, P., 2002. Overexpression of β-carotene hydroxylase enhances stress tolerance in *Arabidopsis*. *Nature,* 418, 203-206.

De Aragão, M. E. F., Guedes, M. M., Otoch, M. L. O., Guedes, M. I. F., de Melo, D. F., Lima, M. G. S., 2005. Differential responses of ribulose-1,5-bisphosphate carboxylase/oxygenase activities of two *Vigna unguiculata* cultivars to salt stress. *Braz. J. Plant Physiol.*, 17, 207-212.

Delfine, S., Alvino, A., Villani, M. C., Loreto, F., 1999. Restrictions to CO_2 conductance and photosynthesis in spinach leaves recovering from salt stress. *Plant Physiol.*, 119, 1101-1106.

Delfine, S., Alvino, A., Zacchini, M., Loreto, F., 1998. Consequences of salt stress on conductance to CO_2 diffusion, Rubisco characteristics and anatomy of spinach leaves. *Aust. J. Plant Physiol.*, 25, 395–402.

Downton, W. J. S., Grant, W. J. R., Robinson, S. P., 1985. Photosynthetic and stomatal response of spinach leaves to salt stress. *Plant Physiol.*, 78, 85-88.

Dubey, R. S., Singh, A. K., 1999. Salinity induces accumulation of soluble sugars and alters activity of sugar metabolizing enzymes in rice plants. *Biol. Plant,* 42, 233-239.

Ellis, R. J., 1979. The most abundant protein in the world. *Trends Biochem. Sci.*, 4, 241–244.

Fageria, N. K., Baligar, V. C., Li, Y. C., 2008. The role of nutrient efficient plants in improving crop yields in the twenty first century. *J. Plant Nutr.*, 31, 1121-1157.

Feller, U., Anders, I., Demirevska, K., 2008. Degradation of Rubisco and other chloroplast proteins under abiotic stress. *Gen. Appl. Plant Physiol.*, 34, 5-18.

Flexas, J., Bota, J., Loreto, F., Cornic, G., Sharkey, T. D., 2004. Diffusive and metabolic limitations to photosynthesis under drought and salinity in C3 plants. *Plant Biol.*, 6, 269-279.

Flowers, T. J., Troke, P. F., Yeo, A. R., 1977. The mechanism of salt tolerance in halophytes. *Annu. Rev. Plant Physiol.*, 28, 89-121.

Food and Agricultural Organization, 2009. Land and plant nutrition management. Service, www.fao.org/wsfs/forum2050/

Food and Agricultural Organization, 2008. Land and plant nutrition management. Service, http://www.fao.org/ag/agl/agll/spush/

Geissler, N., Hussin, S., Koyro, H. W., 2009. Interactive effects of NaCl salinity and elevated atmospheric CO_2 concentration on growth, photosynthesis, water relations and chemical composition of the potential cash crop halophyte *Aster tripolium* L. *Environ. Exp. Bot.*, 65, 220-231.

Genkov, T., Meyer, M., Griffiths, H., and Spreitzer, R. J., 2010. Functional hybrid Rubisco enzymes with plant small subunits and algal large subunits: Engineered rbcS cDNA for expression in *Chlamydomonas. J. Biol. Chem.*, 285, 19833-19841.

Gomathi, R., Rakkiyapan, P., 2011. Comparative lipid peroxidation, leaf membrane thermostability, and antioxidant system in four sugarcane genotypes differing in salt tolerance. *Int. J. Plant Physiol. Biochem.*, 3, 67-74.

Greenway, H., Munns, R., 1980. Mechanism of salt tolerance in non halophytes. *Ann. Rev. Plant Physiol.*, 31, 149-190.

Gutteridge, S., 1990. Limitations of the primary events of CO_2 fixation in photosynthetic organisms: the structure and mechanisms of Rubisco. *Biochem. Biopbys. Acta.*, 1015, 1-14.

Gutteridge, S., Rhoades, D. F., Herrmann, C., 1993. Site-specific mutations in a loop region of the C-terminal domain of the large subunit of ribulose-1,5-bisphosphate carboxylase/oxygenase that influence substrate partitioning. *J. Biol. Chem.*, 268, 7818-7824.

Hajduch, M., Rakwal, R., Agrawal, G. K., Yonekura, M., Pretova, A., 2001. High-resolution two-dimensional electrophoresis separation of proteins from metal-stressed rice (*Oryza sativa* L.) leaves: Drastic reductions/fragmentation of ribulose-1,5-bisphosphate carboxylase/ oxygenase and induction of stress-related proteins. *Electrophoresis,* 22, 2824-2831.

Hartman, F. C., Lee, E. H., 1989. Examination of the function of the active site lysine 329 of ribulose-1,5-bisphosphate carboxylase/oxygenase as revealed by proton exchange reactions. *J. Biol. Chem.*, 246, 11784-11789.

He, Y., Yu, C., Zhou, L., Chen, Y., Liu, A., Jin, J., Hong, J., Qi, Y., Jiang, D., 2014. Rubisco decrease is involved in chloroplast protrusion and Rubisco-containing body formation in soybean (*Glycine max.*) under salt stress. *Plant Physiol. Biochem.*, 74, 118-124.

Heuer, B., 2005. Photosynthetic carbon metabolism of crops under salt stress. In: Pessarakli M (eds.) Handbook of Photosynthesis. Chapter 40. Taylor & Francis Group, LLC.

Hu, T., Yi, H., Hu, L., Fu, J., 2013. Stomatal and metabolic limitations to photosynthesis resulting from NaCl stress in perennial ryegrass genotypes differing in salt tolerance. *J. Am. Soc. Hort. Sci.*, 138, 350-357.

Hudson, G. S., Evans, J. R., von Caemmerer, S., Arvidsson, Y. B. C., Andrews, T. J., 1992. Reduction of ribulose-1,5-bisphosphate carboxylase oxygenase content by antisense RNA reduces photosynthesis in transgenic tobacco plants. *Plant Physiol.*, 98, 294-302.

Hui, H., Xu, X., Li, S., 2004. Possible mechanism of inhibition on photosynthesis of *Lycium barbarum* under salt stress. *Chin. J. Ecol.*, 23, 5-9.

Isaksson, C., Andersson, S., 2008. Oxidative stress does not influence carotenoid mobilization and plumage pigmentation. *Proc. R. Soc. Biol. Sci. Ser. B.,* 275, 309-314.

Jouyban, Z., 2012. The effect of salt stress on plant growth. *Tech. J. Engineering App. Sci.*, 2, 7-10.

Kanevski, I, Maliga, P., 1994. Relocation of the plastid rbcL gene to the nucleus yields functional ribulose-1,5-bisphosphate carboxylase in tobacco chloroplasts. *Proc. Nat. Acad. Sci. USA,* 5, 1969-1973.

Kang, S. M., Titus, J. S., 1989. Increased proteolysis of senescing rice leaves in the presence of NaCl and KCl. *Plant Physiol.*, 91, 1232-1237.

Khan, M. I. R., Asgher, M., Khan, N. A., 2014. Alleviation of salt-induced photosynthesis and growth inhibition by salicylic acid involves glycinebetaine and ethylene in mungbean (*Vigna radiata* L.). *Plant Physiol. Biochem.,* 80, 67-74.

Khan, N. A., Syeed, S., Masood, A., Nazar, R., Iqbal, N., 2010. Application of salicylic acid increases contents of nutrients and antioxidative metabolism in mungbean and alleviates adverse effects of salinity stress. *Int. J. Plant Biol.*, 1, e1.

Khorshidi, M. B., Yarnia, M., Hassanpanah, D., 2009. Salinity effect on nutrients accumulation in alfalfa shoots in hydroponic condition. *J. Food Agric. Environ.*, 7, 787-790.

Kilian, J., Whitehead, D., Horak, J., Wanke, D., Weinl, S., Batistic, O., D'Angelo, C., Bornberg-Bauer, E., Kudia, J., Harter, K., 2007. The AtGenExpress global stress expression data set: protocols, evaluation and model data analysis of UV-B light, drought and cold stress responses. *Plant J.*, 50, 347-363.

Kim, D. W., Rakwal, R., Agrawal, G. K., Jung, Y. H., Shibato, J., Jwa, N. S., Iwahashi, Y., Iwahashi, H., Kim, D. H., Shim, I. S., Usui, K., 2005. A hydroponic rice seedling culture model system for investigating proteome of salt stress in rice leaf. *Electrophoresis,* 26, 4521-4539.

Kingston-Smith, A. H., Foyer, C. H., 2000. Bundle sheath proteins are more sensitive to oxidative damage than those of the mesophyll in maize leaves exposed to paraquat or low temperatures. *J. Exp. Bot.*, 51, 123-130.

Knight, S., Andersson, I., Branden, C. I., 1990. Crystallographic analysis of ribulose1,5-bisphosphate carboxylase from spinach at 2.4 A resolution. *J. Mol. Biol.* 215, 113-160.

Larson, E. M., Larimer, F. W., Hartman, F. C., 1995. Mechanistic insights provided by deletion of a flexible loop at the active site of ribulose-1,5-bisphosphate carboxylase/oxygenase. *Biochemistry,* 34, 4531-4537.

Lea, P. J., Azevedo, R. A., 2006. Nitrogen use efficiency. 1. Uptake of nitrogen from the soil. *Ann. App. Biol.*, 149, 243-247.

Lea, P. J., Azevedo, R. A., 2007. Nitrogen use efficiency. 2. Amino acid metabolism. *Ann. App. Biol.*, 151, 269-275.

Li, X. T., Zhang, Y., Liu, H., Wu, Y. T., Zhang, H. X., 2010. Stable expression of *Arabidopsis* vacuolar Na^+/H^+ antiporter gene *AtNHX1* and salt tolerance in transgenic soybean for over six generations. *Chinese Sci. Bull.*, 55, 1127-1134.

Laing, W. A., Ogren, W. L., Hageman, R. H., 1974. Regulation of soybean net photosynthetic CO_2 fixation by the interaction of CO_2, O_2 and ribulose-1,5-bisphosphate carboxylase. *Plant Physiol.*, 54, 678-685.

Makino, A., 2003. Rubisco and nitrogen relationships in rice. Leaf photosynthesis and plant growth. *Soil Sci Plant Nutr.*, 49, 319-327.

Makino, A., Mae, T., Ohira, K., 1984. Relation between nitrogen and ribulose-1,5-bisphosphate carboxylase in rice leaves from emergence through senescence. *Plant Cell Physiol.*, 25, 429-437.

Makino, A., Nakano, H., Mae, T., 1994. Responses of ribulose-1,5-bisphosphate carboxylase, cytochrome f, sucrose synthesis enzymes in rice leaves to leaf nitrogen and their relationships to photosynthesis. *Plant Physiol.*, 105, 173-179.

Manaa, A., Mimouni, H., Wasti, S., Gharbi, E., Aschi-Smiti, S., Faurobert, M., Ahmed, H. B., 2013. Comparative proteomic analysis of tomato (*Solanum lycopersicum*) leaves under salinity stress. *Plant Omics J.*, 6, 268-277.

Martino, C. D., Delfine, S., Alvino, A., Loreto, F., 1999. Photorespiration rate in spinach leaves under moderate NaCl stress. *Photosynthetica,* 36, 233-242.

Meyer, S., Genty, B., 1998. Mapping intercellular CO_2 mole fraction (Ci) in *Rosa rubiginosa* leaves fed with abscisic acid by using chlorophyll fluorescence imaging: significance of Ci estimated from leaf gas exchange. *Plant Physiol.*, 116, 947-957.

Mishra, S. K., Subrahmanyam, D., Singhal, G. S., 1991. Interactionship between salt and light stress on the primary process of photosynthesis. *J. Plant Physiol.,* 138, 92-96.

Miteva, T. S., Zhelev, N. Z., Popova, L. P., 1992. Effect of salinity on the synthesis of ribulose-1,5-bisphosphate carboxylase/oxygenase in barley leaves. *J. Plant Physiol.,* 140, 46-51.

Miteva, T. S., Zhelev, Z. H., Popova, L. P., 1992. Effect of salinity on the synthesis of ribulose bisphosphate carboxylase/oxygenase in barley leaves. *J. Plant Physiol.*, 140, 46-51.

Monson, R., Baldocchi, D., 2014. Control over metabolic fluxes in Terrestrial Biosphere-atmosphere fluxes. Cambridge university press, UK, chapter 4, 74.

Munns, R., 2002. Comparative physiology of salt and water stress. *Plant Cell Environ.*, 25, 239–250.

Nazar, R., Iqbal, N., Syeed, S., Khan, N. A., 2011. Salicylic acid alleviates decreases in photosynthesis under salt stress by enhancing nitrogen and sulfur assimilation and antioxidant metabolism differentially in two mungbean cultivars. *J. Plant Physiol.,* 168, 807-815.

Newman, A., Gutteridge, S., 1993. The X-ray structure of *Synechoccus* ribulose-1,5-bisphosphate carboxylase/oxygenase activated quaternary complex at 2.2 AÊ resolution. *J. Biol. Chem.*, 268, 25876-25886.

Parida, A. K., Mittra, B., Das, A. B., Das, T. K., Mohanty, P., 2005. High salinity reduces the content of a highly abundant 23-kDa protein of the mangrove *Bruguiera parviflora. Planta,* 221, 135-140.

Parker, R., Flowers, T., Moore, A., Harpham, N., 2006. An accurate and reproducible method for proteome profiling of the effects of salt stress in the rice leaf lamina. *J. Exp. Bot.*, 57, 1109-1118.

Parry, M. A. J., Flexas, J., Medrano, H., 2005. Prospects for crop production under drought. *Ann. App. Biol.*, 147, 211-278.

Parry, M. A. J., Madgwick, P. J., Carvahlo, J. F. C. Andralojc, P. J., 2007. Prospects for increasing photosynthesis by overcoming the limitations of Rubisco. *J. Agric. Sci.,* 145, 31-43.

Parry, M. A. J., Reynolds, M. P., 2007. Improving resource use efficiency. *Ann. App. Biol.,* 151, 133-135.

Parry, M. A. J., Schmidt, C. N. G., Cornelius, M. J., Keys, A. J., Millard, B. N., Burton, S., Gutteridge, S., Dyer, T. A., Keys, A. J., 1987. Variations in properties of ribulose-1,5 bisphosphate carboxylase from species related to differences in amino acid sequences. *J. Exp. Bot.,* 38, 1260-1271.

Pinheiro, H. A., Silva, J. V., Endres, L., Ferreira, V. M., Camara, C. A., Cabral, F. F., Oliveira, J. F., de Carvalho, L. W. T., dos Santos, J. M., dos Santos Filho, B. G., 2008. Leaf gas exchange, chloroplastic pigments and dry matter accumulation in castor bean (*Ricinus communis* L.) seedlings subjected to salt stress conditions. *Ind. Crop. Prod.*, 27, 385-392.

Plaut, Z., 1995. Photosynthesis in plant/crops under water and salt stress. In: Pessarakli, M., (Ed.), Handbook of Plant and Crop Physiology. New York, Marcel Dekker, 587–603.

Portis, A. R., 1992. Regulation of ribulose 1,5-bisphosphate carboxylase/oxygenase activity. *Ann. Rev. Plant Physiol. Plant Mol. Biol.,* 43, 415-437.

Portis, A. R., 1995. The regulation of Rubisco by Rubisco activase. *J. Expl. Bot.,* 46, 1285-1291.

Portis, A. R., 2003. Rubisco activase: Rubisco's catalytic chaperone. *Photosyn. Res.,* 75, 11-27.

Portis, A. R., Parry, M. A. J., 2007. Discoveries in Rubisco (ribulose 1,5-bisphosphate carboxylase/oxygenase): a historical perspective. *Photosyn. Res.,* 94, 121-143.

Primrose, S. B., Twyman, R., 2013. Principles of Gene Manipulation and Genomics. John Wiley & Sons. Science.

Quick, W. P., Schurr, U., Scheibe, R., Schulze, E. D., Rodermel, S. R., Bogorad, L., Stitt, M., 1991. Decreased ribulose 1,5-bisphosphate carboxylase-oxygenase in transgenic tobacco transformed with 'antisense' rbcS. I. Impact on photosynthesis in ambient growth conditions. *Planta,* 183, 542-554.

Rahnama, A., Poustini, K., Tavakkol-Afshari, R., Tavakoli, A., 2010. Growth and stomatal responses of bread wheat genotypes in tolerance to salt stress. *Int. J. Biol. Life Sci.*, 6, 216-221.

Rajmane, N. A., Karadge, B. A., 1986. Photosynthesis and photorespiration in winged bean *Posphocarpus tetragonolobus* L.) grown under saline conditions. *Photosynthetica, 20,* 139-145.

Ranieri, A., Giuntini, D., Ferraro, F., Nali, C., Baldan, B., Lorenzinib, G., Soldatini, G. F., 2001. Chronic ozone fumigation induces alterations in thylakoid functionality and composition in two poplar clones. *Plant Physiol. Bioch.*, 39, 999–1008.

Read, B. A., Tabita, F. R., 1992. A hybrid ribulosebisphosphate carboxylase/oxygenase enzyme exhibiting a substantial increase in substrate specificity factor. *Biochemistry,* 31, 5553–5560.

Rivelli, A. R., Lovelli, S., Perniola, M., 2002. Effects of salinity on gas exchange, water relations and growth of sunflower (*Helianthus annuus*). *Funct. Plant Biol.*, 29, 1405-1415.

Rodermel, S. R., Abbott, M. S., Bogorad, L., 1988. Nuclear organelle interactions: nuclear antisense gene inhibits ribulose bisphosphate carboxylase enzyme levels in transformed tobacco plants. *Cell,* 55, 673-681.

Rokka, A., Zhang, L., Aro, E. M., 2001. Rubisco activase: an enzyme with a temperature-dependent dual function? *The Plant J.,* 25, 463-471.

Roy, H., Andrews, T. J., 2000. Rubisco: Assembly and mechanism. In Photosynthesis: Physiology and Metabolism, Leegood, R. C., Sharkey, T. D., von Caemmerer, S., (Eds.), Dordrecht, The Netherlands, Kluwer Academic Publishers, 53–83.

Saibo, N. J. M., Lourenço, T., Oliveira, M. M., 2009. Transcription factors and regulation of photosynthetic and related metabolism under environmental stresses. *Ann. Bot.*, 103, 609-623.

Salama, S., Trivedi, S., Busheva, M., Arafa, A. A., Garab, G., Erdei, L., 1994. Effects of NaCl salinity on growth, cation accumulation, chloroplast structure and function in wheat cultivars differing in salt tolerance. *J. Plant Physiol.,* 144, 241-247.

Santos, C., Azevedo H., Caldeira, G., 2001. In situ and in vitro senescence induced by KCl stress: nutritional imbalance, lipid peroxidatin and antioxidant metabolism. *J. Exp. Bot.*, 52, 351-360.

Santos, C., Caldeira, G., 1999. Comparative responses of *Helianthus annuus* plants and calli exposed to NaCl: I. Growth rate and osmotic regulation in intact plants and calli. *J. Plant Physiol.*, 155, 769-777.

Santos, C. V., 2004. Regulation of chlorophyll biosynthesis and degradation by salt stress in sunflower leaves. *Sci. Hort.*, 103, 93-99.

Saravanavel, R., Ranganathan, R., Anantharaman, P., 2011. Effect of sodium chloride on photosynthetic pigments and photosynthetic characteristics of *Avicennia officinalis* seedlings. *Recent Res. Sci. Technol.*, 3, 177-180.

Seemann, J. R., Critchley, C., 1985. Effects of salt stress on the growth, ion content, stomatal behaviour and photosynthetic capacity of a salt sensitive species. *Phaseolus vulgaris* L. *Plant Physiol.*, 82, 555-560.

Seemann, J. R., Sharkey, T. D., 1986. Salinity and nitrogen effects on photosynthesis, ribulose- 1,5-bisphosphate carboxylase and metabolic pool size in *Phaseolus vulgaris* L. *Plant Physiol.*, 82, 555-560.

Sharkey, T. D., 1989. Evaluating the role of Rubisco activation in photosynthesis of C_3 plants. *Phil. Trans. R. Soc. (Lond.), Series B.,* 323, 435-448.

Singh, A. K., Dubey, R. S., 1995. Changes in chlorophyll a and b contents and activity of photosystem I and II in rice seedlings induced by NaCl. *Photosynthetica,* 31, 489-499.

Solomon, A., Beer, S., Waisel, Y., Jones, G. P., Paleg, L. G., 1994. Effect of NaCl on the carboxylating activity of Rubisco from *Tamerix Jordonis* in the presence and absence of proline-related compatible solutes. *Physiol. Plant,* 90, 198–204.

Soper, T. S., Larimer, F. W., Mural, R. J., Lee, E. H., Hartman, F. C., 1992. Role of asparagine 111 at the active site of ribulose-1,5-bisphosphate carboxylase/oxygenase from *Rhodospirillum rubrum* as demonstrated by site directed mutagenesis. *J. Biol. Chem.*, 267, 8452-8457.

Spreitzer, R. J., 2003. Role of the small subunit in ribulose-1,5-bisphosphate carboxylase/oxygenase. *Arch. Biochem. Biophys.,* 414, 141-149.

Spreitzer, R. J., Peddi, S. R., Satagopan, S., 2005. Phylogenetic engineering at an interface between large and small subunits imparts land-plant kinetic properties to algal Rubisco. *Proc. Natl. Acad. Sci. USA,* 102, 17225-17230.

Spreitzer, R. J., Salvucci, M. E., 2002. RUBISCO: structure, regulatory interactions, and possibilities for a better enzyme. *Annu. Rev. Plant Physiol. Plant Mol. Biol.,* 53, 449-475.

Stitt, M., Von Schaewen, A., Willmitzer, L., 1991. 'Sink' regulation of photosynthetic metabolism in transgenic tobacco plants expressing yeast invertase in the cell wall involves a decrease of the Calvin-cycle enzymes and an increase of glycolytic enzymes. *Planta,* 183, 40-50.

Suleyman, I. A., Nishiyama, Y., Miyairi, S., Yamamoto, H., Inagaki, N., Kaneasaki, Yu, Murata, M., 2002. Salt stress inhibits the repair of photodamaged photosystem II by suppressing the transcription and translation of psb A genes in *Synechocystis. Plant Physiol.,* 130, 1443-1453.

Suzuki, Y., Makino, A., Mae, T., 2004. Changes in the turnover of Rubisco and levels of mRNAs of rbcL and rbcS in rice leaves from emergence to senescence. *Plant Cell Environ.,* 24, 1353–1360.

Taiz, L., Zeiger, E., 2010. A companion to Plant Physiology. 5th edition. Chapter 8. Sinauer Associates, Incorporated, Science.

Taleisnik, E. L., 1987. Salinity effects on growth and carbon balance in *Lycopersicon esculentum* and *L. pennellii. Physiol. Plant,* 71, 213-218.

Tambussi, A., Bort, J., Araus, J. L., 2007. Water use efficiency in C3 cereals under Mediterranean conditions: a review of physiological aspects. *Ann. App. Biol.,* 150, 307–321.

Tsonev, T. D., Lazova, G. N., Stoinova, Z. G., Popova, L. P., 1998. A possible role of jasmonic acid in adaptation of barley seedlings to salinity stress. *J. Plant Growth Regul.,* 17, 153-159.

Tuberosa, R., Giuliani, S., Parry, M. A. J., Araus, J. L., 2007. Improving water use efficiency in Mediterranean agriculture: what limits the adoption of new technologies? *Ann. App. Biol.,* 150, 157-162.

Verma, S., Mishra, S. N., 2005. Putrescine alleviation of growth in salt stressed *Brassica juncea* by inducing antioxidative defense system. *J. Plant Physiol.,* 162, 669-677.

Volkmar, K. M., Hu, Y., Steppuhn, H., 1998. Physiological responses of plants to salinity: a review. *Can. J. Plant Sci.,* 78, 19-27.

Von Cammerer, S., Farquhar, G. D., 1981. Some relationship between the biochemistry of photosynthesis and the gas exchange of leaves. *Planta,* 153, 376-387.

Wang, Y., Nii, N., 2000. Changes in chlorophyll, ribulose bisphosphate carboxylase-oxygenase, glycine betaine content, photosynthesis and transpiration in *Amaranthus tricolor* leaves during salt stress. *J. Hortic. Sci. Biotechnol.,* 75, 623-627.

Wen-Yuan, W., Xiao-Feng, Y., Ying, J., Bo, Q., Yu-Feng, X., 2012. Effects of Salt Stress on Water Content and Photosynthetic Characteristics in *Iris lactea* Var. Chinensis Seedlings. *Middle East J. Sci. Res.,* 12, 70-74.

Whitney, S. M., Andrews, T. J., 2001. The gene for the ribulose-1,5- bisphosphate carboxylase/oxygenase (Rubisco) small subunit relocated to the plastid genome of tobacco directs the synthesis of small subunits that assemble into Rubisco. *Plant Cell,* 13, 193-205.

Whitney, S. M., Baldet, P., Hudson, G. S., Andrews, T. J., 2001. Form I Rubiscos from non-green algae are expressed abundantly but not assembled in tobacco chloroplasts. *Plant J.*, 26, 535-547.

Whitney, S. M., Sharwood, R. E., 2007. Linked Rubisco subunits can assemble into functional oligomers without impeding catalytic performance. *J. Biol. Chem.,* 282, 3809-3818.

Woodrow, I. E., Berry, J. A., 1988. Enzymatic regulation of photosynthetic CO_2 fixation in C_3 plants. *Ann. Rev. Plant Physiol. Plant Mol. Biol.,* 39, 533-594.

Yan, S. P., Zhang, Q. Y., Tang, Z. C., Su, W. A., Sun, W. N., 2006. Comparative proteomic analysis provides new insights into chilling stress responses in rice. *Mol. Cell Proteomics,* 5, 484-496.

Yang, J. Y., Zheng, W., Tian, Y., Wu, Y., Zhou, D. W., 2011. Effects of various mixed salt-alkaline stresses on growth, photosynthesis, and photosynthetic pigment concentrations of *Medicago ruthenica* seedlings. *Photosynthetica,* 49, 275-284.

Ziska, L. H., Seemann, J. R., DeJong, T. M., 1990. Salinity induced limitations on photosynthesis in *Prunus salicina*, a deciduous tree species. *Plant Physiol.*, 93, 864-870.

Zhang, L., Zhang, Z., Gao, H., Meng, X., Yang, C., Liu, J., Meng, Q., 2011. Mitochondrial alternative oxidase pathway protects plants against photoinhibition by alleviating inhibition of the repair of photodamaged PSII through preventing formation of reactive oxygen species in Rumex K-1 leaves. *Physiol. Plant,* 143, 396-407.

Zhu, X. G., de Sturler, E., Long, S. P., 2007. Optimizing the distribution of resources between enzymes of carbon metabolism can dramatically increase photosynthetic rate: a numerical simulation using an evolutionary algorithm. *Plant Physiol.*, 145, 513-526.

Zhu, X. G., Portis, A. R., Long S. P., 2004. Would transformation of C-3 crop plants with foreign Rubisco increase productivity? A computational analysis extrapolating from kinetic properties to canopy photosynthesis. *Plant Cell Environ.,* 27, 155-165.

Ziaf, K., Amjad, M., Pervez, M. A., Iqbal, Q., Rajwana, I. A., Ayyub, M., 2009. Evaluation of different growth and physiological traits as indices of salt tolerance in hot pepper (*Capsicum annuum* L.). *Pak. J. Bot.,* 41, 1797-1809.

Ziska, L. H., Seemann, J. R., DeJong, T. M., 1990. Salinity induced limitations on photosynthesis in *Prunus salicina*, a deciduous tree species. *Plant Physiol.*, 93, 864-870.

In: Photosynthesis
Editor: Nafees Khan
ISBN: 978-1-63463-304-8

Chapter 9

OXIDATIVE STRESS MANAGEMENT THROUGH SUGAR SIGNALING AND ANTIOXIDANT NETWORK IN PLANTS

Resham Sharma,[1] Renu Bhardwaj,[1,*] Dhriti Kapoor,[1] Sonia,[1] Amandeep Rattan,[1,2] Vinod Kumar,[1] Anket Sharma,[1] A.K. Thukral,[1] Satwinderjeet Kaur[1] and Nitika Arora[2]

[1]Department of Botanical & Environmental Sciences,
Guru Nanak Dev University, Amritsar, Punjab, India;
[2]Hans Raj Mahila Maha Vidyalaya, Jalandhar, Punjab, India

ABSTRACT

Oxidative stress is an imbalance between reactive oxygen species (ROS) generation and scavenging of the reactive intermediates leading to damage and redox disturbances in living cells. It leads to disruption and impairment of cellular sugars, proteins, lipids and nucleic acids. Plants under stress tend to generate and accumulate ROS hindering signal cascades at organelle and cellular levels. Soluble sugars such as fructans, raffinose family oligosaccharides (RFOs), trehalose and polyols in plants have long been characterized to activate antioxidant pathways or themselves act as ROS scavengers under abiotic stress. Herein, we discuss their antioxidant functioning during environmental stress in cell organelles such as chloroplasts, mitochondria and peroxisomes. This physiological sugar-antioxidant interplay exerts a control over various genetic and molecular approaches facilitating favourable strategies in developing plants of agronomic and commercial importance.

Keywords: Antioxidant network, cell organelles, environmental stress, ROS, sugars

---

* Corresponding Author: renu_bhardwaj@rediffmail.com, renubhardwaj82@gmail.com.

INTRODUCTION

All organisms, including plants, deal with a plethora of abiotic (temperature-heat/ freezing, water-deficit/logging, heavy metal and radionuclide toxicity, salinity, overuse of agricultural supplements-insecticides/pesticides/herbicides etc.) and biotic (pathogenic invasions) stresses as a result of sudden or gradual environmental changes and constant human interference (Sharma and Dietz 2009; Miller et al. 2010; Atkinson and Urwin 2012; Suzuki et al. 2014; Khan and Khan 2014a). As a result of these activities, severe oxidative damage hits their finely tuned homeostasis, disrupting their normal morphological, physiological, and genetic functions. Oxidative stress or damage manifests itself in the form of an imbalance between the routine ROS (reactive oxygen species) generation and scavenging process (Møller and Sweetlove 2010; Gill and Tuteja 2010; Baxter et al. 2014). These are normally produced during aerobic metabolism and are confined to cellular compartments i.e., chloroplast, mitochondria, and peroxisomes. However, their sudden surge due to stress in plants results in enzyme/protein damage, nucleic acid and lipid degradation and overall retardation of plant growth and accelerated senescence (Miller et al. 2010; Choudhury et al. 2013). To counter their ill-effects and to balance them out, various antioxidant groups such as antioxidants and antioxidant enzymes (ascorbate, tocopherol, glutathione, ascorbate peroxidase, catalase, glutathione peroxidase, glutathione reductase, superoxide dismutase, phenol and flavanoid compounds etc.) have long been identified (Foyer and Noctor 2005; 2013; Foyer and Shigeoka 2011; Gill et al. 2012; Khan and Khan 2014b; Khan et al. 2014). However, more recently, osmoprotectants (betaines, amino compounds and sugars) have been observed for performing a similar role in boosting plant immunity. Several reports since the past two decades document this group's involvement in a massive signalling scheme to counter excess ROS generation in plants facing stress (Khan et al. 2013; ElSayed et al. 2013; Bhardwaj et al. 2014; Gill et al. 2014; Sharma et al. 2014). Our prime focus in this context remains to explore sucrose metabolism and its involvement in cutting down oxidative damage. Sugars/carbohydrates play a central role in development, stress response , and productivity by controlling protein synthesis, regulation of micro RNAs, transcription factors, and various genes for crosstalk with other hormonal and defense signaling pathways (Bolouri-Moghaddam and Van den Ende 2013; Ruan 2014; Lastdrager et al. 2014). Sugars involved as stress busters include fructans, raffinose family oligosaccharides (RFOs), trehalose and polyols (Gill and Tuteja 2010; Keunen et al. 2013). This chapter aims at elaborating these sugar families and their key role as ROS scavengers besides performing other functions. The activation and up-regulation of several factors have been listed and discussed for this with special context to the chloroplast (as sugars are the major outcome of photosynthesis- the prime function of chloroplasts) and mitochondria (as the powerhouse of energy involved in ROS scavenging). Taking into account the recently discovered significance of sugars for providing protection against stress and damage, the current scenario along with future prospects have also been discussed.

ENVIRONMENTAL STRESSES AND OXIDATIVE BURST IN PLANTS

Abiotic stresses, like salinity, drought, extreme temperatures, toxicity due to chemicals and oxidative stress cause serious risks to agriculture and lead to the worsening of environmental quality (Osakabe et al. 2013). Another limiting root cause of crop loss and productivity is biotic stress due to widespread sporadic/epidemic pathogen and pest attacks, which causes reduction in the average yields of major crop plants by more than 50% (Bray et al. 2000; Oerke 2006). In many regions, where drought and salinity are prevalent, severe salinization in 50% of all arable lands till 2050 has been estimated. Abiotic stress causes morphological, physiological, biochemical and molecular changes, which negatively affect the growth and productivity of plants (Wang et al. 2001). These environmental stresses and oxidative stress are interconnected, causing cellular damage in plants worldwide. Drought/salinization results in the disruption of ion distribution and homeostasis in the cells, and it is mostly manifested as an osmotic imbalance (Zhu 2001a). Denaturation of functional and structural proteins is caused by oxidative stress, which often accompanies salinity and high temperature (Uzilday et al. 2012). As a result, these varied environmental stresses frequently activate the same cell signaling pathways (Shinozaki and Yamaguchi-Shinozaki 2000, Knight and Knight 2001; Zhu 2001b; 2002) and cellular responses, like generation of stress proteins, accumulation of compatible solutes, including sugars and up-regulation of antioxidants (Cushman and Bohnert 2000; Krasensky and Jonak, 2012; Gepstein and Glick 2013).

Different environmental factors also affect the distribution of plants and implement a selective pressure on them, which have an inherent adaptation for the same (Amtmann et al. 2005). Those plants which do not boast of an ability to reproduce will not thrive into a community. Thus, highest selectivity corresponds to reproduction capacity (Song et al. 2005; Mittler 2006). Plants, across an evolutionary time gap, have developed a wide range of adaptative strategies to survive in the presence of various environmental stresses. A specific stress response changes with the genotype; however, certain general reactions are common for all genotypes. Abiotic stresses have an effect on different cellular processes of plants like growth, photosynthesis, protein synthesis, carbon partitioning, osmotic homeostasis, carbohydrate and lipid metabolism and gene expression (Munns 2002; Rosa et al. 2004; Ghosh and Xu 2014). Metabolic activities of plants are also affected by chilling (temperatures which are below optimal but above freezing) as well as freezing temperatures, which further cause osmotic stress (Britt 2004). Also, factors such as decrease in photosynthesis, interference with nutrient availability, and restrictions of metabolic processes due to drought conditions confine plant growth (Hu and Schmidhalter 2005; Bohnert et al. 2006). Growth of plants is also interfered by salinity, where plants are affected due to both physiological drought and ion toxicity (Redondo-Gómez et al. 2007). UVBR leads to DNA damage, production of secondary metabolites like phenolic compounds and inhibition in photosynthesis (Ibañez et al. 2008). Heavy metals stimulate the alteration in mitochondrial respiration and thus lead to oxidative damage (Shanker et al. 2005). Consequentially, oxidative stress generates ROS due to the exposure to these abiotic stresses (Apel and Hirt 2004; Foyer and Noctor 2005; Bohnert et al. 2006).

In the field conditions, plants are usually exposed to a combination of these abiotic stresses (Chaves et al. 2002; Rizhsky et al. 2004). Therefore, plant responses to various types of stresses are very complex phenomena, such as exposure of plants to drought stress seem to affect the leaves, even as the stimuli can be alleged in the leaf itself or in other plant part as roots (Tester and Davenport 2003). Plants possess defence strategies like stress avoidance and tolerance mechanisms, to cope with the mixture of stresses. 'Oxidative Stress' refers to a severe imbalance between the generation and removal of ROS (Uzilday et al. 2012). ROS imbalance involves the overproduction of ROS, especially the superoxide, hydroxyl, hydrogen peroxide, peroxy, alkoxy radicals and singlet oxygen, etc., leading to oxidative burst and, eventually, followed by permanent damage of the cell's biochemical machinery (Dat et al. 2000) as seen in Figure 1. Chloroplast is the most active site of ROS production in plants (Asada 2000). They are also generated in the cell wall apoplasmic continuum and peroxisomes (Dat et al. 2000). ROS are involved in various processes like systemic acquired acclimation (SAA), systemic acquired resistance (SAR) and hypersensitive response (HR). In basic cellular processes, the participation of ROS showed their significant role in signalling systems and regulation of stress, and associated amelioration in plants exposed to various environmental stresses (Bowler and Fluhr 2000).

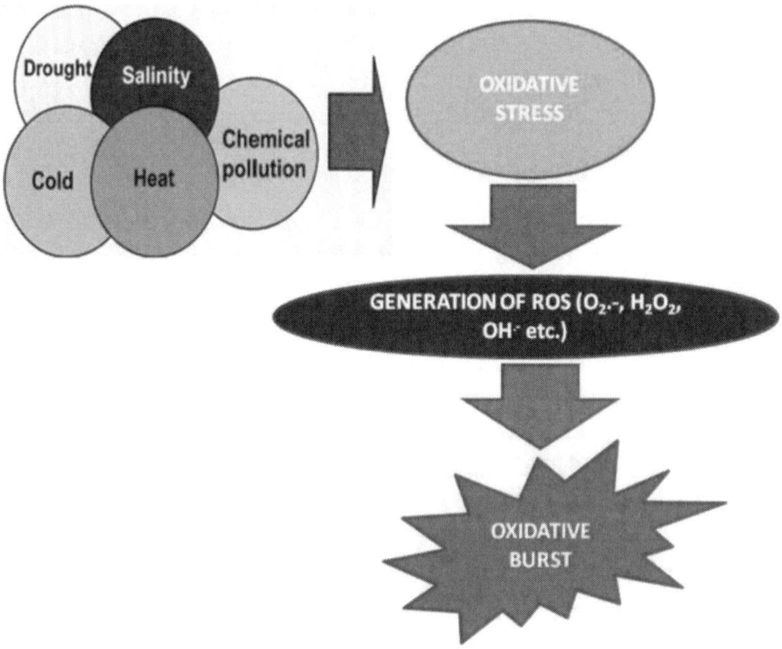

Figure 1. Oxidative burst due to different abiotic stresses through generation of ROS.

Sugar Metabolism and Oxidative Stress Management

Sugars act as both regulatory molecules and nutrients for the whole life of plants. With the interplay amongst various phytohormones, ROS generation, scavenging and signalling systems, a complicated network is established by sugar metabolism and sugar signalling. In plants, normally, metabolic rates are inversely correlated with stress resistance and lifespan during oxidative stress (Halliwell 2006; Valko et al. 2007). High photosynthetic activity in the leaves of plants can stimulate the accumulation of both soluble sugars and ROS. On the other hand, starvation of sugar may also cause ROS accumulation (Coue´e et al. 2006). The defensive role of soluble sugars against oxidative stress i.e., triggering the generation of specific ROS scavengers has been mostly recognized by signalling effects (Ramel et al. 2009). Sugars themselves act as ROS scavengers at higher concentrations (Van den Ende and Valluru 2009). Besides acting as structural constituents and metabolic sources of cells, soluble sugars also play a pivotal role in signal regulating processes, which are associated with plant growth and development (Pego et al. 2000). To modulate metabolic plant responses, sugar signalling pathways interact with stress pathways in an intricate network (Tran et al. 2007). They can also play important roles in cell responses to stress-induced remote signals by acting directly as negative signals or as modulators of plant sensitivity (Ghosh and Xu 2014). Within the mechanisms of stress tolerance, the source-sink partitioning between different organs acts as a key component (Rathinasabapathi 2000). Both glucose and sucrose regulate gene expressions regarding plant metabolism, growth and development of plants and stress resistance. Thus, they are considered as pivotal integrating regulatory molecules (Rolland et al. 2006; Ramon et al. 2008). The processes like sugar sensing and response pathways help the plants to adapt in the changing environment, but due to dual involvement of sugars in nutrition as well as in signalling, makes these mechanisms more intricate (Rolland et al. 2001; 2006). Hexokinase (HXK) acts as the fundamental glucose sensor, as the catalytic activity of mitochondria-associated hexokinase helps in the tracking and stimulation of antioxidant defence factors and synthesis of phenolic compounds. It constantly keeps a tab on the glucose-6-phosphate and ROS levels by doing so (Bolouri-Moghaddam 2010). Snf1-related Kinase 1 and HXK control eukaryotic energy homeostasis, stress resistance, survival and longevity, and they have been recognized as conserved sugar signalling components. Invertase-related sugar signals also appear to be very significant during the defence reactions of plants even during pathogenic invasions (Fotopoulos et al. 2003; Essmann et al. 2008). Conclusively, sugar signals incorporate different endogenous developmental, environmental and metabolic cues, so as to manoeuvre a multifaceted network with the stress related and plant-specific hormone signalling pathways. They operate across cellular and organelle membranes to scavenge ROS, generated as a result of stress exerted on basic processes such as photosynthesis/respiration/detoxification in the chloroplasts/mitochondria/peroxisomes respectively (Keunen et al. 2013). In order to understand these roles, various sugar categories have been discussed with regard to their presence, biosynthesis and functioning in the plant cells.

Fructans

Fructans are poly fructose molecules which function as water soluble carbohydrates, with special regard to plant health and adaptation to changing environmental norms (Huynh et al. 2012). On the basis of glycosidic linkage between fructose units, fructans are classified into various categories such as the insulin containing β (2, 1) linkage fructans mainly present in dicots, the levans containing β (2, 6) linkage mainly occurring in grasses, the graminans containing β (2, 1) and β (2, 6) linkages mainly present in grasses, neolevan and neoinsulin fructans present in monocots such as *Allium* and *Asparagus* (Figure 2).

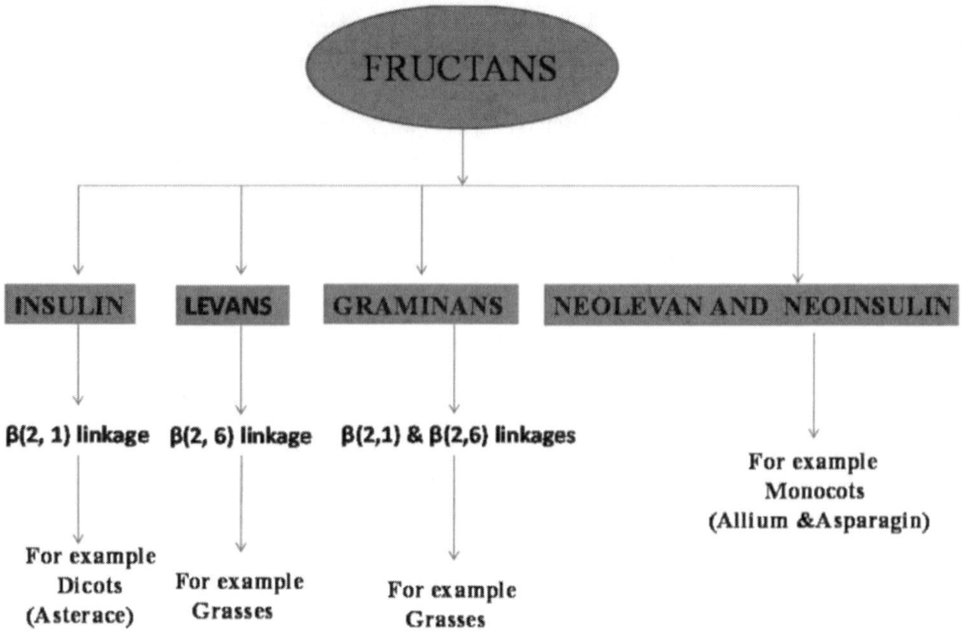

Figure 2. Diagrammatic representation of the classification of Fructans on the basis of glycosidic linkage between the fructose units.

Fructan synthesis takes place in vacuoles, apoplast, xylem and phloem tissues (Livingston and Henson 1998; Wang and Nobel 1998; Van den Ende et al. 2005). Their synthesis from sucrose is catalyzed by a cluster of fructosyltransferases (FTs) which belong to a family of 32 glycoside hydrolases (Duchateau et al. 1995; Ritsema and Smeekens 2003; Chalmers et al. 2005; Altenbach et al. 2009; Van den Ende et al. 2009). The most well studied fructans are the Graminan type fructans which are present mainly in cereals, with beta-2, 6 and beta-2, 1 short linked branched chain (Edelman et al. 1968; Ritsema and Smeekens 2003; Chalmers et al. 2005). The synthesis of beta-2, 6 fructans takes place by continuous action of 1-SST (sucrose: sucrose-1-fructosyltransferase), 6-SFT (sucrose: fructan-1-fructosyl transferase) and 1-FFT (fructan: fructan-1-fructosyltransferase). The cDNA for these enzyme groups has been characterized from wheat (Gupta and Kaur 2005; Kawakami and Yoshida 2002; 2005). The biosynthesis of fructans in wheat relates to these three enzymes. Fructan exo- hydrolases catalyze the degradation of fructan in plants (Van den Ende et al. 2003a; b). 6- SFT and 1-SST are involved in biosynthesis of fructans under cold hardening, whereas 1-FFT and 6-SFT are involved in further elongation of branched graminans (Le Roy et al. 2007; 2013). During drought and cold stress, fructans protect the plant through stabilization of membranes and by regulating the process of osmosis through opening of flowers (Hincha et al. 2003; Le Roy et al. 2007; Ende and El-Esawe 2013). Studies by De Roover et al. (2000) and Valluru and Van den Ende (2008) indicated that fructans play an important role during abiotic stresses such as freezing, chilling, salinity and drought in wheat. Van den Ende (2013) reported that accumulation of fructans in vacuoles takes place under various kinds of stresses such as drought and salinity. Li et al. (2013) studied the effect of nitric oxide on fructan expression in the leaves and seedlings of two wheat varieties namely, spring wheat and winter wheat. The activities of fructansucrases, including 1-SST, 6-SFT, was higher than 1-FFT activity in seedlings of two wheat cultivars. Temporary accumulation of fructans was recorded in the reproductive organs, young kernels and stems of wheat (Lasseur et al. 2009; Ji et al. 2010; Van den Ende et al. 2003 a; 2005). Accumulation of fructans in tuberous fragments of *Gomphrena marginata* was maximum during the dry and quiescent seasons of growth, and minimum during the rainy season (Silva et al. 2013). Recently, accumulation of higher levels of glucose, fructose and sucrose were recorded against pathogens *Xanthomonas oryzae* pv. oryzae and *Magnaporthe oryzae* in rice plants, in response to the induction of the cell wall invertase gene *Grain Incomplete Filling 1* (*GIF1*), suggesting their ROS limiting role (Sun et al. 2014).

RFOs

Raffinose family oligosaccharides (RFOs) are another class of water-soluble carbohydrates which play an important role in generating plant stress responses (Salerno and Curatti 2003). These are α-galactosyl derivatives of sucrose which include raffinose, stachyose and verbascose, involved in formation of carbohydrate portion of seeds (Taji et al. 2002). They serve as anti-stress compounds, membrane stabilizers and store houses of carbohydrates in plants (Peshev and Van den Ende 2013; Van den Ende 2013). RFOs biosynthesis takes place from sucrose by the action of enzyme galactinol synthase (Egert et al. 2013; Elsayed et al. 2013). Galactinol synthase (genes *Gol S1* and *Gol S2*) helps in galactinol biosynthesis from uridine-diphosphate galactose. This galactinol substrate serves as galactosyl donor to produce RFOs like raffinose (Figure 4). Galactinol synthase activity is required as a precondition for biosynthesis of galactinol and raffinose under stressed environment (Panikulangara et al. 2004). The expression and activity of GolS enzyme enhances significantly when plants encounter various environmental stresses (ElSayed et. al. 2013). Transport of raffinose from other organelles to the chloroplast takes place under stress conditions as it is noticed that RFOs are also biosynthesised outside the chloroplast (Schneider and Kelle 2009).

RFOs act as antioxidants in plants under temperature/drought/salinity stress conditions (Mahajan and Tuteja 2005; Nishizawa et al. 2008; Stoyanova et al. 2011; Hernandez-Marin and Martinez 2012; Peshev et al. 2013). These multiple roles have been elaborated in Figure 3. Occurrence of raffinose in chloroplast has been observed (Schneider and Keller 2009; Foyer and Shigeoka 2011) and it is noticed that raffinose helps in stabilizing PS-II in *Arabidopsis* plant under freeze- thaw cycles (Knaupp et al. Q2011). RFO levels are reported to escalate in chloroplasts of plants under stress conditions and it is also observed that raffinose levels are almost equal to that of Ascorbic Acid (AA) and Glutathione (GSH) levels, indicating their role in detoxification of hydroxyl radicals (Nishizawa et al. 2008; Stoyanova et al. 2011). RFOs are reported to gather in thylakoid membranes, where they may act as signalling messengers or as chloroplast detoxifying antioxidants (Nishizawa et al. 2008; Peshev and Van den Ende 2013). RFOs like stachyose, verbascose and ajugose are reported to be present in vacuoles of some plants (Janecek et al. 2011). These sugars along with their metabolic enzymes are involved in ROS signalling pathways (Bolouri-Moghaddam et al. 2010).

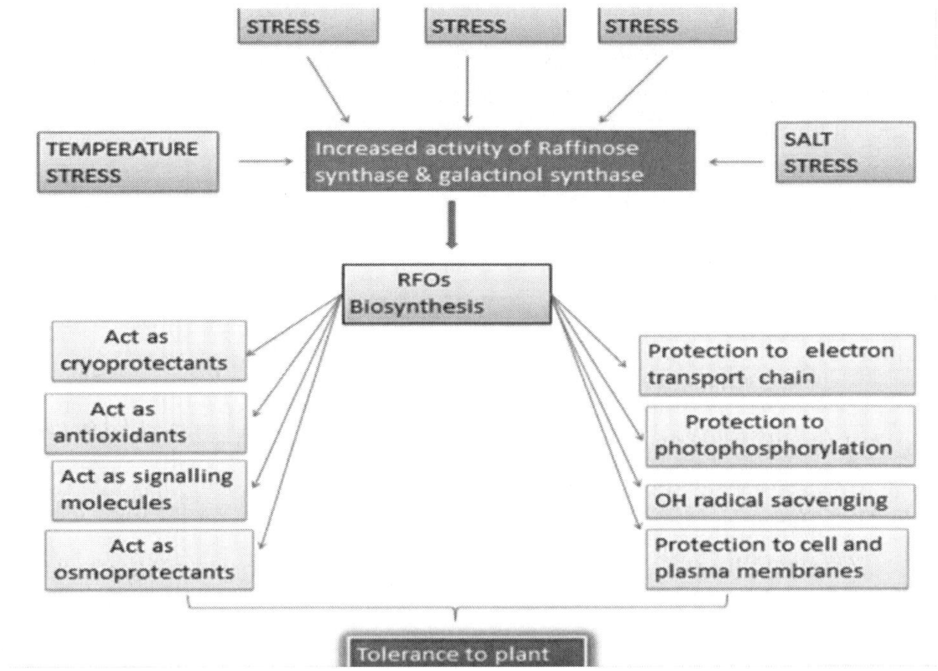

Figure 3. Diagrammatic representation of various roles performed by RFOs under stress conditions in plants.

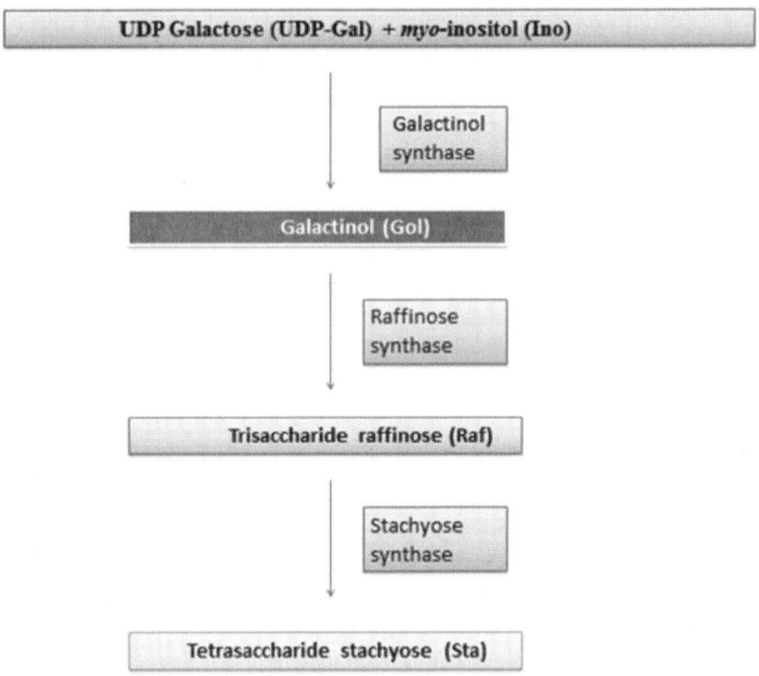

Figure 4. Diagrammatic representation of biosynthesis of raffinose family oligosaccharides (RFOs) under various stress conditions (Modified after Egert et al., 2013).

They are considered as better scavengers for OH radicals as compared to O^{2-} radicals (Stoyanova et al. 2011). In plants, there is no enzyme mediated scavenging of OH radicals; therefore, they are totally dependent upon non-enzymatic antioxidants like RFOs for detoxification of ROS (Buitink et al. 2000; Bailly et al. 2001; Gechev et al. 2006; Lehner et al. 2006). Neutralization of OH radicals is observed in-vitro and raffinose has the same antioxidant potential as that of GSH (Nishizawa et al. 2008). RFOs have another important role in internal transport, carbon storage, cell signalling (Stevenson et al. 2000; Xue et al. 2007), membrane trafficking (Thole and Nielson 2008) and export of mRNA from cell nucleus (Okada and Ye 2009). During drought stress, in order to balance the turgor pressure of cell and cell protein stabilization, RFOs act as osmolytes (Bartels and Sunkar 2005). Elevation of RFO concentration under desiccation stress results in stabilization of phospholipid membranes (Gechev et al. 2012). Galactinol is another signalling molecule under abiotic stress conditions (Valluru and Van den Ende 2011). Also, during pathogen induced biotic stress, RFOs and galactinol act as signalling messengers to enhance the defence mechanism of plants (Kim et al. 2008; Valluru and Van den Ende 2011). Paraquat treatment, salinity and chilling results in increased galactinol and raffinose level in transgenic plants, which lead to tolerance in plants against these abiotic stresses (Shi et al. 2003). Chilling treatment to rice plants for a long time results in elevation of galactinol and raffinose concentrations (Saito and Yoshida 2011). In *Arabidopsis sp.*, it has been observed that the over expression of galactinol synthase (genes *Gol S1, Gol S2, Gol S4*) and raffinose synthase resulted in elevation of galactinol and raffinose levels, which further add on to the ROS detoxification ability and oxidative stress tolerance (Nishizawaet al. 2008). It is reported that Galactinol synthase (GolS) becomes more active under stresses like chilling and drought (Dowrie et al. 2003). In *Arabidopsis,* under chilling stress, the production of raffinose in cytosol plays an important role in protection of plasma membrane (Nagele and Heyer 2013). RFOs are also reported to act as cryoprotectants in plant cells under extreme frozen conditions (Pennycooke et al. 2003). They also serve as mobile carbohydrate moieties through phloem (Ayre et al. 2003). In *Ajuga reptans* L., it is noticed that RFOs increase the frost tolerance in leaves (Peters and Keller 2009). RFOs provide protection to cell membranes of plants (Van den Ende and Valluru 2009). Plant cell organelles like vacuole and chloroplast act as sites for ROS detoxification by RFOs (Van den Ende 2013). These are also described as compounds which act as signal transducing molecules and form antioxidant networks in plants against various abiotic stresses (El Sayed et al. 2013).

Trehalose

Water is a key component for survival; nevertheless, some organisms are adaptative to survival under desiccation, till the occurrence of favourable conditions. These organisms (fungal spores, certain invertebrate species and 'resurrection' plants) accumulate the disaccharide trehalose (Iturriaga et al. 2009). Trehalose (α-D-glucopyranosyl-1, 1-α-D-glucopyranoside) is a naturally occurring non-reducing disaccharide with functionality similar to sucrose but with less sweetness (Figure 5). It contains two glucose molecules linked in a 1, 1-position by an α-glycosidic bond. Although the role of trehalose in vascular plants is still not clear, yet genes involved in trehalose biosynthesis have been elucidated by genome database analysis.

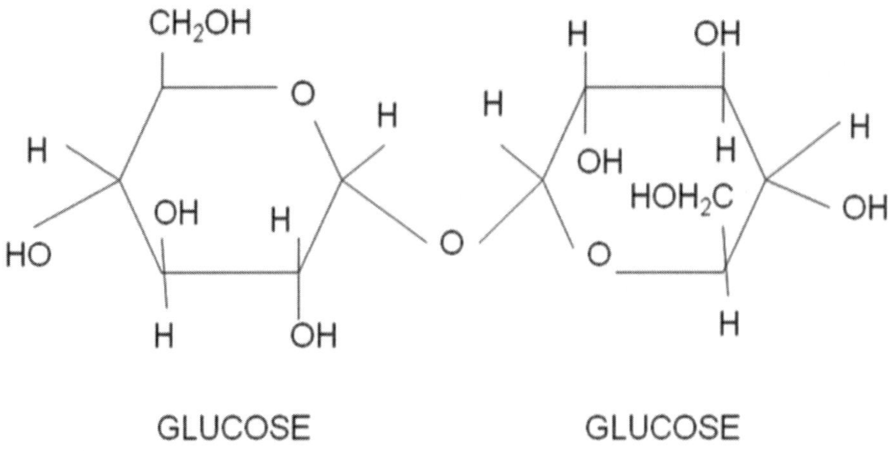

Figure 5. Trehalose Structure: A disaccharide consisting of 2 glucose moieties linked by a glycosidic bond that joins their anomeric carbons.

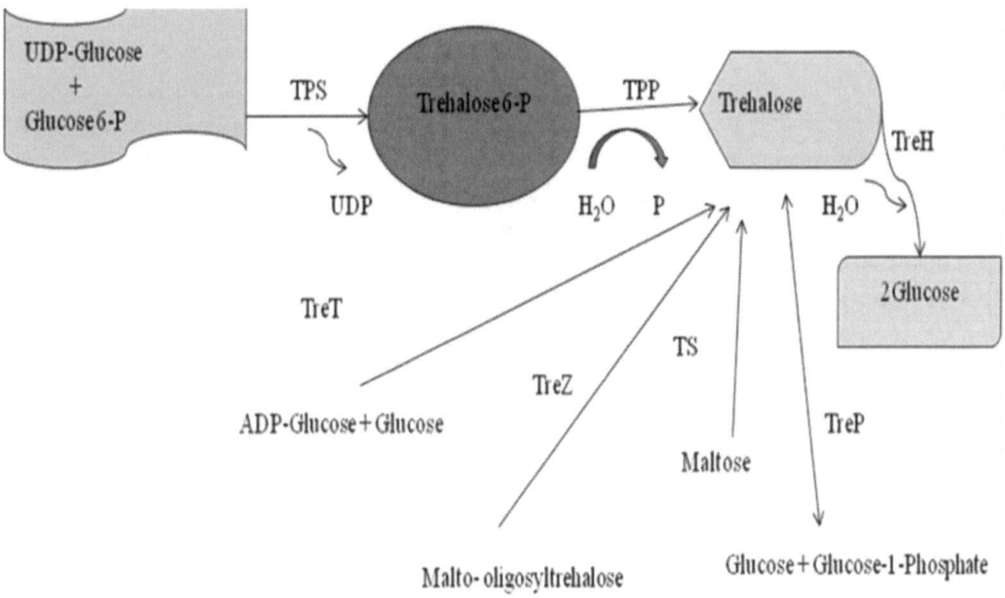

Figure 6. Trehalose metabolic pathway (TPS-TPP pathway present in plants. TreT, TreY- TreZ and TS only described in prokaryotic organisms).

There are at least three different pathways for trehalose biosynthesis, as seen in Figure 6 (Avonce et al. 2005). Trehalose-6-phosphate (T6P) is produced from UDP-glucose and glucose-6-phosphate (G6P) by the enzyme trehalose-6-phosphate synthase (TPS). T6P is dephosphorylated by trehalose-6-phosphate phosphatase (TPP) to produce trehalose. It is further hydrolyzed by the enzyme trehalase into two molecules of glucose (O'Hara et al. 2013). Trehalose biosynthetic pathway is also known as OtsA-OtsB pathway as explained in Figure 6 (Fernandez et al. 2010). In many insects, it is produced as a major blood sugar and acts as a reserve carbohydrate during periods of dehydration and freezing. Trehalose is found in lower concentrations in various consumable foods like mushrooms, bread, fermented beverages and honey. Trehalose was first extracted from ergot of rye, a fungal crop blight; it was identified in several ferns such as *Selaginella lepidophylla* (rose of Jericho) and *Botrychium lunaria*, bryophyte (*Tortula ruralis*), dicots (*Craterostigma plantagineum*, *Boea hygrometrica*) and monocot (*Xerophyta viscose*) (Dinkar and Bartels 2013). The unique feature of trehalose is that it is composed of two hexose rings joined by glycosidic bond, which makes it a stable structure in comparison to sucrose, and it helps in protecting the activity of various thermolabile enzymes like DNA ligases, polymerases and restriction endonucleases. It further helps to increase the shelf life of frozen as well as dried flowers, organs, tissues and cells, and acts as a signalling molecule in many organisms such as yeast, and also as a marker in production of certain transgenic plants. Its role as an effective osmoprotectant is already well known (Garg et al. 2012; Redillas et al. 2012). Trehalose is a potential signal metabolite in plant interactions with pathogenic or symbiotic microorganisms and herbivorous insects. T6P influences the relative amounts of sucrose and starch accumulated in leaves during the day, and regulates the rate of starch degradation at night to match the demand for sucrose. Lunn et al. (2014) proposed that Tr6P influences plant growth and development via inhibition of the SNF1-related protein kinase (SnRK1). The effects of exogenous trehalose (Tre) on salt tolerance of pharmaceutical plant *Catharanthus roseus* and the physiological mechanisms involved were both investigated. It was seen that Tre might act as a signal to make the salt-stressed plants actively increase internal compatible solutes, including soluble sugars and free amino acids, to control water loss, leaf gas exchange and ionic flow at the onset of salt stress. The application of Tre in saline condition also promoted the accumulation of alkaloids (Tsai and Gazzarrini 2014; Chang et al. 2014). Further, enhanced levels of endogenous trehalose were detected in calli exposed to water deficit and trehalose medium. Moreover, water plus Tre treatment showed an additive effect thereby, alleviating the damage induced by water deficit, which was involved in ROS generation, and resulted in elevating cell viability and biomass (Garg et al. 2002; Ma et al. 2013).

Polyols

Polyols are poly-hydroxylated compounds, playing an important role in stress tolerance. The most commonly found polyols (sugar alcohols) in plants are classified into acyclic (mannitol, sorbitol) and cyclic (inositol, pinitol) compounds (Merchant and Richter 2011). These are reduced forms of aldoses or ketoses (Sickler et al. 2007; Nadwodnik and Lohaus 2008). Mannitol and sorbitol are the two main polyols occurring among various plant species. These are compatible solutes which get accumulated to higher concentrations in the cells without adversely affecting cell activity, and their accumulation is associated with the stress tolerance in plants (Williamson et al. 2002; Szabados and Savoure 2010). Mannitol and sorbitol, both are end products of photosynthesis. Mannitol synthesis occurs in mesophyll cells where mannose-6-phosphate isomerase (MPI) converts fructose-6-phosphate to mannose-6- phosphate. Further mannose-6-phosphate reductase (MPR) catalyzes the conversion of mannose-6-phosphate to mannitol-1-phosphate , which has been dephosphorylated by mannitol -1- phosphate phosphatase (MPP) to form mannitol (Loescher and Everard 2000; Merchant and Richter 2011). It is then translocated through the phloem to sink tissue, where it can either be stored or oxidized to mannose and used as a carbon rich energy source (Noiraud et al. 2001; Conde et al. 2008). On the other hand, sorbitol is synthesized in mature leaves where glucose-6-phosphate is converted to sorbitol-6-phosphate by sorbitol-6-phosphate dehydrogenase (SPDH). Afterwards, sorbitol-6-phosphate is dephosphorylated to sorbitol via sorbitol-6-phosphate phosphatase (SPP) (Kanamaru et al. 2004; Liang et al. 2012). The reaction catalyzed by SPDH is the key regulatory step in sorbitol synthesis (Figure 7). It has been reported that in apple plants over expression of SPDH overcomes salt stress by developing tolerance against salinity among plants (Gao et al. 2001). Similarly, inositol (myo-inositol) is synthesized from glucose-6-phosphate in two steps. Firstly, inositol phosphate synthase (INPS) catalyzes the conversion of glucose-6-phosphate to myo-inositol-1-phosphate, which is then dephosphorylated by inositol monophosphatase (IMP). Myo-inositol further converts it to methyl derivatives (D-ononitol, D-pinitol), through the action of inositol methyl transferase (IMT) as seen in Figure VII. Over-expression of INPS and IMT is seen to generate salinity tolerance by enhancing the cyclic polyol levels in tobacco and other halophytes (Majee et al. 2004; Patra et al. 2010). Sengupta et al. (2008) reported the accumulation of myo- inositol and its derivatives D-ononitol and D-pinitol in various halo tolerant plant species under stress. Polyols perform various physiological functions in plants such as enhancement of photosynthesis, storage and translocation of carbon, osmotic adjustments, stress tolerance, protection of enzymesproteins, membranes, liposomes and free radical quenching (Valliyodan and Nguyen 2006; Ortbauer and Popp 2008; Andersen et al. 2011). Mannitol-1-phosphate dehydrogenase (mt1D) catalyzes the reversible conversion of fructose-6-phosphate to mannitol-1-phosphate. Wheat was transformed with the *mtlD* gene of *Escherichia coli* and it showed improved tolerance to water and salt stress. Thus, mannitol accumulation conferred tolerance against stress by overcoming the adverse effects of salinity condition (Abebe et al. 2003). Similarly, over-expression of mannitol-1-phosphate dehydrogenase increased mannitol accumulation and provided protection against chilling stress in *Petunia* (Chiang et al. 2005).Four plant lines were transformed with an *E.coli* gene encoding mtlD. Out of the four plant lines, two lines contained high level of mannitol (M3 and M8) in comparison with non- transformed wild-type plants and the other two had similar levels of mannitol like those of the wild-type plants (M2 and M9). It was observed that M3 and M8 transgenic lines were more tolerant to chilling stress, which favoured the role of mannitol in stress alleviation (Chiang et al. 2005). Mannitol accumulation also developed salinity tolerance in Lemon balm subjected to salt stress. The plants under salt stress showed reduced growth, photosynthetic pigments and mineral content, whereas mannitol accumulation resulted in salinity tolerance by improving the growth, pigment and mineral content (Khalid and Weiming 2011). Mannitol accumulation in transformed tobacco plants stimulated the ROS-mediated plant defence system against fungus *Alternaria alternate* infection through the enhancement of mt1D enzyme expression, which further supported the role of polyols in stress tolerance (Jennings et al. 2002). Similarly, sorbitol offered protection against drought in Nemaguard peach plants by the loss in activity of Sorbitol dehydrogenase (SDH) in sinks (Lo Bianco et al. 2000). Sorbitol protected tissue damage caused by salinity stress, was reported in rice plants, by controlling peroxide radical formation and their timely scavenging, as well as by lipid biosynthesis (Theerakulpisut and Gunnula 2012).

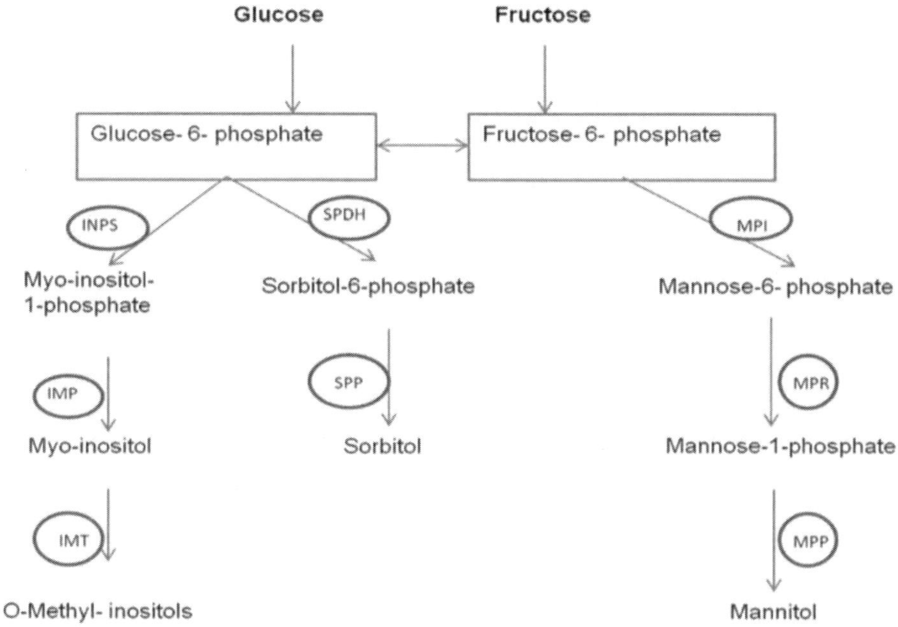

Figure 7. Polyol biosynthesis in plants.

SUGAR METABOLISM AND ANTIOXIDANT NETWORKS

In plants under stressed conditions, ROS generation occurs in the chloroplasts (Suzuki et al. 2012). Hence, to detoxify these harmful ROS, chloroplasts act as sites of sugar metabolism. It is amply clear that soluble sugars act as ROS scavengers in the previous sections to a great extent. Soluble sugars also act as antioxidants and sugar signalling molecules and over express genes involved in oxidative stress management in plants under abiotic stress (Couee et al. 2006). Hexokinase (HXK), which is the first enzyme in glucose metabolism, plays an important role in the regulation of ROS formation in chloroplast under environmental stress (Bolouri-Moghaddam et al. 2010). Figure 8 explains the ROS-Sugar interplay in the cells facing environmental stress. It is noticed that chloroplasts contain maximum ROS content under stress conditions and only minor concentrations of ROS like H_2O_2 have negative impact on the activity of the Calvin cycle enzymes (Miyagawa et al. 2000). Therefore, plants face the need of a strong antioxidant system to detoxify these harmful ROS within the chloroplasts, so that the process of photosynthesis remains unaffected. For this purpose, chloroplasts have a network of enzymatic/ non-enzymatic antioxidants. Enzymatic antioxidants include Mn-SOD, APOX, GPOX and thio-redoxin peroxidase, whereas, non-enzymatic antioxidants include α-tocopherol, β-carotene, AA and GSH (Mittler 2002). It has been widely reported that RFOs produced from galactinol synthase (GolS), raffinose synthase (RafS) and stachyose synthase (StaS) are transported to chloroplast from other organelles (Nagele and Heyer 2013). In Purslane (*Portulaca oleracea* L.), it is observed that there is a significant increase in sugar levels in chloroplast of leaves which provide tolerance to plants under salt stress (Rahdari et al. 2012).

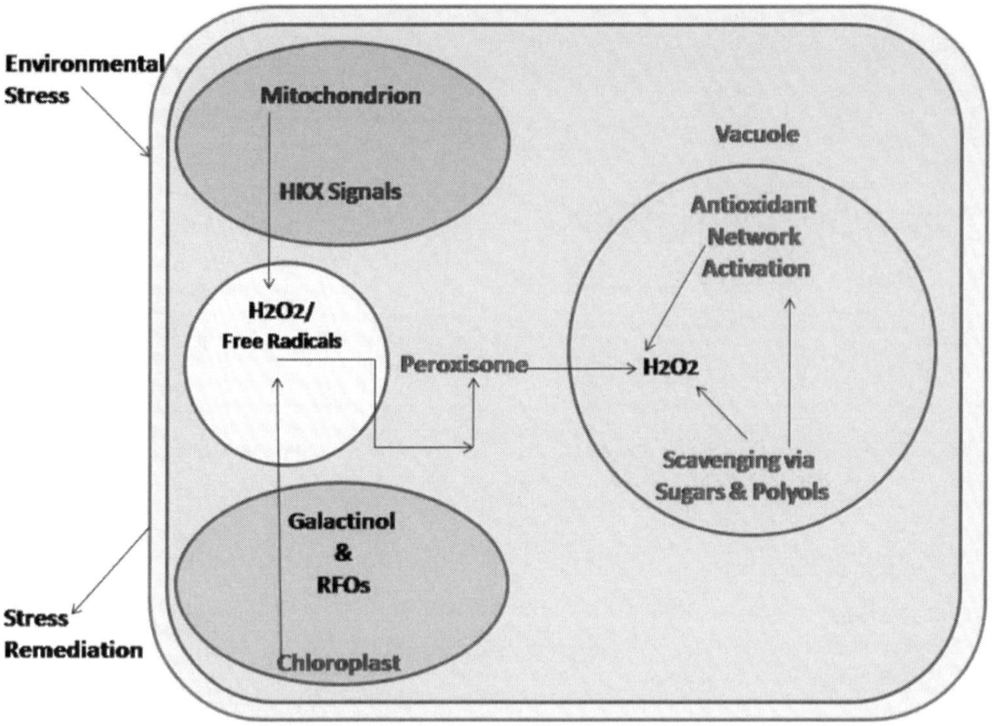

Figure 8. ROS scavenging via sugar interplay in cell organelles.

Raffinose is reported to be present in chloroplast of plants such as *Ajuga reptans*, *Spinacia oleracea* and *Arabidopsis thaliana* during chilling treatment. Also, it is reported that galactinol synthase (GolS) and raffinose synthase (RafS), which are involved in raffinose biosynthesis, are observed to be present in the immediate vicinity of chloroplast. So, in order to transport these enzymes into chloroplast from outside, raffinose transporters are present on the outer covering of chloroplasts (Schneider and Keller 2009). *Arabidopsis thaliana* HXK1 is an important intracellular sensor of glucose. The HXK1 in this case is linked with mitochondria and has also been reported in the nucleus (Chen 2007; Balasubramanian et al. 2007). The HXK localization in mitochondria improves access to ATP molecules that are produced during the complex inter-membrane electron transport chain (ETC) , for consumption by fluxes of active metabolites through cycling of sucrose, sugar nucleotide synthesis and glycolysis. The complete glycolytic metabolism is linked with the outer membrane of mitochondria, providing pyruvate directly to it, which further acts as a substrate for respiration. Glycolytic enzymes have been linked with mitochondria, in response to demand of respiration (Møller 2001; Rontein et al. 2002; Graham et al. 2007; Keunen et al. 2011). These reports suggest that the mitochondrial HXK activity is associated in the regulation of production of ROS in plants and respiration in mitochondria (Da Silva et al. 2004; Rhoads et al. 2006; Noctor et al. 2007). For instance, the potato mitochondrial HXK is more sensitive to inhibition of ADP in micro molar concentrations when glucose is the substrate as compared to fructose, suggesting a varied affinity for these sugars as well (Camacho-Pereira et al. 2008). The mitochondrial HXK plays a significant role in generating ADP to support the process of oxidative phosphorylation and, in turn, avoids the limitation of

ATP generation in respiration and release of hydrogen peroxide in plants (Camacho-Pereira et al. 2008). In the roots of maize, the mt HXK activity is directly associated with the synthesis of NDP-sugar, necessary for the synthesis of phenyl propanoids, celluloses and flavanoids. Its attachment to the mitochondrial membrane is essential for prevention of antioxidant activity in mitochondria of both plants and animals (Sun et al. 2008; Camacho-Pereira et al. 2008; Da Silva et al. 2004). Both glucose-6-phosphate and methyl jasmonates are involved in the detachment of mammalian mt HXK from outer membrane of mitochondria (Ehsani- Zonouz et al. 2001; Goldin et al. 2008; Da Silva et al. 2004). However, glucose-6-phosphate has no effect on mitochondria in plants (Camacho-Pereira et al. 2008). Methyl jasmonate, on the other hand, affects the detachment of mt HXK from mitochondria of plants. The accumulation of methyl jasmonates in plants leads to the accumulation of ROS, changing the movement and morphology of mitochondria and finally causing cell death (Zhang and Xing 2008). Soluble sugars such as sucrose and fructose have a dual role with regard to ROS. These support the oxidative pentose phosphate pathway (PPP), which contributes to the scavenging of ROS. Increased concentration of glucose is associated with pathogenesis and toxicity through enhanced ROS and auto-oxidation of glucose (Russel et al. 2002; Barros et al. 2004). Reduced glucose and calorific limitations strongly decline the production of ROS in mitochondria (Barros et al. 2004). In plant cells, interference of redox equilibrium requires specific antioxidant enzyme activation neutralizing ROS. Antioxidants with low molecular weight play an important role during beginning of oxidative stress and are activated by these soluble sugars. In vitro studies strongly indicate that in vivo capacity of antioxidant sugars might be greatly underestimated. Conclusively, it can be said that RFOs, dissacharides, fructans, sugar alcohols and galactinol function as antioxidants (Nishizawa et al. 2008; Stoyanova et al. 2011; Hernandez-Marin and Martinez 2012; Peshev et al. 2013).

Conclusion and Future Prospects

Environmental stress conditions across the globe disrupt the delicate electron balance in plants. As a result, ROS accumulation occurs as an inevitable physiological phenomenon. However; plants have the right immunity restoring metabolites to tackle the harmful influences exerted by them. These complex antioxidants link together a wide range of metabolic shunt pathways and restrict ROS to cellular compartments such as chloroplasts, mitochondria and peroxisomes. A recent addition to the traditional group of enzymatic and non-enzymatic antioxidants for quenching ROS in these organelles is the sugar clan (fructans, RFOs, trehalose and polyols). Hence, they gain mileage in the current scene for conferring further resistance against abiotic stresses. Enhancing their expression by genetic engineering and exogenous application in food and commercial crops as osmoprotectants and antioxidants provides a promising approach to develop stress resistant and high yielding varieties. Also, sugars are important ingredients in various medicinal and therapeutic concoctions in traditional medicine, and are also used as food additives. What is needed at the current moment is a better understanding of the physiological and molecular mechanisms by which sugars remediate ROS in plants, for both agronomic and economic benefits. Constant changes in sugar levels in plants denote a heightened response to stress, as well as for energy storage. Future work is needed to reveal the role of sugars across a wide range of plant families, for a better grasp of their vitality in generating responses to stress signalling pathways. Also, much work is needed to explore lesser known sugar derivatives such as glycerol, mannitol, sorbitol, etc. as signaling molecules and stress busters in plants.

REFERENCES

Abebe, T., Guenzi, A.C., Martin, B., Cashman, J.C., 2003. Tolerance of mannitol-accumulating transgenic wheat to water stress and salinity. *Plant Physiol.* 131, 1748–1755.

Altenbach, D., Rudino-Pinera, E., Olvera, C., Boller, T., Wiemken, A. and Ritsema, T., 2009. An acceptor-substrate binding site determining glycosyl transfer emerges from mutant analysis of a plant vacuolar invertase and a fructosyltransferase. *Plant Mol. Biol.* 69, 47-56.

Amtmann, A., Bohnert, H.J., Bressan, R.A., 2005. Abiotic stress and plant genome evolution. Search for new models. *Plant Physiol.* 138, 127-30.

Andersen, H.D., Wang, C.H., Arleth, L., Peters, G.H., Westh, P., 2011. Reconciliation of opposing views on membrane–sugar interactions. *Pro. Nat. Acad. Sci. USA*. 108, 1874-1878.

Apel, K., Hirt, H., 2004. Reactive oxygen species: Metabolism, oxidative stress and signal transduction. *Annu. Rev. Plant Biol.* 55, 373-399.

Asada, K., (2000). The water–water cycle as alternative photon and electron sinks, Phil. *Trans. R. Soc. London B*. 335, 1419-1431.

Atkinson, N.J., Urwin, P.E., 2012. The interaction of plant biotic and abiotic stresses: from genes to the field *J. Exp. Bot.* 63, 3523-3543.

Avonce, N., Leyman, B., Thevelein, J., Iturriaga, G., 2005. Trehalose metabolism and glucose sensing in plants. *Biochem. Soc. Trans*. 33, 276-279.

Ayre, B.G., Keller, F., Turgeon, R., 2003. Symplastic continuity between companion cells and the translocation stream. Long distance transport is controlled by retention and retrieval mechanisms in the phloem. *Plant Physiol* 131, 1518-1528.

Bailly, C., Audigier, C., Ladonne, F., Wagner, M.H., Coste, F., Corbineau, F., Coˆme, D., 2001. Changes in oligosaccharide content and antioxidant enzyme activities in developing bean seeds as related to acquisition of drying tolerance and seed quality. *J. Exp. Bot*. 52, 701-708.

Balasubramanian, R., Karve, A., Kandasamy, M., Meagher R.B., Moore, B.D., 2007. A role for F-Actin in hexokinase-mediated glucose signaling. *Plant Physiol*. 145, 1423-1434.

Barros, M.H., Bandy, B., Tahara, E.B., Kowaltowksi, A.J., 2004. Higher respiratory activity decreases mitochondrial reactive oxygen release and increases life span in Saccharomyces cerevisiae. *J. Biol. Chem.* 279, 49883-49888.

Bartels, D., Sunkar, R., 2005. Drought and salt tolerance in plants. *Crit. Rev. Plant Sci.* 24, 23-58.

Baxter, A., Mittler, R., Suzuki, N., 2014. ROS as key players in plant stress signalling *J. Exp. Bot.* 65, 1229-1240.

Bhardwaj, R., Sharma, I., Sharma, R., and Poonam., 2014. Role of Betaines and Related Osmoprotectants in Plant Stress Resistance. In: Plant Adaptation to Environmental

Change: Significance of Amino Acids and their Derivatives. Eds. Naser A. Anjum, Sarvajeet S. Gill and R. Gill, *CABI, Oxfordshire, UK.* 281-299.

Bohnert, H.J., Gong, Q., Li, P., Ma, S., 2006. Unraveling abiotic stress tolerance mechanisms-getting genomics going. *Curr. Opin. Plant Biol.* 9, 180-188.

Bolouri-Moghaddam, M.R., Le Roy, K., Xiang, L., Rolland, F., Van den Ende, W., 2010. Sugar signalling and antioxidant network connections in plant cells. *FEBS J.* 277, 2022-2037.

Bolouri-Moghaddam, M.R., Van den Ende, W., 2013. Sweet immunity in the plant circadian regulatory network. *J. Exp. Bot.* 64, 1439-1449.

Bowler, C., Fluhr, R., 2000. The role of calcium and activated oxygens as signals for controlling cross-tolerance. *Trends Plant Sci.* 5, 241-246.

Bray, E.A., Bailey-Serres, J., Weretilnyk, E., 2000. Responses to abiotic stresses. In: Gruissem W, Buchannan B, Jones R. Biochem. Mol. Biol. Plant. *Amer. Soc. Plant Physiol, Rockville, MD* 1158-1249.

Britt, A.B., (2004). Repair of DNA damage induced by solar UV. *Photosynth. Res.* 81, 105-112.

Buitink, J., Hemminga, M.A., Hoekstra, F.A., 2000. Is there a role for oligosaccharides in seed longevity? An assessment of intracellular glass stability. *Plant Physiol.* 122, 1217-1224.

Camacho-Pereira, J., Meyer, L.E., Machado, L.B., Oliveira, M.F., Galina, A., 2008. Reactive oxygen species production by potato tuber mitochondria is modulated by mitochondrially bound hexokinase activity. *Plant Physiol.* 149, 1099-1110.

Chalmers, J., Lidgett, A., Cummings, N., Cao, Y., Forster, J. and Spangenberg, G., 2005. Molecular genetics of fructan metabolism in perennial ryegrass. *Plant Biotechnol. J.* 3, 459-474.

Chang, B., Yang, L., Cong' W., Zu Y., Tang Z., 2014. The improved resistance to high salinity induced by trehalose is associated with ionic regulation and osmotic adjustment in *Catharanthus roseus. Plant Physiol Biochem.* 77, 140-148.

Choudhury,S., Panda, P., Sahoo, L., Panda, S.K., 2013. Reactive oxygen species signaling in plants under abiotic stress. *Plant Signal Behav.* 8, e23681.

Chaves, M.M., Pereira, J.S., Maroco, J., Rodrígues, M.L., Ricardo, C.P.P., Osorio, M.L., 2002. How plants cope with water stress in the field. *Photosynth. Growth. Ann. Bot.* 89, 907-916.

Chen, J.G., (2007). Sweet sensor, surprising partners. *Sci Stke.* 207, 1-4.

Chiang, Y., Stushnoff, J.C., McSay, A.E., 2005. Over expression of mannitol-1-phosphate dehydrogenase increases mannitol accumulation and adds protection against chilling injury in Petunia. *J. Amer. Soc. Hort. Sci.*130, 605-510.

Conde, C., Delrot, S., Gerós, H., 2008. Physiological, biochemicaland molecular changes occurring during olive development andripening. *J. Plant Physiol.*165, 1545-1562.

Couee, I., Sulmon, C., Gouesbet, G., Amrani, A.E., 2006. Involvement of soluble sugars in reactive oxygen species balance and responses to oxidative stress in plants. *J. Expt. Bot.* 57, 449-459.

Cushman, J.C., Bohnert, H.J., 2000. Genomic approaches to plant stress tolerance. *Curr. Opin. Plant Biol.* 3, 117-124.

Da Silva, W.S., Gomez-Puyou, A., de Gomez-Puyou, M.T., Moreno-Sanchez, R., De Felice, F.G., de Meis, L., Oliveira, M.F., Galina, A., 2004. Mitochondrial bound hexokinase activity as a preventive antioxidant defense: steady-state ADP formation as a regulatory mechanism of membrane potential and reactive oxygen species generation in mitochondria. *J. Biol Chem.* 279, 39846-39855.

Dat, J., Vandenabeele, S., Vranova, E., Van Montagu, M., Inzé, D., Van Breusegem, F., 2000. Dual action of the active oxygen species during plant stress responses. *Cell. Mol. Life Sci.* 57, 779-795.

De Roover, L., Vandenbranden, K., Van Laere, A., Van den Ende, W., 2000. Drought induces fructan synthesis and 1-SST (sucrose: sucrose fructosyltransferase) in roots and leaves of Cichorium seedlings (*Cichorium intybus* L.). *Planta.* 210, 808-814.

Dinkar, C., Bartels, D., 2013. Dessication tolerance in resurrection plants: new insights from transcriptome, proteome, and metablome analysis. *Front. Plant Sci.* 4, 1-14.

Dowrie, B., Gurusinghe, S., Dahal, P., Thacker, R.R., Synhder, J.C., Nonogaki, H., Yim, K., Fukanaga, K., Alvarado, V., Bradford, K.J., 2003. Expression of a *galactinol synthase* gene in tomato seeds is up-regulated before maturation desiccation and again after imbibition whenever radicle protrusion is prevented. *Plant Physiol.* 131, 1347-1359.

Duchateau, N., Bortlik, K., Simmen, U., Wiemken, A., Bancal, R., 1995. Sucrose: fructan 6-fructosyltransfearse, a key enzyme for diverting carbon from sucrose to fructan in barley leaves. *Plant Physiol.* 107, 1249-1255.

Edelman, J., Jefford, T.G., 1968. The mechanism of fructosan metabolism in higher plants as exemplified in *Helianthus tuberosus*. *New Phytol.* 67, 517-531.

Egert, A., Keller, F., Peters, S., 2013. Abiotic stress-induced accumulation of raffinose in *Arabidopsis* leaves is mediated by a single raffinose synthase (RS5, At5g40390). *BMC Plant Biol.* 13, 218.

Ehsani-Zonouz, A., Golestani, A., Nemat-Gorgani, M., 2001. Interaction of hexokinase with the outer mitochondrial membrane and a hydrophobic matrix. *Mol Cell Biochem.* 223, 81-87.

ElSayed, A.I., Rafudeen, M.S., Golldack, D., 2013. Physiological aspects of raffinose family oligosaccharides in plants: protection against abiotic stress. German Botanical Society and The Royal Botanical Society of the Netherlands. *Plant Biol.* 16, 1-8.

Ende, V.W., El-Esawe, K.S., 2013. Sucrose signaling pathways leading to fructan and anthocyanin accumulation: A dual function in abiotic and biotic stress responses. *Environ. Exp. Bot.* doi.org/10.1016/j.envexpbot.2013.09.017

Essmann, J., Schmitz-Thom, I., Schon, H., Sonnewald, S., Weis, E., Scharte, J., 2008. RNA interference-mediated repression of cell wall invertase impairs defense in source leaves of tobacco. *Plant Physiol.* 147, 1288-1299.

Fotopoulos, V., Gilbert, M.J., Pittman, J.K., Marvier, A.C., Buchanan, A.J., Sauer, N., Hall, J.L., Williams, L.E., 2003. The monosaccharide transporter gene, AtSTP4, and the cell-wall invertase, Atbetafruct1, are induced in Arabidopsis during infection with the fungal biotroph *Erysiphe cichoracearum*. *Plant Physiol.* 132, 821-829.

Foyer, C., Noctor, G., 2005. Oxidant and antioxidant signalling in plants: a re-evaluation of the concept of oxidative stress in a physiological context. *Plant Cell Environ.* 28, 56-71.

Foyer, C.H., Shigeoka, S., 2011. Understanding oxidative stress and antioxidant functions to enhance photosynthesis. *Plant Physiol.* 155, 93-100.

Foyer, C.H., Noctor, G., 2013. Redox Signaling in Plants. *Antioxi. Redox Signal.* 18, 2087-2090.

Frenandez, O., Bethencourt, L., Quero, A., Sangwan, R. S., Clement, C., 2010. Trehalose and plant stress responses: friend or foe. *Trends Plant Sci.* 15, 409-417.

Garg, A.K., Kim, J.K., Owens, T.G., Ranwala, A.P., Choi, Y.D., Kochian, L.V., Wu, R.J., 2002. Trehalose accumulation in rice plants confers high tolerance levels to different abiotic stresses. *Proc. Natl. Acad. Sci.* USA. 99, 15898-15903.

Garg, A.K., Kim, J.K., Owens, T.G., Ranwala, A.P., Choi, Y.D., Kochian, L.V., 2012. Trehalose accumulation in rice plants confers high tolerance levels to different abiotic stresses. *Proc. Natl. Acad. Sci.* USA 99, 15898-15903.

Gao, M., Tao, R., Miura, K., Dandekar, A. M. Sugiura, A., 2001.Transformation of Japanese persimmon (*Diospyros kaki* Thunb)with apple cDNA encoding NADP-dependent sorbitol-6-phosphate dehydrogenase. *Plant Sci.*160, 837–845.

Gechev, T.S., Van Breusegem, F., Stone, J.M., Denev, I., Laloi, C., 2006. Reactive oxygen species as signals that modulate plant stress responses and programmed cell death. *Bio Essays.* 28, 1091-1101.

Gechev, T.S., Dinakar, C., Benina, M., Toneva, V., Bartels, D., 2012. Molecular mechanisms of desiccation tolerance in resurrection plants. *Cell Mol. Life Sci.* 69, 3175-86.

Gepstein, S., Glick, B.R., 2013. Strategies to ameliorate abiotic stress-induced plant senescence. *Plant Mol. Bio.* 82, 623-633.

Ghosh, D., Xu, Z., 2014. Abiotic stress responses in plant roots: a proteomics perspective. *Plant Sci.* doi: 10.3389/fpls.2014.00006.

Gill, S.S., Nafees, A., Tutejab, N., 2012. Cadmium at high dose perturbs growth, photosynthesis and nitrogen metabolism while at low dose it up regulates sulfur assimilation and antioxidant machinery in garden cress (*Lepidium sativum* L.). *Plant Sci.* 182, 112-120.

Gill, S. S.; Anjum, N. A.; Gill, R., 2014. Target osmoprotectants for abiotic stress tolerance in crop plants - glycine betaine and proline. In: Plant adaptation to environmental change: significance of amino acids and their derivatives. Eds: Ed. Naser A. Anjum, Sarvajeet S. Gill and R. Gill. *CABI, Oxfordshire, UK.* 97-108.

Gill,S. S., Tuteja, N., 2010. Reactive oxygen species and antioxidant machinery in abiotic stress tolerance in crop plants. *Plant Physiol Biochem.* 48, 909-930.

Goldin, N., Arzoine, L., Heyfets, A., Israelson, A., Zaslavsky, Z., Bravman, T., Bronner, V., Notcovich, A., Shoshan- Barmatz, V., Flescher, E., 2008. Methyl jasmonate binds to and detaches mitochondria-bound hexokinase. *Oncogene* 27, 4633-4635.

Graham, J.W., Williams, T.C., Morgan, M., Fernie, A.R., Ratclive, R.G., Sweetlove, L.J., 2007. Glycolytic enzymes associate dynamically with mitochondria in response to respiratory demand and support substrate channeling. *Plant Cell* 19, 3723-3738.

Gupta, A K., Kaur, N., 2005. Sugar signalling and gene expression in relation to carbohydrate metabolism under abiotic stresses in plants. *J. Biosci.* 30, 761-776.

Halliwell, B., (2006). Reactive species and antioxidants. Redox biology is a fundamental theme of aerobic life. *Plant Physiol.* 141, 312–322.

Hernandez-Marin, E., Martínez, A., 2012. Carbohydrates and their free radical scavenging capability: a theoretical study. *J. Phys. Chem.* 116, 9668-9675.

Hincha, D.K., Zuther, E., Heyer, A.G., 2003. The preservation of liposomes by raffinose family oligosaccharides during drying is mediated by effects on fusion and lipid phase transitions. *Biochim. Biophys. Acta. Biomemb.* 1612, 172-177.

Hu, Y.C., Schmidhalter, U., 2005. Drought and salinity: a comparison of their effects on mineral nutrition of plants. *J. Plant Nutr. Soil Sci.* 168, 541-549.

Huynh, L.B., Mather, E.D., Schreiber,W.A., Toubia, J., Baumann, U., Shoaei, Z., Stein, N., Ariyadasa, R., Stangoulis, R.C.J., Edwards, J., Shirley, N., Langridge, P., Fleury, D., 2012 .Clusters of genes encoding fructan biosynthesizing enzymes in wheat and barley. *Plant Mol Biol.* 80, 299-314.

Ibañez, S., Rosa, M., Hilal, M., González, J.A., Prado, F.E., 2008. Leaves of *Citrus aurantifolia* exhibit a different sensibility to solar UV-B radiation according to development stage in relation to photosynthetic pigments and UV-B absorbing compounds production. *J. Photochem. Photobiol.* 90, 163-169.

Iturriaga, G., Suarez, R., Nova-Franco, B., 2009. Trehalose metabolism: from osmoprotection to signalling. *Int. J. Mol. Sci.* 10: 3793-3810.

Janecek, S., Lanta, V., Klimesova, J., Dolezal, J., 2011. Effect of abandonment and plant classification on carbohydrate reserves of meadow plants. *Plant Biol.* 13, 243-251.

Jennings, D.B., Daub, M.E, Pharr, D.M., Williamson, J.D., 2002. Constitutive expression of a celery mannitol dehydrogenase in tobacco enhances resistance to the mannitol-secreting fungal pathogen *Alternaria alternata*. *Plant J.* 32, 41-49.

Ji, X., Shiran, B., Wan, J., Lewis, D.C., Jenkins, C.L., Condon, A.G., Richards, R.A., Dolferus, R., 2010. Importance of pre-anthesis anther sink strength for maintenance of grain number during reproductive stage water stress in wheat. *Plant Cell Environ.* 33, 926-942.

Kanamaru, N., Ito, Y., Komori, S., Saito, M., Kato, H., Takahashi, S., Omura, M., Soejima, J., Shiratake,, K., Yamada K., Yamaki, S., 2004. Transgenic apple transformed by sorbitol-6-phosphate dehydrogenase cDNA Switch between sorbitol and sucrose supply due to its gene expression. *Plant Sci.*167, 55-61.

Kawakami, A., Yoshida, M., 2002. Molecular characterization of sucrose: sucrose 1 fructosyltransferase and sucrose: fructan 6-fructosyltransferase associated with fructan accumulation in winter wheat during cold hardening. *Biosci. Biotechnol.Biochem.* 66, 2297-2305.

Kawakami, A., Yoshida, M., 2005. Fructan:fructan 1-fructosyltransferase, a key enzyme for biosynthesis of graminan oligomers in hardened wheat. *Planta* 223, 90-104.

Keunen, E., Remans, T., Bohler, S., Vangronsveld, J., Cuypers, A., 2011. Metal induced oxidative stress and plant mitochondria. *Int. J. Mol. Sci.* 12, 6894-6918.

Keunen, E., Peshev, D., Vangronsveld, J., Van Den Ende, W., Cuypers, A., 2013. Plant sugars are crucial players in the oxidative challenge during abiotic stress: extending the traditional concept. *Plant Cell Environ.* 36, 1242-1255.

Khalid, K.A., Weiming C., 2011. The effects of mannitol and salinity stresses on growth and biochemical accumulations in lemon balm. *Acta Ecol. Sinica*. 31, 112-120.

Khan, M.I.R., Khan, N.A., 2014a. Salicylic acid and jasmonates: Approaches in abiotic stress tolerance. *J. Plant Biochem. Physiol.* 1:4 doi.org/10.4172/2329-9029.1000e113

Khan, M.I.R., Khan, N.A., 2014b. Ethylene reverses photosynthetic inhibition by nickel and zinc in mustard through changes in PS II activity, photosynthetic nitrogen use efficiency, and antioxidant metabolism. *Protoplasma*. doi.org/10.1007/s00709-014-0610-7.

Khan, M.I.R., Asgher, M., Khan, N.A., 2014. Alleviation of salt-induced photosynthesis and growth inhibition by salicylic acid involves glycinebetaine and ethylene in mungbean (*Vigna radiata* L.). *Plant Physiol. Biochem.* 80, 67-74.

Khan, M.I.R., Iqbal, N., Masood, A., Per, T.S., Khan, N.A., 2013. Salicylic acid alleviates adverse effects of heat stress on photosynthesis through changes in proline production and ethylene formation. *Plant Signal. Behav.* 8, e26374.

Kim, M.S., Cho, S.M., Kang, E.Y., Im, Y.J., Hwangbo, H., Kim, Y.C., Ryu, Ch.-M., Yang, K.Y., Chung, G.Ch., Cho, B.H., 2008. Galactinol is a signaling component of the induced systemic resistance caused by *Pseudomonas chlororaphis* O6 root colonization. *Mol. Plant Microb. In.* 12, 1643-1653.

Knaupp, M., Mishra, K.B., Nedbal, L., Heyer, A.G., 2011. Evidence for a role of raffinose in stabilizing photosystem II during freeze–thaw cycles. *Planta.* 234, 477-486.

Knight, H., Knight, M.R., 2001. Abiotic stress signalling pathways: specificity and cross-talk. *Trends Plant Sci.* 6, 262–267.

Krasensky, J., Jonak, C., 2012. Drought, salt, and temperature stress-induced metabolic rearrangements and regulatory networks. *J. Exp. Bot.* 63, 1593-1608.

Lastdrager, J., Hanson , J. Smeekens, S., 2014. Sugar signals and the control of plant growth and development. *J. Exp. Bot.* 65, 799-807.

Lasseur, B., Schroeven, L., Lammens, W., Le Roy, K., Spangenberg, G., Manduzio, H.,Vergauwen, R., Lothier, J., Prud'homme, M.-P., Van den Ende, W., 2009. Transforming a fructan:fructan 6G-fructosyltransferase from perennial ryegrass into a sucrose:sucrose 1-fructosyltransferase. *Plant Physiol.* 149, 327-339.

Le Roy, K., Vergauwen, R., Cammaer, V., Yoshida, M., Kawakami, A., Van Laere, A., Van den Ende, W., 2007. Fructan 1-exohydrolase is associated with flower opening in *Campanula rapunculoides. Funct. Plant Biol.* 34, 972-983.

Le Roy, K., Vergauwen, R., Struyf, T., Yuan, S., Lammens, W., Mátrai, J., De Maeyer, M., Van den Ende, W., 2013. Understanding the role of defective invertases in plants: tobacco Nin88 fails to degrade sucrose. *Plant Physiol.* 161, 1670-1681.

Lehner, A., Bailly, C., Flechel, B., Poels, P., Cume, D., Corbineau, F., 2006. Changes in wheat seed germination ability, soluble carbohydrate and antioxidant enzyme activities in the embryo during the desiccation phase of maturation. *J. Cereal Sci.* 43, 175-182.

Li, C., Li, T., Zhang, D., Jiang, L., Shao, Y., 2013. Exogenous nitric oxide effect on fructan accumulation and FBEs expression in chilling-sensitive and chilling-resistant wheat. *J. Exp. Bot.* 86, 2-8.

Liang, D., Cui, M., Wu, S., Ma, F.W., 2012. Genomic structure, sub-cellular localization, and promoter analysis of the gene encoding sorbitol-6–phosphate dehydrogenase from apple. *Plant Mol. Biol.* 30, 904-914.

Livingston, D.P., Henson, C.A., 1998. Apoplastic sugars, fructans, fructan exohydrolase, and invertase in winter oat: responses to second phase cold hardening. *Plant Physiol.* 116, 403-408.

Lo Bianco, R., Rieger, M., Sung, S.S., 2000. Effect of drought on sorbitol and sucrosemetabolism in sinks and sources of peach. *Physiol Plant.* 108, 71-78.

Loescher, W., Everard, J.D., 2000 Regulation of sugar alcohol biosynthesis. In storage carbohydrates in vascular plants. Ed. D.H. Lewis. *Kluwer Academic: Dordrecht, Netherlands.* 275-299.

Lunn, J.E., Delorge, I., Figueroa, C.M., Van Dijck, P., Stitt, M., 2014. Trehalose metabolism in plants. *The Plant J..* doi: 10.1111/tpj.12509.

Ma, C., Wang, Z., Kong, B., Lin, T., 2013. Exogenous trehalose differentially modulates antioxidant defense system in wheat callus during water deficit and subsequent recovery. *Plant Growth Regul.* 70, 275-285.

Mahajan, S., Tuteja, N., 2005. Cold, salinity and drought stresses: an overview. *Arch. Biochem. Biophys.* 444, 139-158.

Majee, M., Maitra, S., Dastidar, K.G., Pattnaik S., Chatterjee, A., Hait, N.C., Das,.K.P.Majumder, A.L., 2004. A novel salt-tolerant L-myo-inositol-1-phosphate synthase from *Porteresia coarctata* (Roxb.) Tateoka, a halophytic wild rice: molecular cloning, bacterial over expression, characterization, and functional introgression into tobacco-conferring salt tolerance phenotype. *J. Biol. Chem.* 279, 28539-28552.

Merchant, A., Richter, A.A., 2011. Polyols as biomarkers and bioindicators for 21st centuary plant breeding. *Funct. Plant Bio.* 38, 934-940.

Miller G., Suzuki N., Ciftci-Yilmaz S. & Mittler R., 2010. Reactive oxygen species homeostasis and signalling during drought and salinity stresses. *Plant Cell Environ.* 33, 453-467.

Mittler, R., (2002). Oxidative stress, antioxidants and stress tolerance. *Trends Plant Sci.* 9, 405-410.

Mittler, R., (2006). Abiotic stress, the field environment and stress combination. *Trends Plant Sci.* 11, 15-19.

Miyagawa, Y., Tamoi, M., Shigeoka, S., 2000. Evaluation of the defense system in chloroplasts to photooxidative stress caused by paraquat using transgenic tobacco plants expressing catalase from *Escherichia coli. Plant Cell Physiol.* 41, 311-320.

Møller, I.M., (2001). Plant mitochondria and oxidative stress: electron transport, NADPH turnover, and metabolism of reactive oxygen species. *Ann. Rev. Plant Physiol. Plant Mol. Biol.* 52, 561-591.

Møller, I.M., Sweetlove, L.J., 2010. ROS signalling-specificity is required. *Trends Plant Sci.* 15, 370-374.

Munns, R., (2002). Comparative physiology of salt and water stress. *Plant Cell Environ.* 25, 239-250.

Nadwodnik, J., Lohaus, G., 2008. Subcellular concentrations of sugar alcohols and sugars in relation to phloem translocation in *Plantago major*, *Plantago maritima*, *Prunus persica* and *Apium graveolens*. *Planta.* 227, 1079–1089.

Nägele, T., Heyer, A. G., 2013. Approximating subcellular organisation of carbohydrate metabolism during cold acclimation in different natural accessions of *Arabidopsis thaliana*. *New Phytol.* 198, 777–787.

Nishizawa, A., Yabuta, Y., Shigeoka, S., 2008. Galactinol and raffinose constitute a novel function to protect plants from oxidative damage. *Plant Physiol.* 147, 1251-1263.

Noctor, G., De Paepe, R., Foyer, C.H., 2007. Mitochondrial redox biology and homeostasis in plants. *Trends Plant Sci.* 12, 125–134.

Noiraud, N., Maurousset, L., Lemoine, R., 2001. Transport of polyols in higher plants. *Plant Physiol and Biochem.*39, 717-728.

Oerke, E.C., (2006). Crop losses to pests. *J Agr Sci.* 144, 31-43.

O'Hara, L.E., Paul, M.J., Wingler, A., 2013. How Do Sugars Regulate Plant Growth and Development? New Insight into the Role of Trehalose-6-Phosphate. *Mol. Plant.* 6, 261-274.

Okada, M., Ye, K., 2009. Nuclear phosphoinositide signaling regulates messenger RNA export. *RNA Biol.* 6, 12-16.

Ortbauer, M., Popp, M., 2008. Functional role of polyhydroxy compounds onprotein structure and thermal stability studied by circular dichroismspectroscopy. *Plant Physiol. Biochem.* 46, 428-434.

Osakabe, Y., Osakabe, K., Shinozaki, K., 2013. Plant environmental stress responses for survival and biomass enhancement. In: Climate Change and Plant Abiotic Stress Tolerance. Eds: N, Tuteja, S.S. Gill. *Wiley-VCH Verlag GmbH and Co. KGaA, Weinheim, Germany.*

Panikulangara, T.J., Eggers-Schumacher, G., Wunderlich, W., Stransky, H., Schöffl, F., 2004. *Galactinol synthase1*.A novel heat shock factor target gene responsible for heat-induced synthesis of raffinose family oligosaccharides in Arabidopsis. *Plant Physiol.* 136, 3148-3158.

Patra, B., Ray, S., Richter, A., Majumder, A.L., 2010. Enhanced salt tolerance of transgenic tobacco plants by co-expression of PcINO1 and McIMT1 is accompanied by increased level of myo-inositol and methylated inositol. *Protoplasma* 245,143-152.

Pego, J.V., Kortstee, A.J., Huijser, C., Smeekens, S.C.M., 2000. Photosynthesis, sugars and the regulation of gene expression. *J. Exp. Bot.* 51, 407-16.

Pennycooke, J.C., Jones, M.L., Stushnoff, C., 2003. Down-regulating a-galactosidase enhances freezing tolerance in transgenic petunia. *Plant Physiol.* 133, 901–909.

Peshev D., Vergauwen R., Moglia A., Hideg E. Van den Ende W., 2013.Towards understanding vacuolar antioxidant mechanisms: a role for fructans. *J. Exp. Bot.* 64, 1025-38.

Peshev, D., Van den Ende, W., 2013. Sugars as antioxidants in plants. In: Crop Improvement under Adverse Conditions Eds N. Tuteja and S.S. Gill. *Springer-Verlag, Berlin, Heidelberg, Germany* 285-308.

Peters, S., Keller, F., 2009. Frost tolerance in excised leaves of the common bugle (*Ajuga reptans* L.) correlates positively with the concentrations of raffinose family oligosaccharides (RFOs). *Plant Cell Environ.* 32, 1099-1107.

Rahdari, P., Tavakoli, S., Hosseini, S.M., 2012. Studying of salinity stress effect on germination, proline, sugar, protein, lipid and chlorophyll content in Purslane (*Portulaca oleracea* L.) Leaves. *J. Stress Physiol. Biochem. 8, 182-193.*

Ramel, F., Sulmon, C., Bogard, M., Coue´e, I., Gouesbet, G., 2009. Differential patterns of reactive oxygen species and antioxidative mechanisms during atrazine injury and sucrose-induced tolerance in Arabidopsis thaliana plantlets. *BMC Plant Biol.* 9, 28.

Ramon, M., Rolland, F., Sheen, J., 2008. Sugar sensing and signaling. The Arabidopsis Book. *Amer. Soc. Plant Biol.* 6: e0117.

Rathinasabapathi, B., (2000). Metabolic engineering for stress tolerance: installing osmoprotectant synthesis pathways. *Ann Bot.* 86, 709-716.

Redillas, M.C.F.R., Park, S.-H., Lee, J.W., Kim, Y.S., Jeong, J.S., Jung, H., 2012. Accumulation of trehalose increases soluble sugar contents in rice plants conferring tolerance to drought and salt stress. *Plant Biotechnol. Rep.* 6, 89-96.

Redondo-Gómez, S., Mateos-Naranjo, E., Davy, A.J., Fernández-Muñoz, F., Castellanos, E.M., Luque, T., Figueroa, M.E., 2007. Growth and photosynthetic responses to salinity of the saltmarsh shrub *Atriplex portulacoides*. *Ann. Bot.* 100, 555-563.

Rhoads, D.M., Umbach, A.L., Subbaiah, C.C., Siedow, J.N., 2006. Mitochondrial reactive oxygen species. Contribution to oxidative stress and interorganellar signaling. *Plant Physiol.* 141, 357–366.

Ritsema, T. and Smeekens, S., 2003. Fructans: beneficial for plants and humans. *Curr. Opin. Plant Biol.* 6, 223–230.

Rizhsky, L., Liang, H., Shuman, J., Shulaev, V., Davletova, S., Mittler, R., 2004. When defense pathways collide. The response of Arabidopsis to a combination of drought and heat stress. *Plant Physiol.* 134, 1683-1696.

Rolland, F., Baena-Gonzalez, E., Sheen, J., 2006. Sugar sensing and signaling in plants: conserved and novel mechanisms. *Annu. Rev. Plant Biol.* 57, 675-709.

Rolland, F., Winderickx, J., Thevelein, J.M., 2001. Glucose sensing mechanisms in eukaryotic cells. *Trends Biochem. Sci.* 26, 310-317.

Rontein, D., Dieuaide-Noubhani, M., Dufourc, E.J., Raymond, P., Rolin, D., 2002. The metabolic architecture of plant cells. Stability of central metabolism and flexibility of anabolic pathways during the growth cycle of tomato cells. *J Biol Chem.* 277, 43948–43960.

Rosa, M., Hilal, M., González, J.A., Prado, F.E., 2004. Changes in soluble carbohydrates and related enzymes induced by low temperature during early developmental stages of quinoa (*Chenopodium quinoa*) seedlings. *J. Plant Physiol.* 161, 683-689.

Ruan, Y.L., (2014). Sucrose metabolism: Gateway to diverse carbon use and sugar signaling. *Ann. Rev. Plant Biol.* 65: 33-67.

Russell, J.W., Golovoy, D., Vincent, A.M., Mahendru, P., Olzmann, J.A., Mentzer, A., Feldmann EL., 2002. High glucose-induced oxidative stress and mitochondrial dysfunction in neurons. *FASEB J.* 16, 1738-1748.

Saito, M., Yoshida, M., 2011. Expression analysis of the gene family associated with raffinose accumulation in rice seedlings under cold stress. *J. Plant Physiol.* 168, 2268-2271.

Salerno, G.L., Curatti L., 2003. Origin of sucrose metabolism in higher plants: when, how and why? *Trends Plant Sci.* 8, 63-69.

Schneider, T., Keller, F., 2009. Raffinose in chloroplasts is synthesized in the cytosol and transported across the chloroplast envelope. *Plant Cell Physiol.* 50, 2174-2182.

Sengupta, S., Patra, B., Ray, S., Majumder, A.L., 2008. Inositol methyl transferase from a alophytic wildrice, *Porteresia coarctata* Roxb. (Tateoka): regulation of pinitol synthesis under abiotic stress. *Plant Cell Environ.* 31, 1442-1459.

Shanker, A.K., Cervantes C, Loza-Tavera H, Avudainayagam S., 2005. Chromium toxicity in plants. *Environ. Intern.* 31, 739-753.

Sharma S.S., Dietz K.J., 2009. The relationship between metal toxicity and cellular redox imbalance. *Trends Plant Sci.* 14, 43-50.

Sharma, R., Bhardwaj, R., Thukral, A.K., Handa, N., Kaur, R. and Kumar, V., 2014. Osmolyte Dynamics: New Strategies for Crop Tolerance to Abiotic Stress Signals. In: Emerging Technologies and Management of Crop Stress Tolerance, Volume 2 Ed: Parvaiz Ahmad *Elsevier, New York.* DOI: doi.org/10.1016/B978-0-12-800875-1.00017-X.

Shi, J., Wang, H., Wu, Y., Hazebroek, J., Meeley, R.B., Ertl, D.S., 2003. The maize low phytic acid mutant lpa2 is caused by mutation in an inositol phosphate kinase gene. *Plant Physiol.* 131, 507-515.

Shinozaki, K., Yamaguchi-Shinozaki, K., 2000. Molecular responses to dehydration and low temperature: differences and cross-talk between two stress signalling pathways. *Curr. Opin. Plant Biol.* 3, 217-223.

Sickler, C.M., Edwards, G.E., Kiirats, O., Gao, Z.F., Loescher, W., 2007. Response of mannitol-producingArabidopsis thalianato abiotic stress. *Funct. Plant Biol. 34*, 382–391

Silva, D.G.F., Cangussu, B.M.L., Paula, D.A.L.S., Melo, A.G., Silva, A.E., 2013. Seasonal changes in fructan accumulation in the underground organs of *Gomphrena marginata* Seub. (Amaranthaceae) underrock-field conditions. *Theor Exp Plant Physiol.* doi.org/10.1590/S2197-00252013000100006.

Song, J., Feng, G, Tian C., Zhang, F., 2005. Strategies for adaptation of *Suaeda physophora, Haloxylon ammodendron* and *Haloxylon persicum* to a saline environment during seed germination stage. *Ann. Bot.* 96, 309-405.

Stevenson, J.M., Perera, I.Y., Heilmann, I., Persson, S., Boss, W.F., 2000. Inositol signalling and plant growth. *Trends Plant Sci.* 5, 252-258.

Stoyanova, S., Geuns, J., Hideg, E., Van den Ende, W., 2011. The food additives inulin and stevioside counteract oxidative stress. *Int. J. Food Sci. Nurt.* 62, 207-214.

Sun, L., Shukair, S., Naik, T.J., Moazed, F., Ardehali, H., 2008. Glucose phosphorylation and mitochondrial binding are required for the protective effects of hexokinases I and II. *Mol Cell Biol.* 28, 1007-1017.

Sun, L., Yang, D.., Kong, Y., Chen, Y., Li, X. Z., Zeng, L.J., Li, Q., Wang, E.T., He, Z. H., 2014. Sugar homeostasis mediated by cell wall invertase GRAIN INCOMPLETE FILLING 1 (GIF1) plays a role in pre-existing and induced defence in rice. *Mol. Plant Pathol.* 15, 161-173.

Suzuki, N., Koussevitzky, S., Mittler, R., Miller, G., 2012. ROS and redox signalling in the response of plants to abiotic stress. *Plant Cell Environ.* 35, 259-270.

Suzuki, N., Rivero, R. M., Shulaev, V., Blumwald, E. and Mittler, R., 2014. Abiotic and biotic stress combinations. *New Phytol.* doi: 10.1111/nph.12797.

Szabados, L., Savoure, A., 2010. Proline: A multifunctional amino acid. *Trends Plant Sci.*15, 89-97.

Taji, T., Qhsumi, C., Iuchi, S., Seki, M., Kasuga, M., Kobayashi, M., Yamaguchi-Shinozaki, K., Shinozaki, K., 2002. Important roles of drought-and cold-inducible genes for galactinol synthase in stress tolerance in *Arabidopsis thaliana*. *Plant J.* 29, 417-426.

Tester, M., Davenport, R., 2003. Na^+ tolerance and Na^+ transport in higher plants. *Ann. Bot.* 91, 503-527.

Theerakulpisut, P., Gunnula, W., 2012. Exogenous Sorbitol and Trehalose Mitigated Salt Stress Damage in Salt-sensitive but not Salt-tolerant Rice Seedlings. *Am. J. Crop Sci.* 4, 165-170.

Thole, J.M., Nielsen, E., 2008. Phosphoinositides in plants: novel functions in membrane trafficking. *Curr. Opin. Plant Biol.* 11, 620-631.

Tran, L.S., Nakashima, K., Shinozaki, K., Yamaguchi Shinozaki, K., 2007. Plant gene networks in osmotic stress response: from genes to regulatory networks. *Methods Enzymol* 28, 109-128.

Tsai, A.Y.L., Gazzarrini, S., 2014. Trehalose-6-phosphate and SnRK1 kinases in plant development and signaling: the emerging picture. *Front Plant Sci.* doi: 10.3389/fpls. 2014.00119.

Uzilday, B., Turkana, I., Sekmena, A.H., Ozgura, R., Karakayab, H.C., 2012. Comparison of ROS formation and antioxidant enzymes in Cleome gynandra (C4) and Cleome spinosa (C3) under drought stress. *Plant Sci.* 182, 59-70.

Valko, M., Leibfritz, D., Moncol, J., Cronin, M.T.D., Mazur, M., Telser, J., 2007. Free radicals and antioxidants in normal physiological functions and human disease. *Int. J. Biochem. Cell Biol.* 39, 44-84.

Valliyodan, B., Nguyen, H.T., 2006. Understanding regulatory networksand engineering for enhanced drought tolerance in plants.*Curr.Opin. Plant Biol.* 9,189-195.

Valluru, R., Van den Ende, W., 2008. Plant fructans in stress environments: emerging concepts and future prospects. *J. Exp. Bot.* 59, 2905-2916.

Valluru, R., Van den Ende, W., 2011. Myo-inositol and beyond–Emerging networks under stress. *Plant Sci.* 181, 387-400.

Van den Ende, W., 2013. Multifunctional fructans and raffinose family oligosaccharides. *Fron plant sci.* 4, 247.

Van den Ende, W., Clerens, S., Vergauwen, R., Van Riet, L., Van Laere, A., Yoshida, M., Kawakami, A., 2003. Fructan 1-exohydrolase: (2,1) trimmers during graminan biosynthesis in stems of wheat (*Triticum aestivum* L.): purification, characterization, mass mapping and cloning of two 1-FEH isoforms. *Plant Physiol.* 131, 621-631.

Van den Ende, W., Clerens, S., Vergauwen, R., Van Riet, L., Van laere, A., 2003.Graminan biosynthesis in stems of wheat? Purification, characterization, massmapping, and cloning of two fructan 1-exohydrolase isoforms. *Plant Physiol.*131, 621-631.

Van den Ende, W., Lammens, W., Van Laere, A., Schroeven, L. and Le Roy, K., 2009. Donor and acceptor substrate selectivity among plant glycoside hydrolase family 32 enzymes. *FEBS J.* 276, 5788-5798.

Van den Ende, W., Valluru, R., 2009. Sucrose, sucrosyl oligosaccharides, and oxidative stress: scavenging and salvaging. *J. Exp Bot.* 60, 9-18.

Van den Ende, W., Yoshida, M., Clerens, S., Vergauwen, R., Kawakami, A., 2005. Cloning, characterization and functional analysis of novel 6-kestose exohydrolases (6-KEHs) from wheat (*Triticum aestivum* L.). *New Phytol.* 166, 917-932.

Wang, N., Nobel, P.S., 1998. Phloem transport of fructans in the Crassulacean acid metabolism species Agave deserti. *Plant Physiol.* 116, 709-714.

Wang, W.X., Vinocur, B., Shoseyov, O., Altman, A., 2001. Biotechnology of plant osmotic stress tolerance: physiological and molecular considerations. *Acta. Hort.* 560, 285-292.

Williamson, J.D., Jennings, D.B., Guo, W.W., Pharr, D.M., Ehrenshaft, M., 2002. Sugar alcohols, salt stress, and fungal resistance: polyols–multifunctional plant protection? *J. Am. Soc. Horti. Sci.* 127, 467-473.

Xue, H., Chen, X., Li, G., 2007. Involvement of phospholipid signaling in plant growth and hormone effects. *Curr. Opin. Plant Biol.* 10, 483-489.

Zhang, L., Xing, D., 2008. Methyl jasmonate induces production of reactive oxygen species and alterations in mitochondrial dynamics that precede photosynthetic dysfunction and subsequent cell death. *Plant Cell Physiol.* 49, 1092–1111.

Zhu, J.K., (2001a). Plant salt tolerance. *Trends Plant Sci.* 6, 66-71.

Zhu, J.K., (2001b). Cell signaling under salt, water and cold stresses. *Curr. Opin. Plant Biol.* 4, 401-406.

Zhu, J.K., (2002). Salt and drought stress signal transduction in plants. *Annu. Rev. Plant Biol.* 53, 247-273.

In: Photosynthesis
Editor: Nafees Khan
ISBN: 978-1-63463-304-8

Chapter 10

THE EFFECT OF SILICON ON PLANT PHOTOSYNTHESIS DURING PATHOGENS INFECTION

Fabrício de Ávila Rodrigues*[*], *Wilka Messner da Silva Bispo and Carlos Eduardo Aucique-Pérez
Viçosa Federal University, Department of Plant Pathology,
Laboratory of Host-Pathogen Interaction, Viçosa,
Minas Gerais State, Brazil

ABSTRACT

Plants are constantly exposed to pathogens infection and mechanisms of defense must be triggered to avoid their deleterious effect on host physiology. Alterations on host secondary metabolism as a result of pathogens infection have been investigated by many researchers; however, studies reporting their effects on primary metabolism are yet poorly characterized. Defense induction generates energetic costs for their hosts, which require, for instance, modifications on photosynthetic and respiratory performance besides changes on source-sink patterns, leading to reductions in plant fitness then translated in production and profit losses. It is invaluable the importance of photosynthesis in resistance against pathogens, so the development of strategies that favors plant physiology and enhance the activation of plant defense responses can provide a powerful tool for disease control, delaying or reducing the use of fungicides or pesticides since the demand for sustainable plant-protection strategies is urgent. The application of silicon as a broad plant-protection strategy has been currently evidenced; despite virtually nothing is known about the molecular basis and regulation of silicon-mediated disease resistance. The aim of this Chapter is to introduce to the readers some aspects of silicon involvement on plant physiology and thus encourage the discussion and investigation on this intriguing subject.

Keywords: Photosynthesis, plant physiology, primary metabolism, silicon

---

[*] Corresponding Author address: Email: fabricio@ufv.br

INTRODUCTION

According to fossil evidence, the establishment of the first land plants occurred approximately 480 million years ago; however, recent studies have demonstrated molecular-clock early colonization estimates of approximately 700 million years (Heckman et al., 2001). Regardless of the inconsistencies related to the temporal locations of colonization in the terrestrial environment, evidence suggests that the establishment of eukaryotes in this changing environment has most likely been facilitated by symbiosis between a fungus and a photosynthetic organism. Thus, this evidence suggests that plants have been in contact with microorganisms since becoming established on land and that plants have coevolved and developed dynamic interactions with microorganisms (Gehrig et al., 1996, Heckman et al., 2001; Chisholm et al., 2006; Jones and Dangl, 2006). Thus, because plants are exposed to a plethora of environmental conditions and organisms and are sessile, they possess no mechanisms for escaping from environmental challenges. In addition, the abilities of plants to identify and defend against potentially pathogenic microorganisms have been, and remain, essential to the development of existing plants (Chisholm et al., 2006).

Plants employ different defense mechanisms before and after the attack of pathogens. These defense responses are intrinsically linked to increasing demands for energy, reducing equivalents and carbon skeletons, which are products of the major metabolic pathways (Bolton et al., 2009). However, this increase in demand for assimilates and energy is not one-sided because the invading microorganisms require nutrients for their growth and reproduction (Korves and Bergelson, 2003; Berger et al., 2007). Photoassimilates become targets of a deadly and relentless tug-of-war between the host and its "tenant". Pathogen recognition and the responses of downstream resistance have been extensively studied. However, the roles of primary metabolism on the production of substrates and resistance responses require greater emphasis (Bolton et al., 2009) to increase knowledge regarding the mechanisms involved in plant defense against pathogen and herbivore attack (Freeman and Beattie, 2008) and to discover methods or compounds involved in potentiating plant metabolism against diseases (such as the use of silicon in agriculture).

This chapter seeks to gather information about the involvement of primary metabolism in plant defense mechanisms, to provide an overview of the photosynthesis alterations provoked by plant pathogens and to highlight the role of silicon as a potentiating agent for primary metabolism and plant defenses. Interest regarding this mineral element in the agricultural sector is currently evident and is growing.

STRESS INSTALLATION

Without a mechanism for reallocating, plants must face adversities and maintain their vital functions. Specifically, plants must be capable of coping with a wide range of environmental conditions, including high and low irradiances, temperature and humidity, changes in water availability, CO_2, and nutrient contents (Berger et al., 2004),and the presence of pathogenic microorganisms. Nevertheless, nearly all organisms have organic limits for tolerating large deviations from normal conditions. If these limits are exceeded, severe damages to the structure and function of the plant cells and organs can occur.

Deviations from the normal metabolic status may be characterized as stress states and generate specific responses at a cellular and organism level that result in the development of protective responses that aim to restore homeostasis (Roháček et al., 2008). Nevertheless, when stress continuously interferes with the plant physiology and leads to the abnormal performance of vital functions, such as photosynthesis and the absorption and transport of water and minerals, a disease condition occurs that results in unbalanced energy availability and reduces the plant *fitness* (Freeman and Beattie, 2008).

According to Lichtenthaler (1998) and as deeply discussed by Roháček et al. (2008), the stress responses of plants can be divided into four phases, response, restitution, end and regeneration. The decline in metabolic processes observed at the begging of stress induction enhances the signaling cascades, which culminate in the responses that comprise the restitution phase. During this phase, adaptation and repair processes begin and the plant resistance increases, which may favor the containment of the disease and the restoration of the host homeostasis. However, if long-term stress causes damages that are large enough to overcharge the plants adaptation capacity, the "end" phase (or an exhaustion stage) occurs. In contrast, when smooth damages occur and the stressor agent is removed, a regeneration phase occurs with the partial or full regeneration of the physiological functions.

Pathogen Attack and Triggered Immunity

To be considered pathogenic, a microorganism must be able to reach inside its host, either directly or through natural openings (such as stomata and lenticels), and initiate the colonization process in the host tissues by reproducing and by feeding on the host's tissues (Chisholm et al., 2006). Using the analogy proposed by Freeman and Beattie (2008), many plant pathogens act as 'silent thieves' and use specialized tools that are designed to disable the security system of the host to avoid detection while establishing close associations with their hosts. Pathogens that keep their hosts alive and feed on their living tissues are called biotrophics, such as the pathogens that cause rusts. Other pathogens act as 'thieves who use explosives' when entering and inhabiting the host and are named necrotrophics. These pathogens are known to produce non-selective toxins and lytic enzymes that degrade the host tissues and promote the rapid release of nutrients. Pathogens with this life style include the fungus *Botrytis cinerea,* which causes gray mold, and the bacteria *Pectobacterium carotovorum* subsp. *carotovorum*, which provokes soft rot. Moreover, some pathogens use both strategies initially and act as biotrophic pathogens, which initially establish intimate and silent contact with the plant cells and later become necrotrophic during the later stages of the disease. Those pathogens are called hemibiotrophic, and the *Colletotrichum* species is a great example of this class of pathogens.

Regardless of the mode of tissue penetration, the pathogen colonization process begins after the plant tissue has been gouged, which assures its nutrition and survival. Although plant cells are the main targets of the pathogen, the barrier imposed by the rigid cell walls must be overcome to expose the plasma membrane and to allow intimate interactions between the cells and pathogens structures (Chisholm et al., 2006). Because cell exposure is required for the acquisition of resources by pathogens, the plants recognize the pathogens and trigger defense responses. The perception of the pathogen upon contact with the plasmalemma occurs through the recognition of the molecular patterns that are associated with their structures (P/MAMPs, respectively for pathogen or microorganism-associated molecular patterns) or the structures derived from the mechanical damage of the host cells (DAMPs) by the surface receptors (PRRs) (Medzhitov and Janeway, 1997; Spoel and Dong, 2012). P/MAMPs are highly conserved structures that are absent in hosts and are essential to the *fitness* of the organism. These MAMPs (more general term) include peptides such as bacterial flagellin and elongation factor Tu; oligosaccharides, such as glucan and chitin derivatives from fungal cell walls. Furthermore, these MAMPs can include more complex epitopes, such as oomycetes glycoproteins, bacterial lipopolysaccharides or peptidoglycans (Postel and Kemmerling, 2009; Zhang and Zhou, 2010). DAMPs, may be exemplified by oligogalacturonides or oligomers of alpha-1,4-linked galacturonosyl residues and be released from plant cell walls upon partial degradation of homogalacturonans (Aziz et al., 2004; Vorwerk et al., 2004). This recognition pattern comprises the first layer of defense from the innate immune system of the plants, which is called PTI or PAMP-triggered immunity (Nürnberger et al., 2004; Mazzota and Kimmerling, 2011).

However, due to a long co-evolutive process that leads to a continuous 'arms race' (Boller and He, 2009) between the host and pathogen, pathogenic microorganisms have evolved successful methods for challenging the PTI. These mechanisms are employed through the interference in the recognition process on the plasma membrane or by the secretion of effector proteins that alter the patterns of signaling cascades that result from initial recognition (Muthamilarasan and Prasad, 2013). For example, pathogenic bacteria can release approximately 15-30 type III secreted protein effectors in host cells. These proteins contribute to pathogenicity by acting as transcription factors or by acting on the host's transcription factors. In addition, these proteins can interfere in the folding process of histone proteins and in the configuration of chromatin (Feng and Zhou, 2012). When considering the coevolutionary dynamics, because microorganisms have acquired the ability to suppress their primary defenses, plants have developed specialized mechanisms to recognize them. Specifically, plants have formed effector-triggered immunity (ETI), which involves the direct or indirect recognition of microbial effectors that are used to circumvent PTI by plant proteins (Chisholm et al., 2006; Zipfel, 2008). The effector proteins are recognized in a very specific way by the host disease resistance (R) genes that have been shown to control the outcomes of plant-pathogen interactions in a wide variety of hosts against a long list of pathogens (Bent and Mackey, 2007). As the 'arms race' continues, the pathogens seem to have adapted effectors that interfere with the ETI (Chisholm et al., 2006). Thus, because the majority of pathogens are retained in the first 'sieve' of the immune system (PTI), the ability to introduce pathogen effector proteins directly into the host cell to alter plant defense became a unifying feature among the considered pathogenic microorganisms (Chisholm et al. 2006).

Structural and Chemical Defenses: A Brief Review

Faced with all of this evolutionary history, plants have developed a wide variety of structures and defense mechanisms that are classified as constitutive and induced mechanisms. Constitutive mechanisms comprise the performance barriers that are very well exemplified by projections (like trichomes and prickles) (Glover, 2010). For example, the waxy material and cuticle that are located on the walls of the epidermal cells not only prevent desiccation of the plant shoots but also prevent the direct contact of pathogens with the host cells (Samuels, 2008; Glover, 2010). In addition, a polysaccharide-rich cell wall can function as a skeleton for the plant body (Doblin et al., 2010) and shield it from microorganisms (Underwood, 2012). Furthermore, the barks of trees, which are impregnated with lignin and suberin, have great mechanical resistance (Kotula et al., 2009). Such barriers can stop pathogen invasions before the onset of the disease and can eliminate the need for more complex defense responses (Underwood, 2012).

In addition, induced barriers are only activated in response to the presence of pathogens and comprise several factors, including alterations in the cell wall composition, the production of toxic substances and enzymes that are involved in the degradation of pathogen structures and the induction of cell death (Freeman and Beattie 2008). Plant defense responses at the cell wall level have been widely studied, and efforts have focused on characterizing callose deposition and the formation of papillae by using ultrastructural observations or histochemical and immunohistochemical analyses (Chen and Kim, 2009; Underwood, 2012). Callose is an amorphous, high-molecular weight β-(1,3)-glucan polymer that provides a substrate for the deposition of antimicrobial compounds at the cellular sites of attack, and the deposition of this polymer in *Arabidopsis* emerged as a popular model for quantifying and assessing the responses of the immune system of plants (Luna et al., 2011). Ellinger et al. (2013) observed that the early deposition of callose on *Arabidopsis* cell walls results in complete resistance to the penetration of *Golovinomyces cichoracearum*, which causes powdery mildew. Similar results were found by Blumke et al. (2013), who observed that the transient expression of the *Arabidopsis thaliana* callose synthase PMR4 increased the penetration resistance to *Blumeria graminis* f.sp. *hordei* in barley. Although similar results were also observed for other host species (Fernández-Aparicio et al., 2009), the number of conflicting reports regarding the regulation of pathogen-induced callose (Fofana et al., 2005) has increased. To address this controversy, Luna et al. (2011) examined callose deposition under different growth conditions and in response to two different pathogen-associated molecular patterns. These authors made a broader conclusion that the formation of callose in response to stress is a multifaceted defense reaction that is controlled by distinct signaling pathways and its 'feat' depends on the environmental conditions and the challenging pathogen-associated molecular pattern.

Another well-studied defense response against pathogens that can be constitutive or induced is the production of secondary metabolites. Their roles in the plant's interaction with its environment are widely recognized (Selvaraj and Fofana, 2012), and these compounds usually belong to one of three large chemical classes, terpenoids, phenolics, and alkaloids (Freeman and Beattie, 2008). Terpenoids (terpenes) occur in virtually all plants and represent the largest class of secondary metabolites with over 22,000 described compounds (Freeman and Beattie, 2008). Many of these compounds are the primary components of essential oils, which may be highly volatile compounds that contribute to the fragrance of plants and are toxic to insects (some commercially available insecticides are synthetic analogues of pyrethrins), and protectors against fungal or bacterial attack (Freeman and Beattie, 2008). In addition, alkaloids are a large class of nitrogenous compounds that are derived from the aspartate, lysine, tyrosine, and tryptophan amino acids that are found in many vascular plants. Among these amino acids, caffeine, cocaine, morphine, and nicotine, and many others have powerful effects on animal physiology and on plant pathogens (Freeman and Beattie, 2008;

Kim et al., 2011; Joosten and van Veen, 2011). The widely studied class of secondary metabolites includes phenylpropanoid pathway derivatives or phenolic compounds. These derivatives and compounds are synthesized primarily through the shikimic and malonic acids pathways in plants, and many defense-related compounds, such as flavonoids, anthocyanins, phytoalexins, tannins, lignin, and furanocoumarins, are part of this group (Freeman and Beattie, 2008). The involvement of these types of metabolites in plant defense has been evaluated for decades, and many reviews covering the subject are available (Matern and Kneusel, 1988; Nicholson and Hammerschimidt, 1992; Lattanzio et al., 2006; Jabeen et al., 2009; Bhattacharya et al., 2010; Prasad et al., 2010). For example, Petkovsek et al. (2008) observed that an increase in the phenolics concentrations in apple leaves reduced the scab symptoms caused by *Venturia inaequalis*. Likewise, Ewané et al. (2012), found a possible effect of phenolic compounds on the diversity of responses associated with banana resistance to crown rot. These authors suggested that these compounds represent a potential target for developing alternatives to the chemical control of the disease. The phenolic compounds are also employed for various other purposes during plant metabolism, which directly influences the plants defense against pathogens and may potentiate physiological traits that may confer support for mounting defenses, such as UV sunscreens, signal compounds, pigments, internal physiological regulators or chemical messengers (Veberic et al., 2005; Latanzzio et al., 2006).

How Costly Are Defenses for Plants?

Although constitutive defenses may significantly contribute to the protection and the rigidity of plants, their maintenance over the plant life cycle is costly in terms of metabolic energy. The *Arabidopsis* mutant *defense and no death* (*dnd1*) is an example of an energy expenditure that results from setting up constitutive defenses. The mutation induces a constant systemic resistance that increases the cellular concentration of salicylic acid and favors resistance against a broad range of pathogens, such as bacteria, viruses and fungi, and contributes to the formation of dwarf plants (Genger et al., 2008; Bolton, 2009). Another example is the *cpr* mutant of *Arabidopsis*, which increases the salicylic acid concentration and the expression of pathogenesis-related genes that result in increased resistance to pathogens, such as *Peronospora parasitica*. However, these plants were reduced in size when compared with the control plants (Clarke et al., 1998). Based on these previous results, it was proposed that the consumption of carbon for defenses induces the plant to adjust its growth under conditions with low-carbon availability (Smith and Stitt, 2007). Conversely, induced defenses, although costly, do not act when they are not needed (*i.e.,* they are not induced in the absence of the invading microorganisms), which reduces the metabolic burden and is advantageous when compared to the pre-formed defenses. In addition, this behavior allows the plants to become phenotypically plastic and reduces the pathogens chances of adapting to their defenses, which makes their defenses less predictable (Steppuhn et al., 2004; Roda et al., 2004).

As discussed by Berger et al. (2007), who reviewed the physiological changes that result from plant-pathogen interactions, there are obvious reasons why contact with pathogens can alter the primary metabolism of the plant and leads to changes in metabolic energy expenditure. The primary metabolites that result from photosynthesis are channeled to different metabolic pathways for the synthesis of the cell wall 'makeup' such as heightened cellulose, lignin and callose depositions. In addition, secondary metabolites are produced and several compounds are modified and stored to cope with abiotic and biotic stresses.

After pathogens reach successful infection and use the nutrients in the host tissues for their own nutrition and for the production of their own defenses, the pathogens demand photoassimilates and attempt to manipulate the carbohydrate metabolism of the plants to their own favor (Berger et al., 2007), which reduces the plant growth and *fitness* (Bolton et al., 2009). However, according to Bolton et al. (2009), although it should not be assumed that direct competition for resources and reduced plant *fitness* always result from carbon limitations, there is little doubt that plant defense is costly in terms of energy. In addition, the activation of defense responses provides a high metabolic requirement at the infected region because carbon skeletons and energy are required for the synthesis of new molecules. Thus because the defense response may change the magnitude of the concentrations of a number of metabolic intermediates, the photosynthetic performance of the infected plant is liable to undergo changes to adapt to the new cell demands.

Most studies that have involved assessments of plant physiology during infection by a particular pathogen have shown that photosynthesis rates are dramatically reduced upon pathogen ingress (Chou et al., 2000; Berger et al., 2004; Scharte et al., 2005; Zou et al., 2005; Bonfig et al., 2006; Swarbrick et al., 2006; Berger et al., 2007; Doehlemann et al., 2008; Horst et al., 2008; Bolton et al., 2009). In addition, other studies have indicated that photosynthesis locally increases in areas adjacent to lesion spots, which occurs for foliar pathogens (Berger et al., 2004; Scharte et al., 2005). The observed decreases in photosynthetic performance can be associated with several factors. One factor is the reduction in the area of healthy green tissues that are capable of absorbing and utilizing the energy in the photochemical process. This reduction results from increases in the areas of necrotic lesions that arise from the actions of reactive oxygen species. In addition, plats appear to turn off photosynthesis and other assimilatory metabolism processes to initiate other processes, such as respiration, that are required for defense. The down-regulation of transcripts related to genes involved in the photosynthetic process, such as the small subunit of ribulose-bisphosphate carboxylase/oxygenase (RbcS) and the chlorophyll a/b-binding proteins (Cab), during the early stages of the defense reactions are a common trait (Scharte et al., 2005; Berger et al., 2007; Bolton et al., 2009; Magnin-Robert et al., 2011). This reduction in the transcription of genes and the photosynthetic activity can also function to protect the photosynthetic apparatus against oxidative damage or may result from oxidative damage (Niyogi, 2000; Blokhina et al., 2003; Bolton et al., 2009).

The localized reduction in photosynthetic metabolism in addition to increased demands during cell resistance response often triggers changes in carbon metabolism (Roitsch et al., 2003). It is generally observed, especially in the case of pathogens that affect leaves, that an increase in the activity of the enzyme invertase occurs, especially regarding the extracellular invertases that are important for apoplastic phloem unloading and key enzymes in determining sink strength (Roitsch et al., 2003; Berger et al., 2004; Essmann et al., 2008; Hyun et al., 2011). Consequently, in addition to this increase, an increase in the hexose concentration occurs because invertase acts on sucrose and produces glucose and fructose. These hexoses are transported into the cell by hexose transporters, which provide substrates for energy production requirements and reduce resistance responses. Therefore, these hexoses increase the availability of nutrients for the pathogens and favor signaling for the expression of genes related to defense and the induction of cell death because sugars are signals that can regulate some genes expression (Koch, 1996). Unfortunately, increases in hexose concentrations may also act on the retro inhibition of the photosynthetic process (Sinha et al., 2002; Roitsch et al., 2003; Fotopoulos et al., 2003; Kocal et al., 2008; Bolton et al., 2009). Although the repression of photosynthesis and changes in carbon metabolism appear to be general responses to pathogenic infection, their effects on sugar concentrations vary considerably between the different organisms involved in the different plant-pathogen interactions (Berger et al., 2007).

Enhancers of Plant Physiology

Because the photosynthesis is important in the resistance of plants against pathogens, it is important to develop strategies that promote plant physiology when the plants are under attack by a certain pathogen. In addition, it is important that the activation of plant defense responses are enhanced against diseases and pests to provide a powerful tool for disease control that delays or reduces the use of pesticides because the demands for sustainable plant-protection strategies by society are urgent (Van Bockhaven et al., 2013).

In fact, the induction of host resistance by applying inducers of resistance has been studied for decades with significant results for both laboratory and field conditions. Activators are compounds that are capable of activating plant defense responses, enhancing photosynthetic performance and favoring the growth and development of plants (von Rad et al., 2005; Cavalcanti et al., 2006). However, to be recognized and classified as an activator, a compound must be capable of inducing identical resistance responses in a plant relative to a broad spectrum of different pathogens. In addition, both the activator and its metabolites cannot have any direct antimicrobial activities (Kessmann et al., 1994). Plant activators include a wide spectrum of compounds, such as salicylic acid (SA), benzo-(1,2,3)-thiadiazole-7-carbotioic acid S-methyl ester (BTH), 2,6-dichloroisonicotinic acid (INA), jasmonates (JA), ethylene (ET) and abscisic acid (ABA), which are natural or synthetic plant activators (von Rad et al., 2005).

Many efforts have been made to uncover new natural or synthetic activator compounds (as observed on studies from the RIKEN Plant Science Center (PSC) and reviewed by Picket et al. (2006)) because the demand for food and the need to avoid pre and post-harvest (mainly due to plant disease) losses are continuously increasing. The SA, JA and ET are archetypal defense hormones, and their importance in signaling cascades of the innate immune system are well established in plants (De Vleesschauwer et al., 2013; Van Bockhaven et al., 2013). Other hormones, such as ABA, gibberellins, cytokinins, auxins and brassinosteroids, have emerged as modulators of plant-microbe interactions (Asselbergh et al., 2008; Robert-Seilaniantz et al., 2011; Van Bockhaven et al., 2013). Most of these hormones are differentially active and show several modes of action. Due to this great variability, the activation of defenses against one type of pathogen often increases the plants susceptibility to another type of pathogen (Pieterse et al., 2009; Van Bockhaven et al., 2013). This condition led to significant progress in engineering plants that are resistant to specific classes of pathogens. However, examples of treated plants that present a broad-spectrum disease resistance are scarce (Van Bockhaven et al., 2013).

Evidence has been presented regarding the application of silicon (Si) as a broad plant-protection strategy; however, virtually nothing is known about the molecular basis and regulation of silicon-mediated disease resistance. Si is one of the most abundant elements on the earth's crust, and its role in plant protection against a suite of pathogens with different lifestyles and modes of infection is of great interest for agriculture (Ma et al., 2006; Epstein, 2009; Guntzer et al., 2012; Van Bockhaven et al., 2013). Accordingly, novel studies are required regarding the physiological alterations that arise from using Si amendments and the regulatory mechanisms that should help us understand the abilities of this element to protect plants from a multitude of stresses. Therefore, in the next section, several recent studies will be addressed that focus on the physiological changes that occur in plants that are supplied with Si and are attacked by different types of plant pathogens. In addition, the effects of this element on the enhancement of plant defense responses will be investigated. Furthermore, the role of Si will be addressed as a stimulator or enhancer of plant primary metabolism, thereby facilitating the maintenance of plant homeostasis under stress and the production of defense components as previously outlined.

Silicon: A Powerful Tool

Si is the second most abundant element in soils (oxygen is the most abundant) because most mineral soil are composed of silicates or aluminosilicates (Epstein and Bloom, 2005). These minerals present great variability regarding their solubility, which results in variable concentrations in the soil solution that range from 0.1 to 0.6 mM and are of the same magnitude as potassium, calcium, and other major plant nutrients (Epstein, 1994). Si is mainly found in the soil as monosilicic acid (H_4SiO_4), and because it forms part of the soil solution, the H_4SiO_4 is promptly absorbed by terrestrial plants, which contain appreciable concentrations of H_4SiO_4 (from fractions of 1 dag/kg of the dry matter to several percent) (Epstein, 1994). Despite its ubiquity in plants, Si is not considered as an "essentials" element, with the exception of certain algae species and members of Equisitaceae family. Thus, Si is not used in the formulation of nutrient solutions and is usually not identified as a primary target of investigations that involve plant nutrition and behavior (Epstein, 1994; 1999; Epstein and Bloom, 2005).

Although Si is not yet considered an essential mineral element for plant nutrition, its benefits for plant growth and development for many plant species must be considered and evaluated because these attributes may be related to the beneficial roles of Si for enhancing to the resistance of plants to various biotic and abiotic stresses (Epstein, 1994; 1999; Epstein and Bloom, 2005). Evidence of the control of plant diseases by silicon, especially those caused by fungi and bacteria, have been conducted recently an include the studies of Datnoff et al. (1997), Kim et al. (2002), Ma et al. (2006) and Zhang et al. (2006).

Since Si was found to control plant diseases, its role as a physical barrier has traditionally been used to explain the enhancement of plant resistance (Cai et al., 2009). When absorbed by a given plant, Si is readily transported to the plant shoots by the transpiration stream. In addition, supersaturation of Si occurs as water flows out from the leaves, which leads to the polymerization and of Si and to the formation of amorphous silica bodies, which are called opal or opal phytoliths (Epstein and Bloom, 2005). Si can accumulate and deposit beneath the cuticle to form a cuticle-Si double layer and can lend rigidity and roughness to the wall to afford the plant protection against the penetration of pathogens (Samuels et al., 1991; Fauteux et al., 2005; Epstein and Bloom, 2005; Cooke and Leishman, 2011). Kwon et al. (1974) showed increased resistance in rice plants that were associated with more silicified cells. Later, many researchers continued to discuss the roles of the Si-cell wall associations in blocking the ingress of many pathogens (Seebold et al., 2001; Kim et al., 2002; Hayasaka et al., 2008). However, as discussed and exposed by Cai et al. (2009), early and recent studies have only partially support the hypothesis that cell silicification and papilla formation explain the enhanced resistance of plants against pathogens. The sole attribution of these mechanisms in disease resistance is increasingly contested. Meanwhile, in the last 20 years, several studies have shown an intimate association between Si and other plant mechanisms, such as the activation of enzymatic defense complexes (Liang et al., 2005) and the production of compounds with antifungal activities, such as phenolics and phytoalexins (Rodrigues et al., 2003, 2004, 2005).

Among other factors, biochemical analysis allowed us to hypothesize that the reduction of disease severity upon Si amendment resulted from an intricate pattern of responses that were associated with the enzymes and compounds that are involved in defense and antioxidant metabolism (Table 1). For example, in diseases caused by *Fusarium oxysporum* f.sp. *cubense*, *Coletotrichum lindemuthianum*, *Meloidogyne exigua*, *Pythium* spp. and *Ramularia areola,* which affect banana, beans, coffee, cucumber and cotton, respectively, phenylalanine ammonia lyase (PAL), a key enzyme from the phenylpropanoid pathway, had greater activity when plants were supplied with Si. This behavior may be related to the enhanced production of phenolic compounds and lignin that occurs to cope with the progressing disease. This hypothesis was supported by the largest increase in the lignin-thioglycolic acid (LTGA) derivatives, which are important compounds for increasing the penetration resistance of pathogens (Chérif et al., 1994, Silva et al., 2010, Fortunato et al., 2012, Pollanco et al., 2012, Curvelo et al., 2013). Moreover, in rice, cotton, cucumber and wheat plants that were supplied with Si and were inoculated with *Bipolaris oryzae*, *Ramularia areola, Podosphaera xantii* and *Bipolaris sorokiniana*, respectively, showed an increase in the activities of interspecific peroxidases (POX) and chitinases (CHI) (Liang et al., 2005; Domiciano et al., 2010; Dallagnol et al., 2011; Curvelo et al., 2013). Regarding anti-oxidative mechanisms, researchers must consider that the production of reactive oxygen species (ROS), including superoxide anions ($O_2^{\bullet-}$), hydroxyl radicals (OH^-) and hydrogen peroxide (H_2O_2), are imminent under stress. In addition, the formation and actions of these radicals may damage the essential macromolecules of the plant cells, including the pigments, proteins and nucleic acids (Mandal et al., 2008; Debona et al., 2012; Resende et al., 2012). During the evaluation of the interactions in banana-*Fusarium oxysporum* f.sp. *cubense*, rice-*Magnaporthe grisea*, wheat-*Pyricularia oryzae* and sorghum-*Colletotrichum sublineolun* that were mediated by Si, the lipid peroxidation was reduced in the stressed plants that were supplied with Si when compared with the inoculated plants that were grown in the absence of Si. This response is related to increased activities of the enzymes that are involved in the metabolization of ROS (Table 1) (Sun et al., 2010; Fortunato et al., 2012; Resende et al., 2012; Debona et al., 2013). Moreover, the transcriptomic analysis indicated that the interactions of *Aradidopsis thaliana-Erysiphe cichoracearum* that were mediated by Si activated nearly 3,970 genes.

Table 1. Response of ascorbate peroxidase (APX), catalase (CAT), chitinase (CHI), *β*-1-3 glucanase (GLU), glutathione reductase (GR), malondialdehyde (MDA), lipoxygenase (LOX), lignin-thioglycolic acid (LTGA) derivatives, phenylalanine ammonia-lyase (PAL), peroxidase (POX), polyphenoloxidase (PPO), superoxide dismutase (SOD) and total soluble phenolics (TSP) in the different crops that were grown and supplied with silicon and infected by different pathogens

Crops	Pathogens	APX[1]	CAT[1]	CHI[1]	GLU[1]	GR[1]	MDA[2]	LOX[1]	LTGA[2]	PAL[1]	POX[1]	PPO[1]	SOD[1]	TSP[2]	References
Banana	*Fusarium oxysporum* f.sp. *cubense*	*	*	+	+	*	-	*	+	+	+	+	*	+	Fortunato et al., 2012
Bean	*Colletotrichum lindemuthianum*	*	*	-	-	*	*	+	+	+	-	-	*	+	Polanco et al., 2012
Coffee	*Meloidogyne exigua*	*	*	*	*	*	*	*	+	+	+	+	*	*	Silva et al., 2010
Cotton	*Ramularia areola*	*	*	+	+	*	*	*	+	+	+	+	*	*	Curvelo et al., 2013
Cucumber	*Pythium* spp.	*	*	*	*	*	*	*	*	+	*	+	*	*	Chérif et al., 1994
Rice	*Magnaporhte oryzae*	*	-	*	*	*	-	+	*	*	*	*	*	*	Sun et al., 2010
Sorghum	*Colletotrichum sublineolum*	+	+	*	*	+	-	*	*	*	*	*	+	*	Resende et al., 2012
Wheat	*Pyricularia oryzae*	-	+	+	*	+	-	*	+	*	-	-	-	+	Xavier-Filha et al., 2011. Debona et al., 2013
Wheat	*Xanthomonas transulens* pv. *undulosa*	*	*	-	*	*	*	*	-	*	-	+	*	+	Silva et al., 2010

[1] Enzymatic activity. [2] Concentration of the compound. The symbols + and - indicate higher and lower response, respectively, when plants supplied with silicon are compared with plants non-supplied with silicon. * = non-determined response.

Furthermore, 60% of up-regulation activity was closely associated with the expression of the resistance genes and the biosynthesis of SA, ET, JA, and ROS scavengers and antifungal compounds. The remaining 40% were strongly associated with primary metabolism. Specifically genes 17 and 11 were related to photosynthesis and energy metabolism, respectively (Fauteux et al., 2005).

The studies conducted with rice-*Bipolaris oryzae*, sorghum-*Colletotrichum sublineolum* and wheat-*Pyricularia oryzae* demonstrated that the production of defense compounds increased and the defense-related enzymes and the enhancement of the antioxidant metabolism decreased the percentage of affected tissues and favored the exchange of gases (CO_2 and H_2O). These changes resulted in higher rates of CO_2 fixation, and were associated with moderate transpiration rates and lower stomatal limitations in the plants that were supplied with Si when compared to the plants without Si. Thus, these plants presented pronounced declines in photosynthetic performance of 50% and alterations in carbohydrate synthesis (Resende et al., 2012; Dallagnol et al., 2013; Aucique-Pérez et al., 2013). In addition, studies that evaluated rice plants showed that the supply of Si induced higher crop yields, even in the absence of stressors. The absence of stresses increased the stalk strength, which exerted a feed-forward mesophyll conductance-associated effect on photosynthesis. Furthermore, the relationship between carbon and nitrogen metabolism was investigated and demonstrated that Si nutrition could modulate the carbon/nitrogen balance in unstressed rice plants (Detmann et al., 2012, 2013).

Meanwhile, photochemical parameters, such as the quantum yield of the PSII photochemistry (F_v/F_m), have shown improved performance when plants that were supplied with Si were subjected to biotic or abiotic stressful conditions (Sun et al., 2010; Aucique-Pérez et al., 2013). When accompanied by NPQ reductions, increments in q_P and ETR in wheat plants supplied with Si and inoculated with *P. oryzae* resulted in an increased flow of electrons along the electron transport chain of the chloroplasts, which allowed for the formation of NADPH and facilitated the electrochemical proton gradient that was required for the formation of ATP (Aucique-Pérez et al., 2013). The chlorophyll concentration in wheat plants that were supplied with Si and infected with *P. oryzae* was greater than in the inoculated plants that were not supplied with Si (Aucique-Pérez et al., 2013). In several studies, the photorespiratory process has been shown to be an excellent mechanism of resistance to various pathogen infections. However, studies that have focused on the possible effects of Si on this process are scarce and constitute a new horizon of research (Von Bonkhaven et al., 2013). Therefore, evidence shows that Si nutrition improves physiological plant traits by assisting in several vital processes that are directly or indirectly linked to photosynthesis, which leads to increased growth and favors a substantial gain in crop yields (Aucique-Pérez et al., 2013).

The effects of Si on photosynthetic parameters and soluble sugar concentrations were determined in leaves of rice (cv. Oochikara) and in mutant Oochikara plants that were defective in active Si uptake [*low silicon 1* (*lsi1*)] (Dallagnol et al., 2013). Plants were grown in hydroponic cultures that were amended with 0 (-Si) or 2 mM Si (+Si) and with or without inoculation with Bipolaris oryzae, the causal agent of brown spot in rice. Plant inoculation reduced the photosynthetic parameters that were measured, regardless of plant material or Si supply. A decrease in the net carbon assimilation rate (A) of the inoculated plants compared with that of the non-inoculated plants was associated with damage in the photosynthetic apparatus. However, for +Si cv. Oochikara, stomatal restriction [low water vapor conductance

(g_s)] contributed to A reduction. Damage to the photosynthetic apparatus of the inoculated plants was the main reason for the reduction in A for the +Si and -Si *lsi1* mutant plants. In addition, for -Si cv. Oochikara, a reduction in g_s contributed to the reduced A. However, for +Si cv. Oochikara, g_s was the limiting factor for A. Inoculated plants of +Si cv. Oochikara had higher A values than the +Si *lsi1* mutant plants. Soluble sugars were not detected in the leaf tissues of the plants. In addition, Si improved the hexose concentrations in the non-inoculated plants at 144 h after inoculation (hai) for the *lsi1* mutant plants and from 96 hai onwards for the cv. Oochikara compared with the -Si plants. However, plant inoculation reduced the hexose concentration relative to the non-inoculated plants, mainly in the +Si plants, regardless of the plant material. The sucrose concentration increased in the cv. Oochikara leaves in the presence of Si regardless of whether the plant was inoculated. For the +Si *lsi1* mutant plants, the sucrose concentrations only increased at 48 hai compared with the -Si plants regardless of inoculation. According to Dallagnol et al. (2013), a minimum Si concentration is needed in rice plant leaf tissues to avoid the negative impacts of B. oryzae infection on photosynthesis and sugar concentration. High leaf Si concentrations resulted in an increased soluble sugar concentration, and the soluble sugar and Si contents separately reduced the severity of brown spot in the rice.

The evidence recounted above suggests that Si plays a positive role in plants that are attacked by pathogens due to a physical barrier that reduces the chances of penetration and colonization of the pathogens and reduces the amount of affected tissue. The green leaf tissue that remains is important for atmospheric carbon fixation and is necessary for the carbohydrate metabolism that will be used as a defense mechanism. Moreover, Si may be involved in stimulating the synthesis of antimicrobial compounds and in the activation of antioxidative mechanisms and strategies that the plant can use to react against pathogen infection. In approaching years, it is expected that research at the physiological, biochemical and molecular levels may help to clarify the roles of Si in plant-pathogen interactions and explain the stimulation of primary metabolism and the activation of defense mechanisms. In addition, further research is expected to include Si as part of the fertilization program, which would permit a reduction in the environmental impacts of exaggerated fungicide use and result in more sustainable agricultural production methods.

References

Aucique-Perez, C. E.; Rodrigues, F. A.; Moreira, W. R.; DaMatta, F. M. Leaf gas exchange and chlorophyll *a* fluorescence in wheat plants supplied with silicon and infected with *Pyricularia oryzae*. *Phytopathology* 2013. DOI: 10.1094/PHYTO-06-13-0163-R.

Altenbach, D.; Robatzek, S. Pattern recognition receptors: from the cell surface to intracellular dynamics. *Molecular Plant-Microbe Interactions* 2007, *20* (9), 1031-1039.

Asai, T.; Tena, G.; Plotnikova, J.; Willma, M. R.; Chiu, W.; Gomez-Gomez, L.; Boller, T.; Ausubel, F. M.; Sheen, J. MAP kinase signalling cascade in *Arabidopsis* innate immunity. *Nature* 2002, *415*, 977-983.

Asselbergh, D.; De Vleesschauwer, D.; Hofte, M. Global switches and fine-tuning: ABA modulates plant-pathogen defense. *Molecular Plant-Microbe Interactions* 2008, *21*, 709-719.

Aziz A.; Heyraud, A.; Lambert, B. Oligogalacturonide signal transduction, induction of defense-related responses and protection of grapevine against *Botrytis cinerea*. *Planta* 2004, *218*, 767-774.

Bent, A. F.; Mackey, D. Elicitors, effectors, and R genes: the new paradigm and a lifetime supply of questions. *Annual Review of Phytopathology* 2007, *45*, 399-436.

Berger, S.; Papadopoulos, M.; Schreiber, U.; Kaiser, W.; Roitsch, T. Complex regulation of gene expression, photosynthesis and sugar levels by pathogen infection in tomato. *Physiologia Plantarum* 2004, *122*, 419-428.

Berger, S.; Sinha, A. K.; Roitsch, T. Plant physiology meets phytopathology: plant primary metabolism and plant-pathogen interactions. *Journal of Experimental Botany* 2007, *58*, 4019-4026.

Bhattacharya, A.; Sood, P.; Citovsky, V. The roles of plant phenolics in defence and communication during *Agrobacterium* and *Rhizobium* infection. *Molecular Plant Pathology* 2010, *11* (5), 705-719.

Blümke, A.; Somerville, S. C.; Voigt, C. A. Transient expression of the *Arabidopsis* thaliana callose synthase PMR4 increases penetration resistance to powdery mildew in barley. *Advances in Bioscience and Biotechnology* 2013, *4*, 810-813.

Boller, T.; Felix, G. A Renaissance of elicitors: perception of microbe-associated molecular patterns and danger signals by pattern-recognition receptors. *Annual Review of Plant Biology* 2009, *60*, 379-406.

Boller, T.; He, S. P. Innate immunity in plants: An arms race between pattern recognition receptors in plants and effectors in microbial pathogens. *Science* 2009, *324* (5928), 742-744.

Bolton, M. D. Primary metabolism and plant defense-fuel for the fire. *Molecular Plant-Microbe Interactions* 2009, *22* (5), 487-497.

Bonfig, K. B.; Gabler, A.; Simon, U. K.; Luschin-Ebengreuth, N.; Hatz, M.; Berger, S.; Roitsch, T. Post-translational derepression of invertase activity in source leaves via down-regulation of invertase inhibitor expression is part of the plant defense response. *Molecular Plant* 2010, *3* (6), 1037-1048.

Cai, K.; Gao, D.; Chen, J.; Luo, S. Probing the mechanisms of silicon-mediated pathogen resistance. *Plant Signaling & Behavior* 2009, 4(1), 1-3.

Cavalcanti, F. R.; Resende, M. L. V.; Pereira, R. B.; Costa, J. C. B. C.; Carvalho, C. P. S. Atividades de quitinase e *β*-1,3-glucanase após eliciação das defesas do tomateiro contra a mancha-bacteriana. *Pesquisa Agropecuária Brasileira* 2006, *41*, 1721-1730.

Chen, X.; Kim, J. Callose synthesis in higher plants. *Plant Signaling & Behavior* 2009, *4* (6), 489-492.

Chérif, M.; Asselin, A.; Bélanger. R. R. Defense responses induced by soluble silicon in cucumber root infected by *Pythium* spp. *Phytopathology* 1994, *84*, 236-242.

Chisholm, S. T.; Coaker, G.; Day, B.; Staskawicz, B. J. Host-microbe interactions: shaping the evolution of the plant immune response. *Plant Cell* 2006, *124*, 803-814.

Chou, H-M.; Bundock, N.; Rolfe, S. A.; Scholes, J. D. Infection of Arabidopsis thaliana leaves with *Albugo candida* (white blister rust) causes a reprogramming of host metabolism. *Molecular Plant Pathology* 2000, *1*, 99-113.

Clarke, J. D.; Liu, Y.; Klessig, D. F.; Dong, X. Uncoupling PR gene expression from NPR1 and bacterial resistance: Characterization of the dominant *Arabidopsis* cpr 6-1 mutant. *Plant Cell* 1998, *10*, 557-570.

Cooke J.; Leishman M. R. Is plant ecology more siliceous than we realise? *Trends in Plant Science* 2011, *16*, 61-68.

Curvelo, C. R. S.; Rodrigues, F. A.; Silva, L. C.; Nascimento, K. J. T.; Berger, P. G. Mecanismos bioquímicos da defesa do algodoeiro à mancha de ramulária mediados pelo silício. *Bragantia* 2013, *72*, 41-51.

Dallagnol, L. J.; Rodrigues, F. A.; DaMatta, F. M.; Mielli, M. V. B.; Pereira, S. C. Deficiency in silicon uptake affects cytological, physiological, and biochemical events in the rice-*Bipolaris oryzae* interaction. *Phytopathology* 2011, *101*, 92-104.

Dallagnol, L. J.; Rodrigues, F. A.; Vale, F. X. R.; Chaves, A. R.; DaMatta, F. M. Photosynthesis and sugar concentration are impaired by the defective active silicon uptake in rice plants infected with *Bipolaris oryzae*. *Plant Pathology* 2013, *62*, 120-129.

Datnoff, L. E.; Deren, C. W.; Snyder, G. H. Silicon fertilization for disease management of rice in Florida. *Crop Protection* 1997, *16*, 525-31.

De Vleesschauwer, D.; Gheysen, G.; Höfte, M. Hormone defense networking in rice: tales from a different world. *Trends in Plant Science* 2013, *18*, 555-565.

Debona, D.; Rodrigues, F. A.; Rios, J. A.; Nascimento, K. J. T. Biochemical changes in the leaves of wheat plants infected by *Pyricularia oryzae*. *Phytopathology* 2012, *102*, 1121-1129.

Debona, D.; Rodrigues, F. A.; Rios, J. A.; Nascimento, K. J. T.; Silva, L. C. The effect of silicon on antioxidant metabolism of wheat leaves infected by *Pyricularia oryzae*. *Plant Pathology* 2013, DOI: 10.1111/ppa.12119.

Detmann, K. C.; Araújo, W. L.; Martins, S. C. V.; Fernie, A. R., DaMatta, F. M. Metabolic alterations triggered by silicon nutrition: Is there a signaling role for silicon?. *Plant Signaling & Behavior* 2013, *8*, 1-4.

Detmann, K. C.; Araújo, W. L.; Martins, S. C. V.; Sanglard, L. M. V. P.; Reis, J. V.; Detmann, E.; Rodrigues, F. A.; Nunes-Nesi, A.; Fernie, A. R.; DaMatta, F. M. Silicon nutrition increases grain yield, which, in turn, exerts a feed- forward stimulation of photosynthetic rates via enhanced mesophyll conductance and alters primary metabolism in rice. *New Phytologist* 2013, *196*, 752-762.

Dodds, P. N.; Rathjen, J. P. Plant immunity: towards an integrated view of plant-pathogen interactions. *Nature* 2010, *11* (8), 539-548.

Doehlemann, G.; Wahl, R.; Horst, R. J.; Voll, L. M.; Poree, F.; Stitt, M.; Kahmann, R. Reprogramming a maize plant: transcriptional and metabolic changes induced by the fungal biotroph *Ustilago maydis*. *The Plant Journal* 2008, *56*, 181-195.

Domiciano, G. P.; Rodrigues, F. A.; Vale, F. X. R.; Xavier Filha, M. S.; Moreira, W. R.; Andrade, C. C. L.; Pereira, S. C. Wheat resistance to spot blotch potentiated by silicon. *Journal of Phytopathology* 2010, *158*, 334-343.

Ellinger, D.; Naumann, M.; Falter, C.; Zwikowics, C.; Jamrow, T.; Manisseri, C.; Voigt, C. A. Elevated early callose deposition results in complete penetration resistance to powdery mildew. *Plant Physiology* 2013, *161*, 1433-1444.

Epstein, E. The anomaly of silicon in plant biology. *Proceedings of the National Academy of Sciences of the United States of America* 1994, 91, 11-17.

Epstein, E. Silicon. *Annual Review of Plant Physiology and Plant Molecular Biology* 1999, *50*, 641-664.

Epstein E. Silicon: its manifold roles in plants. *Annals of Applied Biology* 2009, *155*, 155-160.

Epstein, E., Bloom, A.J., 2005. Mineral Nutrition of Plants: Principles and Perspectives, 2nd Ed. Sinauer, Sunderland, MA. 2005, p. 48-49.

Essmann, J.; Schmitz-Thom, I.; Schoen, H.; Sonnewald, S.;Weis, E.; Scharte, J. RNA interference-mediated repression of cell wall invertase impairs defense in source leaves of tobacco. *Plant Physiology* 2008, *147*, 1288-1299.

Ewané, C. A.; Lepoivre, P.; Bellaire, L. D. L. De. Involvement of phenolic compounds in the susceptibility of bananas to crown rot. *Biotechnology, Agronomy, Society and Environment* 2012, *16* (3), 393-404.

Fauteux, F.; Chain, F.; Belzile, F.; Menzies, J. G.; Bélanger, R. R. The protective role of silicon in the *Arabidopsis*-powdery mildew pathosystem. *PNAS* 2006, *103*, 17554-17559.

Fauteux, F.; Rémus-Borel, W.; Menzies, J. G.; Bélanger, R. R. Silicon and plant disease resistance against pathogenic fungi. *FEMS Microbiology Letters* 2005, *249*, 1-6.

Feng, F.; Zhou, J. M. Plant-bacterial interactions mediated by type III effectors. *Current Opinion in Plant Biology* 2012, 15, 469-476.

Fernández-Aparicio, M.; Prats, E.; Emeran, A. A.; Rubiales, D. Characterization of resistance mechanisms to powdery mildew (*Erysiphe betae*) in beet (*Beta vulgaris*). *Phytophatology* 2009, *99* (4), 385-389.

Fortunato, A. A.; Rodrigues, F. A.; Nascimento, K. J. T. Physiological and biochemical aspects of the resistance of banana plants to *Fusarium* wilt potentiated by silicon. *Phytopathology* 2012, *102*, 957-966.

Fotopoulos, V.; Gilbert, M. J.; Pittman, J. K.; Marvier, A. C.; Buchanan, A. J.; Sauer, N.; Williams, L. E. The monosaccharide transporter gene, *AtSTP4*, and the cell-wall invertase, *Atβfruct1*, are induced in *Arabidopsis* during infection with the fungal biotroph *Erysiphe cichoracearum*. *Plant Physiology* 2003, *132*, 821-829.

Freeman, B.C.; G.A. Beattie. An overview of plant defenses against pathogens and herbivores. 2008. The Plant Health Instructor. DOI: 10.1094/PHI-I-2008-0226-01. https://www.apsnet.org/edcenter/intropp/topics/Pages/OverviewOfPlantDiseases.aspx (accessed Nov 2013).

Gehrig, H.; Schussler, A.; Kluge, M. Geosiphon pyriforme, a fungus forming endocytobiosis with Nostoc (cyanobacteria), is an ancestral member of the Glomales: evidence by SSU rRNA analysis. *Journal of Molecular Evolution* 1996, *43*, 71-81.

Genger, R. K., Jurkowski, G. I., McDowell, J. M., Lu, H., Jung, H. W., Greenberg, J. T., and Bent, A. F. Signaling pathways that regulate the enhanced disease resistance of *Arabidopsis* "Defense, No Death" mutants. *Molecular Plant-Microbe Interaction* 2008, *21*, 1285-1296.

Glover, B. J. Epidermis: outer cell layer of the plant. In: eLS. John Wiley & Sons Ltd, Chichester [e-book], 2010, <Go to>: http://www.els.net [doi: 10.1002/9780470015902.a0002072.pub2] (accessed Nov 2013).

Guntzer, F.; Keller, C.; Meunier, J. D. Benefits of plant silicon for crops: a review. *Agronomy for Sustainable Development* 2012, *32*, 201-213.

Hayasaka, T.; Fujii, H.; Ishiguro, K. The role of silicon in preventing appressorial penetration by the rice blast fungus. *Phytopathology* 2008, *98*, 1038-1044.

Heckman, D.S.; Geiser, D. M.; Eidell,B. R.; Stauffer, R. L.; Kardos, N. L.; Hedges, B. Molecular evidence for the early colonization of land by fungi and plants. *Science* 2001, *293*, 1129-1133.

Henry, G.; Thonart, P.; Ongena, M. PAMPs, MAMPs, DAMPs and others: an update on the diversity of plant immunity elicitors. *Biotechnology, Agronomy, Society and Environment* 2012, *16* (2), 257-268.

Hofius, D.; Tsitsigiannis, D. I.; Jones, J. D. G.; Mundya, J. Inducible cell death in plant immunity. *Seminars in Cancer Biology* 2007, 17, 166-187.

Horst, R. J.; Engelsdorf, T.; Sonnewald, U.; Voll, L. M. Infection of maize leaves with *Ustilago maydis* prevents establishment of C_4 photosynthesis. *Journal of Plant Physiology* 2008, *165*, 19-28.

Hyun, T. K.; Eom, S. H.;Rim, Y.; Kim, J. Alteration of the expression and activation of tomato invertases during *Botrytis cinerea* infection. *Plant Omics Journal* 2011, *4* (7), 413-417.

Jabeen, N.; Ahmed, N.; Ghani, M. Y.; Sofi, P. A. Role of phenolic compounds in resistance to chilli wilt. *Communications in Biometry and Crop Science* 2009, *4* (2), 52-61.

Jones, J. D; Dangl, J. L. The plant immune system. *Nature* 2006, *444*, 323-329.

Joosten, L.; van Veen, J. A. Defensive properties of pyrrolizidine alkaloids against microorganisms. *Phytochemistry Review* 2011, *10*, 127-136.

Kim, S. G.; Kim, K. W.; Park, E. W.; Choi, D. Silicon-induced cell wall fortification of rice leaves: A possible cellular mechanism of enhanced host resistance to blast. *Phytopathology* 2002, *92*, 1095-103.

Kim, Y.; Lim, S.; Yoda, H.; Choi, C.; Choi1, Y.; Sano, H. Simultaneous activation of salicylate production and fungal resistance in transgenic chrysanthemum producing caffeine. *Plant Signaling & Behavior* 2011, *6* (3), 409-412.

Kocal, N.; Sonnewald, U.; Sonnewald, S. Cell wall-bound invertase imits sucrose export and is involved in symptom development and inhibition of photosynthesis during compatible interaction between tomato and *Xanthomonas campestris* pv. *vesicatoria*. *Plant Physiology* 2008, 148, 1523-1536.

Koch K. E. Carbohydrate-modulated gene expression in plants. *Annual Review of Plant Physiology and Plant Molecular Biology* 1996, 47, 509-540.

Korves, T. M.; Bergelson, J. A developmental response to pathogen infection in *Arabidopsis*. *Plant Physiology* 2003, *133*, 339-47.

Kotula, L.; Ranathunge, K.; Schreiber, L.; Steudle, E. Functional and chemical comparison of apoplastic barriers to radial oxygen loss in roots of rice (*Oryza sativa* L.) grown in aerated or deoxygenated solution. *Journal of Experimental Botany* 2009, *60* (7), 2155-2167.

Lattanzio, V.; Lattanzio, V. M. T.; Cardinali, A.; Amendola, V. Role of phenolics in the resistance mechanisms of plants against fungal pathogens and insects. In: Imperato, F. *Phytochemistry: Advances in Research*, 2nd Ed, Research Signspot, Kerala, India 2006, p 23-67.

Liang Y.C., Sun W.C., Si J., Römheld V. Effects of foliar and root applied silicon on the enhancement of induced resistance to powdery mildew in *Cucumis sativus*. *Plant Pathology* 2005, *54*,678-685.

Lichtenthaler, H.K. *Stress of Life: From Molecules to Man*, P. Csermely (Ed.), Annals of the N.Y. Acad. Sci., 851, CDP/PCP, New York, 1998.

Luna, E.; Pastor, V.; Robert, J.; Flors, V.; Mauch-mani, B.; Ton, J. Callose deposition: a multifaceted plant defense response. *Molecular Plant-Microbe Interactions* 2011, *24* (2), 183-193.

Ma, J. F.; Yamaji, N. Silicon uptake and accumulation in higher plants. *Trends in Plant Science* 2006, *11*, 392-7.

Ma, J. F.; Yamaji, N.; Mitani-Ueno, N. Transport of silicon from roots to panicles in plants. *Proceedings of the Japan Academy Series B-Physical and Biological Sciences* 2011, *87*, 377-385.

Magnin-Robert, M.; Letousey, P.; Spagnolo, A.; Rabenoelina, F.; Jacquens, L.; Mercier, L.; Clément, C.; Fontaine, F. Leaf stripe form of esca induces alteration of photosynthesis and defence reactions in presymptomatic leaves. *Functional Plant Biology* 2011, *38*, 856-866.

Mandal, S.; Mitra, A.; Mallick, N. Biochemical characterization of oxidative burst during interaction between *Solanum lycopersicum* and *Fusarium oxysporum* f.sp. *lycopersici*. *Physiological and Molecular Plant Pathology* 2008, *72*, 56-61.

Matern, U.; Kneusel, R. E. Phenolic compounds in plant disease resistance. *Phytoparasitica* 1988, *16* (2), 153-170.

Medzhitov, R.; Janeway, C. A. Innate Immunity: The virtues of a nonclonal system of recognition. *Plant Cell* 1997, *91*, 295-298.

Mazzotta, S.; Kemmerling, B. Pattern recognition in plant innate immunity. *Journal of Plant Pathology* 2011, *93* (1), 7-17.

Muthamilarasan, M.; Prasad, M. Plant innate immunity: An updated insight into defense mechanism. *Journal of Bioscience* 2013, *38*, 433-449.

Nicholson, R. L.; Hammerschmidt, R. Phenolic compounds and their role in disease resistance. *Annual Review of Phytopathology* 1992, *30*, 369-389.

Nürnberger T.; Brunner F.; Kemmerling B.; Piater L. Innate immunity in plants and animals: striking similarities and obvious differences. *Immunological Reviews* 2004, *198*, 249-266.

Pickett, J. A.; Bruce, T. J. A.; Chamberlain, K.; Hassanali, A.; Khan, Z. R.; Matthes, M. C.; Napier, J. A.; Smart, L. E.; Wadhams, L. J.; Woodcock, C. M. Plant volatiles yielding new ways to exploit plant defence. In: Dicke, M.; Takken, W. (eds.). *Chemical ecology: from gene to ecosystem*. Netherlands 2006, p 161-173.

Pieterse, C. M. J.; Leon-Reyes, A.; Van der Ent, S..; Van Wees, S. C. M. Networking by small-molecule hormones in plant immunity. *Nature Chemical Biology* 2009, *5*, 308-316.

Polanco, L. R.; Rodrigues, F. A.; Nascimento, K. J. T.; Shulman, P.; Silva, L. C.; Neves, F. W.; Vale, F. X. R. Biochemical aspects of bean resistance to anthracnose mediated by silicon. *Annual Review of Applied Biology* 2012, *161*, 140-150.

Postel, S.; Kemmerling, B. Plant systems for recognition of pathogen-associated molecular patterns. *Seminars in Cell & Developmental Biology* 2009, 20, 1025-1031.

Petkovs, M. M.; Stampar, F.; Veberic, R. Increased phenolic content in apple leaves infected with the apple scab pathogen. *Journal of Plant Pathology* 2008, *90* (1), 49-55.

Prasad, P. V. V, Staggenborg, S. A., & Ristic, Z. Impacts of dought and/or heat stress on physiological, developmental, growth, and yield processes of crop plants. In: Response of crops to limited water: Understanding and modeling water stress effects on plant growth processes. *Advances in Agricultural Systems Modeling Series 1*, Madison, USA 2008, p 301-356.

Resende, R. S.; Rodrigues, F. A.; Cavatte, P. C.; Martins, S. C. V.; Moreira, W. R.; Chaves, A. R. M.; DaMatta, F. M. Leaf gas exchange and oxidative stress in sorghum plants supplied with silicon and infected by *Colletotrichum sublineolum*. *Phytopathology* 2012, *102*, 892-898.

Robert-Seilaniantz, A.; Grant, M.; Jones, J. D. G. Hormone crosstalk in plant disease and defense: more than just jasmonate- salicylate antagonism. *Annual Review of Phytopathology* 2011, *49*, 317-343.

Roda A.; Halitschke R.; Steppuhn A.; Baldwin, I. T. Individual variability in herbivore-specific elicitors from the plant's perspective. *Molecular Ecology* 2004, *13*, 2421-2433.

Rodrigues, F. A.; Jurick, W. M.; Datnoff, L. E.; Jones, J. B.; Rollins, J. A. Silicon influences cytological and molecular events in compatible and incompatible rice-*Magnaporthe grisea* interactions. *Physiological and Molecular Plant Pathology* 2005, *66*, 144-159.

Rodrigues F.A., McNally D.J., Datnoff L.E., Jones J.B., Labbe C., Benhamou N., Menzies J.G., Bélanger R.R. Silicon enhances the accumulation of diterpenoid phytoalexins in rice: a potential mechanism for blast resistance. *Phytopathology* 2004, *94*, 177-183.

Rodrigues, F. A.; Vale, F. X. R.; Korndorfer, G. H.; Prabhu, A .S.; Datnoff L. E.; Oliveira A. M. A.; Zambolim L. Influence of silicon on sheath blight of rice in Brazil. *Crop Protection* 2003, *22*, 23-29.

Rohácek, K.; Soukupová, J.; Barták, M. Chlorophyll fluorescence: A wonderful tool to study plant physiology and plant stress. In: Schoefs, B. (ed.). *Plant Cell Compartments - Selected Topics*, 64 Kerala, India 2008, p 41-104.

Roitsch, T.; Balibrea, M. E.; Hofmann, M.; Proels, R.; Sinha, A. K. Extracellular invertase: key metabolic enzyme and PR protein. *Journal of Experimental Botany* 2011, *54* (382), 513-524.

Samuels, L.; Kunst, L.; Jetter, R. Sealing plant surfaces: cuticular wax formation by epidermal cells. *Annual Review of Plant Biology* 2008, *59*, 683-707.

Scharte, J.; Schon, H.; Weis, E. Photosynthesis and carbohydrate metabolism in tobacco leaves during an incompatible interaction with *Phytophthora nicotianae. Plant Cell and Environment* 2005, *28*, 1421-1435.

Seebold, K. W.; Kucharek, T. A.; Datnoff, L. E.; Correa-Victoria, F. J.; Marchetti, M. A. The influence of silicon on components of resistance to blast in susceptible, partially resistant and resistant cultivars of rice. *Phytopathology* 2001, *91*, 63-69.

Selvaraj, K.; Fofana, B. An Overview of Plant Photosynthesis Modulation by Pathogen Attacks, Advances in Photosynthesis-Fundamental Aspects, Dr Mohammad Najafpour (Ed.), [Online], 2010, p. 465-486 http://www.intechopen.com/books/advances-in-photosynthesis-fundamental-aspects/an-overview-of-plant-photosynthesis-modulation-by-pathogen-attacks

Silva R. V., Oliveira, R. D. L.; Nascimento K. J. T.; Rodrigues F. A. Biochemical responses of coffee resistance against *Meloidogyne exigua* mediated by silicon. *Plant Pathology* 2010, *59*, 586-593.

Smith A. M.; Stitt M. Coordination of carbon supply and plant growth. *Plant, Cell and Environment* 2007, *30*, 1126-1149.

Spoel, S. H.; Dong, X. How do plants achieve immunity? Defence without specialized immune cells. *Nature Reviews Immunology* 2012, *12* (2), 89-100.

Steppuhn, A.; Gase, K.; Krock, B.; Halitschke, R.; Baldwin, I. T. Nicotine's defensive function in nature. *PLoS Biology* 2004, *2*, 1074-1080.

Sun, W., Zhang, J., Fan, Q., Xue, G., Li, Z., Liang, Y. Silicon-enhanced resistance to rice blast is attributed to silicon-mediated defence resistance and its role as physical barrier. *European Journal of Plant Pathology* 2010, *128*, 39-49.

Swarbrick, P. J.; Schulze-Lefert, P.; Scholes, J. D. Metabolic consequences of susceptibility and resistance (race-specific and broad-spectrum) in barley leaves challenged with powdery mildew. *Plant Cell and Environment* 2006, *29*, 1061-1076.

Tsuda, K.; Katagiri, F. Comparing signaling mechanisms engaged in pattern-triggered and effector-triggered immunity. *Current Opinion in Plant Biology* 2010, *13* (4), 459-465.

Underwood, W. The plant cell wall: a dynamic barrier against pathogen invasion. *Plan Physiology* 2012, *3*, 1-6.

Van Bockhaven, J.; Vleesschauwer, D.; Höfte, M. Towards establishing broad-spectrum disease resistance in plants: silicon leads the way. *Journal of Experimental Botany* 2013, *64* (5), 1281-1293.

Veberic R.; Trobec M.; Herbinger K.; Hofer M.; Grill D.; Stampar F. Phenolic compounds in some apple (*Malus domestica* Borkh) cultivars of organic and integrated production. *Journal of the Science of Food and Agriculture* 2005, *85*, 1687-1694.

Von Rad U.; Mueller M. J.; Durner J. Evaluation of natural and synthetic stimulants of plant immunity by microarray technology. *New Phytologist* 2005, *165*, 191-202.

Vorwerk, S.; Somerville, S.; Somerville, C. The role of plant cell wall polysaccharide composition in disease resistance. *Trends in Plant Science* 2004, *9* (4), 203-209.

Xavier Filha M. S., Rodrigues F. A., Domiciano G. P., Oliveira H. V., Silveira P. R., Moreira W. R. Wheat resistance to leaf blast mediated by silicon. *Australasian Plant Pathology* 2011, *40*, 28-38.

Zhang, G. L.; Dai, Q. G.; Zhang, H. C. Silicon application enhances resistance to sheath blight (*Rhizoctonia solani*) in rice. *Journal of Plant Physiology and Molecular Biology* 2006, *32*, 600-606.

Zhang, J.; Zhou, J. Plant immunity triggered by microbial molecular signatures. *Molecular Plant* 2010, *3* (5), 783-793.

Zipfel, C. Pattern-recognition receptors in plant innate immunity. *Current Opinion in Immunology* 2008, *20*, 10-16.

Zipfel, C.; Robatzek, S. Pathogen-associated molecular pattern-triggered immunity: Veni, Vidi?. *Plant Physiology* 2010, *154*, 551-554.

Zou, J.; Rodriguez-Zas, S.; Aldea, M.; Li, M.; Zhu, J.; Gonzalez, D. O.; Vodkin, L. O.; DeLucia, E.; Clough, S. J. Expression profiling soybean response to *Pseudomonas syringae* reveals new defense-related genes and rapid HR-specific downregulation of photosynthesis. *Molecular Plant-Microbe Interactions* 2005, *18*, 1161-1174.

In: Photosynthesis
Editor: Nafees Khan

ISBN: 978-1-63463-304-8

Chapter 11

NATURAL PRODUCTS AND THEIR SYNTHETIC ANALOGUES AS NEW ACTIVE PRINCIPLES TARGETING THE PHOTOSYNTHETIC MACHINERY

Róbson Ricardo Teixeira*[*1], Angélica Faleiros da Silva Maia[1], Wagner Luiz Pereira[1], Ângela Maria Almeida Lima[1] and Giuseppe Forlani[2]

[1]Departamento de Química, Universidade Federal de Viçosa, Viçosa, Minas Gerais, Brazil
[2]Department of Life Science and Biotechnology, University of Ferrara, Ferrara, Italy

ABSTRACT

The search of new inhibitors of the photosynthetic process through the screening of natural products is described in this chapter. Additionally, synthetic endeavors toward the development of more effective analogues are also analyzed.

Keywords: Photosynthetic inhibitors; natural products; photosynthesis; herbicides

INTRODUCTION

Photosynthesis is a fundamental process in which plants, algae and certain bacteria are capable of using energy from sunlight to drive the synthesis of organic compounds [1-3]. The photosynthetic process involves a complex series of electron transfer reactions. It provides energy and reduced carbon required for the survival of virtually all life on Earth. Most photosynthetic organisms use water as the electron donor, resulting in the release of molecular oxygen and the removal of carbon dioxide from the atmosphere [4-6].

---

[*] Corresponding Author address: Email: robsonr.teixeira@ufv.br.

Various compounds can interfere with the photosynthetic electron transport. This fact has been early explored by agrochemical companies to develop an assortment of herbicides to control weeds; some of them are shown in Figure 1 [7]. The ability of crops to survive their application usually relies upon the presence of detoxifying enzymes. However, because of a high selective pressure for weeds, a rapid diffusion of resistant biotypes occurred within a few years after their introduction, dramatically reducing the commercial life of most active principles. As a consequence, there is a constant need for the discovery of new active scaffolds to manage weed resistance [8-13].

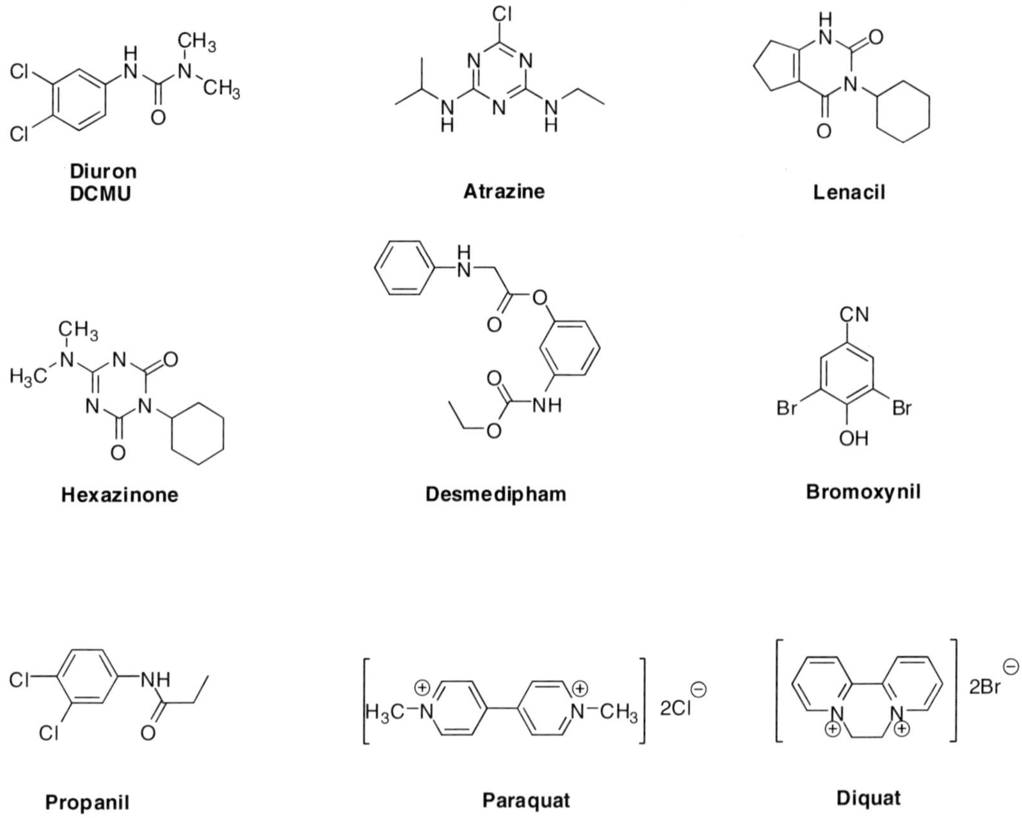

Figure 1. Examples of active ingredients of commercial herbicides that inhibit photosynthesis.

Recently, the vast reservoir of secondary metabolites provided by nature has been intensively explored in search of new inhibitors of the photosynthetic process, yet only a small fraction has been examined. This approach, which can be exploited not only for finding new photosynthetic inhibitors but also for searching new agrochemicals [14-29], is interesting since it may afford new active ingredients possessing low toxicity to non-target organisms, showing high specifity, and that can be applied at low rates. In addition, natural products may serve as lead compounds (prototype bioactive molecules) that can be modified by either semisynthesis or synthesis to afford new analogues with improved activity.

In this chapter, the search of new inhibitors of the photosynthetic process through the screening of natural products will be reviewed. Additionally, synthetic endeavors toward the development of more effective analogues will be analyzed, with special emphasis on recent achievements from our laboratories. Finally, some details about the different modes of action of these compounds will be presented. Although not exhaustive, this review will provide the readers with a concise and updated reference for recent advances in terms of photosynthetic inhibitors.

TERPENES

Terpenes are a group of at least 30,000 secondary metabolites that present various biological activities and a great variety of structure types [30-34]. These structures derive from the repetition of a C_5H_8 unit, and based on the number of carbons in the molecules, the terpenes are classified in monoterpenes (ten carbons), sesquiterpenes (fifteen carbons), diterpenes (twenty carbons), triterpenes (thirty carbons), and tetraterpenes or carotenoids (fourty carbons). Several compounds belonging to this group of substances (Figure 2) have been evaluated as inhibitors of photosynthesis.

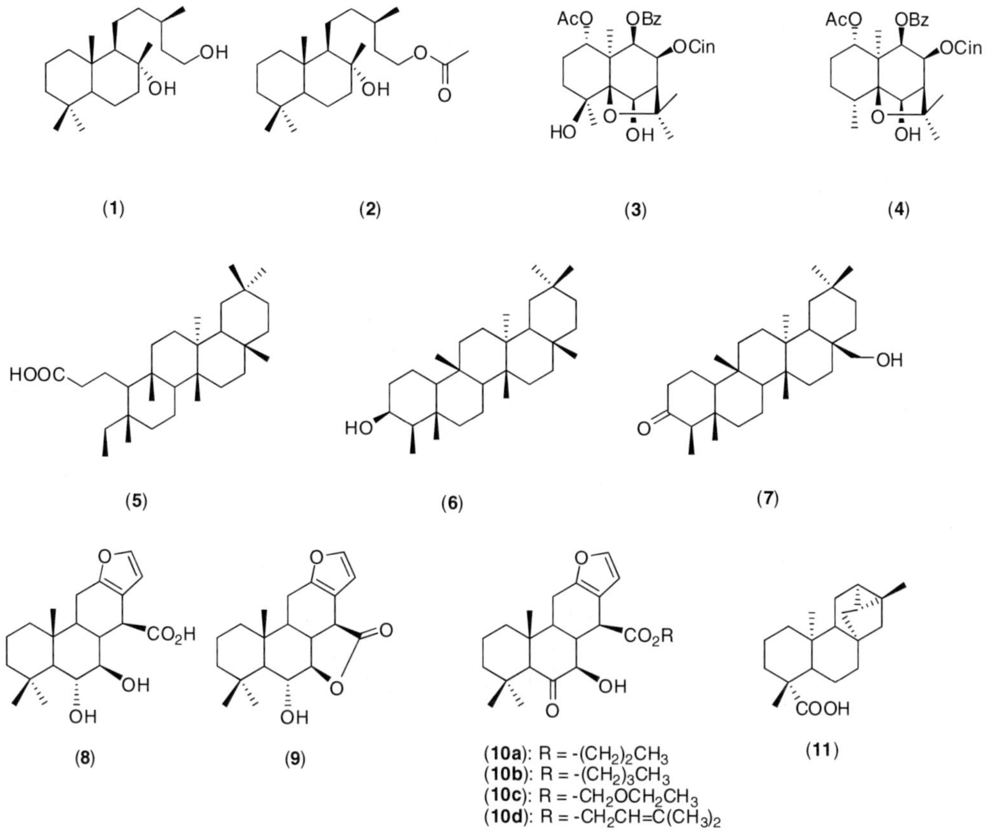

Figure 2. Terpenes evaluated as photosynthetic inhibitors.

The fractioning of hexan extracts from *Croton ciliatoglandulifer* stems resulted in the isolation of diterpene labdane-8α,15-diol (**1**) and its acetyl derivative (**2**) (Figure 2). Biological assays carried out with intact spinach chloroplasts showed that these compounds are capable of interfering with ATP synthesis [35].

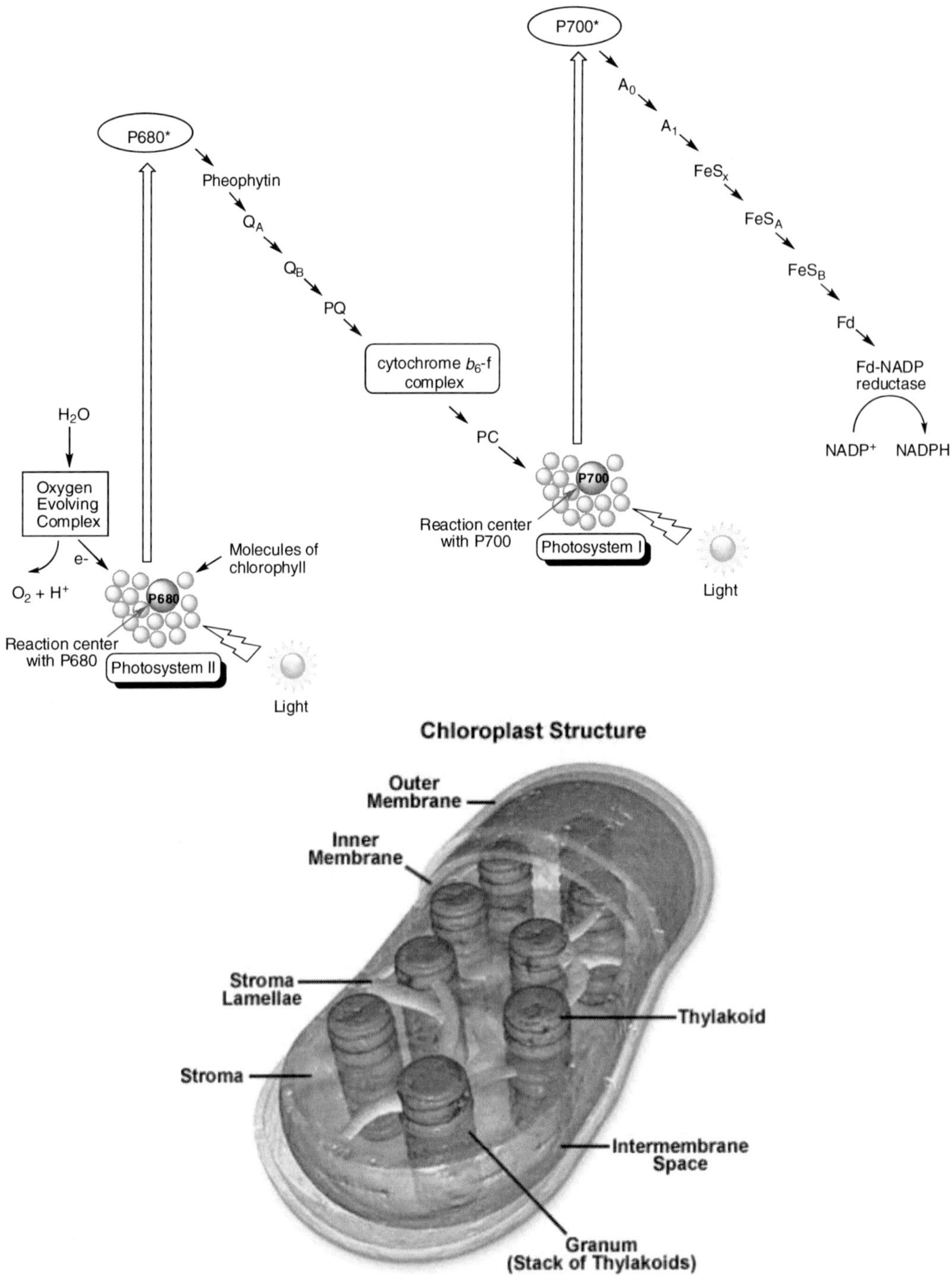

Figure 3. The Z scheme and the electron transfer processes involved in oxygenic photosynthesis inside the chloroplast.

During oxygenic photosynthesis the transfer of electrons from water to $NADP^+$ involves two different reaction centers, known as photosystem (PS) II and PSI. In the presence of light, electrons are transferred from PSII to PSI by intermediate carriers that, when plotted according to their midpoint redox potentials, form the so-called Z scheme (Figure 3). The net reaction is the transfer of electrons from water to $NADP^+$. In higher plants this sequence of reactions occurs inside the chloroplast (Figure 3).

The electron transport from water to $NADP^+$ also builds up an electrochemical gradient of hydrogen ions (protons) across the thylakoidal membrane. This fact occurs due to (i) the release of protons in the lumen of chloroplast during the oxidation of water by the oxygen-evolving complex in PSII and (ii) the translocation of protons from the outer aqueous phase (stroma) to the inner aqueous phase (lumen) by the so-called Q-cycle, in which plastoquinones are reduced to plastohydroquinones taking two protons from the stroma side, and then they are oxidized back to plastoquinone releasing protons into the lumen (Figure 4).

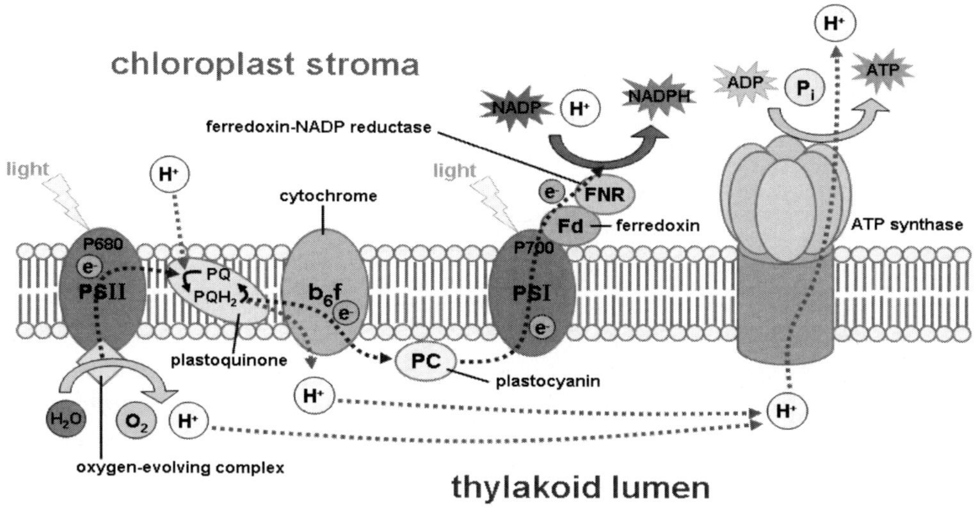

Figure 4. Photosynthetic electron transport yielding the formation of the proton gradient.

The whole process leads to the accumulation in the lumen of no less than 10 protons for four electrons transferred from PSII to PSI [36], and it creates a proton gradient between the stroma (basic) and the lumen (acidic). The sum of the proton chemical potential and the transmembrane electric potential (the so-called *proton motive force*) is used to catalyse ATP formation according to the chemiosmotic hypothesis proposed for the first time by Peter Mitchell [37]. Protons move back to the stroma through the integral membrane portion of ATP synthase, and the energy released allows the formation of ATP from ADP and inorganic phosphate (Figure 4). Such a process is called photophosphorylation.

ATP formation can be measured *in vitro,* using functionally intact chloroplasts by determining the basicity of the medium. In the presence of an artificial electron receptor such as methyl viologen and under continuous actinic illumination, the pH rise continues to increase linearly with time corresponding to the steady state rate of ATP formation. Under appropriate conditions [35], the rate of ATP formation can be determined by back titration with hydrochloric acid using a microelectrode attached to a potentiometer.

Diterpenes (**1**) and (**2**) (Figure 2) were found to inhibit ATP synthesis coupled to electron transport from water to methylviologen in spinach chloroplasts in a concentration-dependent manner. The concentrations causing 50% inhibition *in vitro* (IC_{50}) were determined equal to 72 μmol L^{-1} for compound (**1**) and 10 μmol L^{-1} for compound (**2**) [35]. Inhibition of ATP synthesis can be explained in terms of (a) interruption of the electron transport chain within the thylakoid membrane, (b) dissipation of the H^+ gradient, that is, uncoupling of the ATP synthesis process from the electron transport, and (c) direct inhibition of the H^+-ATPase complex. The block of the electron transport avoids ATP synthesis because the transmembrane electrochemical gradient is not formed. Chemicals that increase proton permeability of thylakoid membranes uncouple phosphorylation from electron flow. Such uncoupling agents inhibit ATP synthesis by decreasing the proton gradient, but allow electron transport to occur at high rates. In contrast, direct inhibitors of photophosphorylation block both ATP synthesis and the portion of electron transport that is a consequence of proton efflux linked to phosphorylation.

In order to elucidate the mechanism of action of (**1**) and (**2**) on photosynthesis, their effects on the electron transport from water to methylviologen were investigated under basal, phosphorylating, and uncoupling conditions. It was found that as concentration of these compounds increased up to 300 μmol L^{-1}, the rates of basal, phosphorylating, and uncoupled electron transport were inhibited, basal and uncoupled electron rates being more affected. The IC_{50} found for compound (**1**) were 200 μmol L^{-1} (basal) and 76.0 μmol L^{-1} (uncoupled) while for (**2**) IC_{50} were 31.0 and 71.5 μmol L^{-1} for basal and uncoupled, respectively.

Further details about the mode of action of natural products (**1**) and (**2**) on the photosynthetic apparatus were obtained by investigating their effects on PSI and PSII. This is possible by using specific artificial electron donors and acceptors [35, 38, 39] (Figure 5).

It was determined that terpene (**1**) inhibits the partial reaction from water do DCPIP. Inhibition of this partial reaction indicates that the compound inhibits the reactions of PSII electron transport chain from water to Q_B (Figure 5), because oxidized DCPIP accepts electrons at this level. It was also found that (**1**) inhibits the PSII partial reaction from water to sodium silicomolybdate (SiMo). Inhibition of this partial reaction means that (**1**) affects the partial reaction from water to quinone Q_A since SiMo is used as electron acceptor at or before Q_A site [40]. The effect of compound (**1**) on the electron flow from DPC to DCPIP was also evaluated. In this case, DCPIP donates electrons at the P_{680} level and the evaluation is carried out with Tris-treated chloroplasts (Figure 5). Such a treatment inhibits the oxygen evolving complex activity. Compound (**1**) did not inhibit this reaction. Overall, results proved that this natural product interacts with the photosynthetic apparatus at the level of the oxygen evolving complex.

Further investigations carried out with terpene (**1**) showed that it is also able to inhibit the partial reaction from reduced PMS to methylviologen. While PMS donates electrons at the P_{700} level, methylviologen accepts electrons at F_x level in photosystem I. This assay was carried out with chloroplasts poisoned with KCN in the presence of DBMIB, an inhibitor of the cytochrome b_6f complex [35]. Therefore, it was established that the second site of interaction of compound (**1**) is located between P_{700} and F_x of PSI.

The effect of (**2**) on the above-mentioned partial reactions was also evaluated. It inhibited the partial reactions from water to DCPIP, from water to SiMo, and from DPC to DCPIP. These results indicate that the compound inhibits the electron flow in photosystem II in the evolving complex system as well as in the path between P_{680} and Q_A or at least at P_{680} site. The labdane (**2**) also inhibited the partial reaction from reduced PMS to oxidized methylviologen [35].

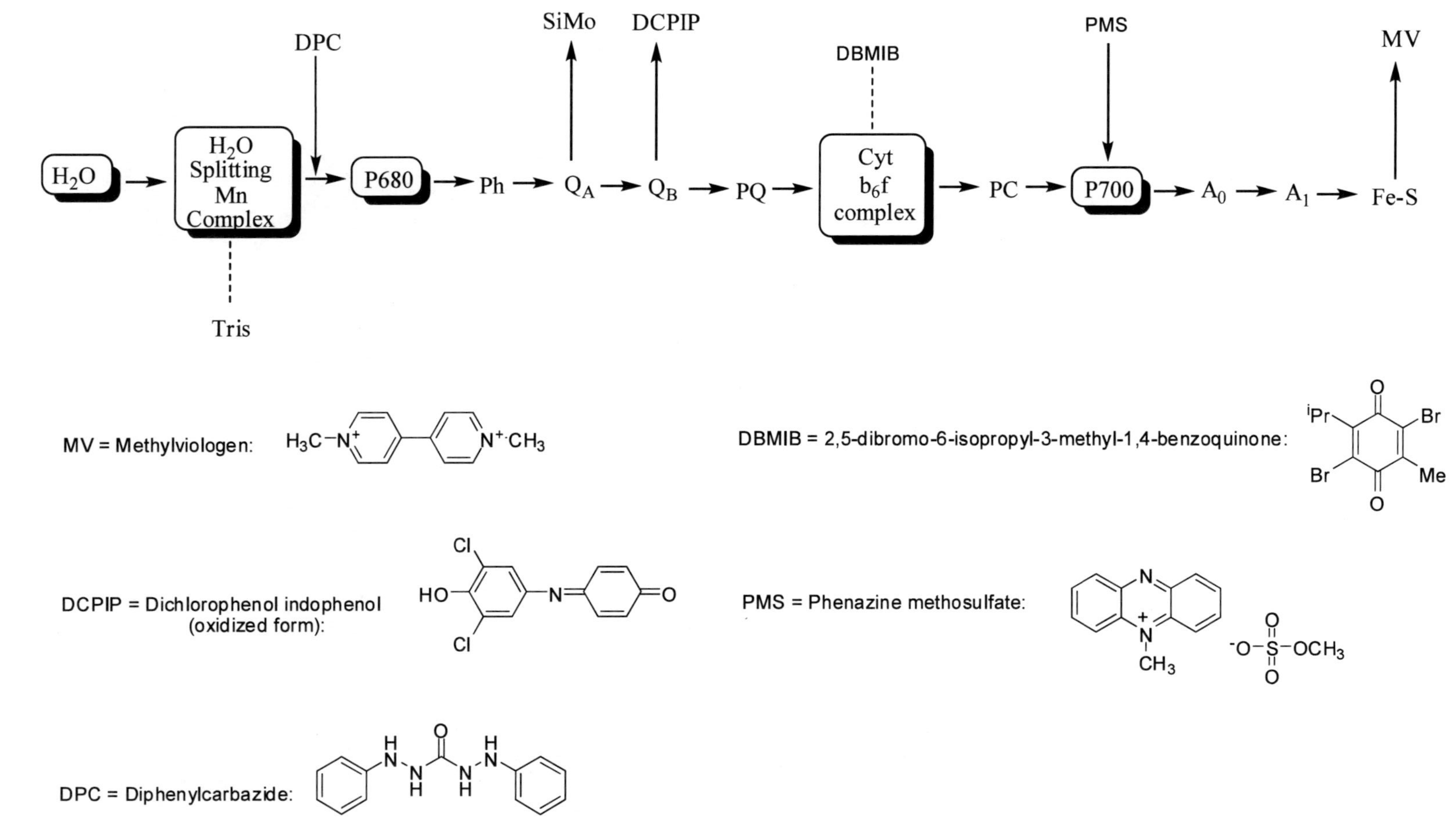

Figure 5. Inhibition sites (broken lines) and donor and acceptor systems (arrows) in the photosynthetic electron transport chain in isolated chloroplasts.

Other terpenes (Figure 2) have been evaluated as potential inhibitors of photosynthesis. For instance, the sesquiterpenoids (**3**) and (**4**) were isolated from the dichloromethane extract of *Celastrus vulcanicola* leaves [40]. Both compounds were capable of inhibiting the photosynthetic photophosphorylation in isolated spinach chloroplasts by interfering with the electron transfer from water to methyl viologen (IC_{50} = 25 μmol L^{-1} for **3** and IC_{50} = 44 μmol L^{-1} for **4**). At low concentrations (up to 50 μmol L^{-1}) compound (**3**) similarly inhibited basal and phosphorylating electron transport rates (approximately 83% and 79%, respectively); however, at higher concentrations (at 300 μmmol L^{-1}) rates were 32% and 68% under basal and phosphorylating conditions, respectively. Moreover, this compound also inhibited the uncoupled electron transport rate. Further investigations showed that terpenoid (**3**) has two sites of interaction: one is located at the oxygen-evolving complex, the other at the H^+-ATPase complex. Compound (**4**) behaves likewise to (**3**) albeit being less active [41].

The seco-3,4-triterpenoid (**5**) was isolated from the hexane extract of air-dried leaves of *Maytenus imbricata*. It inhibited the formation of ATP coupled to electron transport from water to methyl viologen, presenting an IC_{50} value of 148 μmol L^{-1}. Evaluation of the effect of increasing concentrations of (**5**) on basal, phosphorylating, and uncoupled electron flow rates showed that the phosphorylating and uncoupled electron flows are inhibited up to 300 μmol L^{-1}; over this concentration both activities are enhanced. Experimental results support the fact that this triterpenoid acts on the oxygen evolving complex since it inhibits the electron flow from water to sodium silicomolybdate. Furthermore, it behaves as uncoupler activating the Mg^{+2}-ATPase [42]. It is well stablished that uncouplers (e.g., NH_4Cl) stimulate the activity of the Mg^{+2}-ATPase [43].

Epifriedelinol (**6**) and canophyllol (**7**) (Figure 2) are friedelane triterpenoids isolated from the stems of *Celastrus vulcanicola* [44]. Both inhibit ATP synthesis coupled to electron transport from water to methyl viologen showing IC_{50} = 82 μmol L^{-1} for compound (**6**) and 124 μmol L^{-1} for compound (**7**). The evaluation of their effects on noncyclic electron transport from water to methyl viologen under basal, phosphorylating, and uncoupled conditions showed that epifriedelinol (**6**) does not affect these processes. On the other hand, compound (**7**) partially inhibited the electron transport rates at different degrees. Both compounds moderately enhanced the light-activated Mg^{+2}-ATPase activity. Therefore, the friedelane triterpenoids (**6**) and (**7**) act by decreasing the pH variation, by blocking ATP formation, and by interacting with the Mg^{+2}-ATPase in thylakoids. It was also found that compound (**7**) is capable of inhibiting the electron flow from water to the electron acceptor 2,5-dichloro-1,4-benzoquinone (DCBQ), thus acting as Hill reaction inhibitor [44]. A Hill reaction inhibitor can be defined as an electron acceptor that is photoreduced at the expense of water, which is oxidized to molecular oxygen.

The natural diterpene α,7β-dihydroyvouacapan-17β-oic acid (**9**) was isolated from *Pterodon polygalaeflorus* [45]. Treament of this compound with acetic anhydride afforded the lactone derivative (**9**), as shown in Figure 6 [46].

When both compounds were evaluated as possible inhibitors of ATP synthesis in isolated spinach chloroplasts, compound (**8**) was completely inactive. On the contrary, tha lactone derivative (**9**) inhibited this process in a concentration-dependent manner (IC_{50} = 90 μmol L^{-1}). Lactone (**9**) did not affect PSI, but it inhibited the electron transport at the PSII level by targeting the oxygen evolving complex as well as the electron transport between P_{680} and Q_A [45].

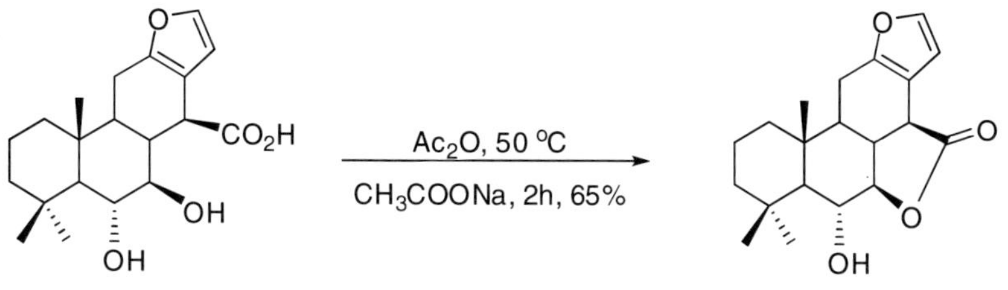

(8) (9)

Figure 6. Conversion of natural product (**8**) into lactone (**9**).

Figure 7 depicts the synthesis of esters (**10a**-**10d**) which were obtained employing lactone (**8**) as starting material [47]. Initial conversion of carboxylic acid (**8**) into lactone **9** followed by Swern oxidation and opening of the lactone ring by sodium hydroxide afforded esters (**10a**-**10d**).

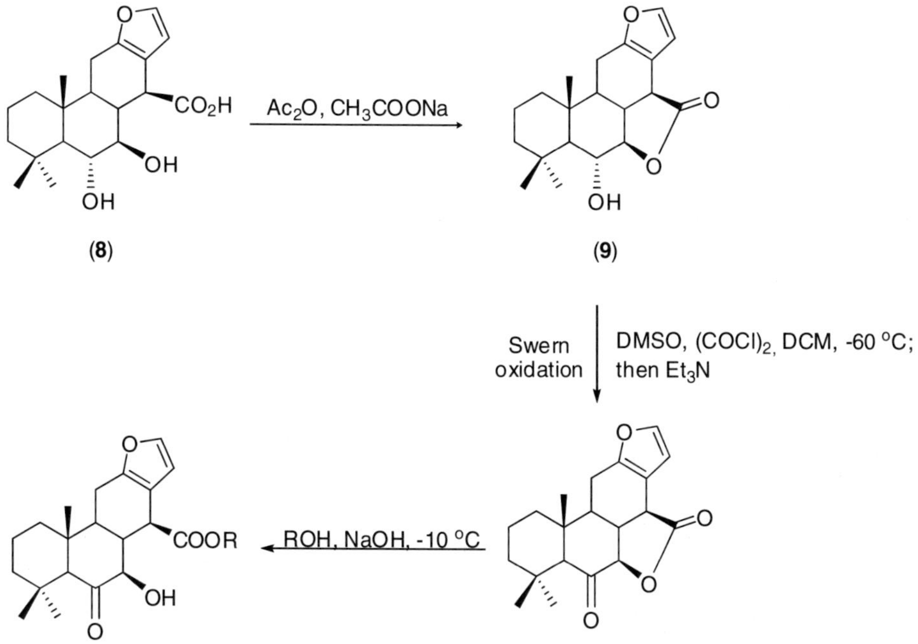

(10a): R = $-(CH_2)_2CH_3$
(10b): R = $-(CH_2)_3CH_3$
(10c): R = $-CH_2OCH_2CH_3$
(10d): R = $-CH_2CH{=}C(CH_3)_2$

Figure 7. Synthetic route used in the synthesis of esters (**10a**-**10d**).

In vitro evaluation of compounds (**10a**-**10d**) showed that all of them were capable of inhibiting ATP synthesis as well as the uncoupled electron transport from water to the artificial electron acceptor methylviologen. Therefore, these synthetic esters act as Hill reaction inhibitors. Compared to compounds (**8**) and (**9**) (Figures 2 and 6), ATP synthesis inhibition by (**10a**-**10d**) is 4 to 9 fold higher, being compound (**10a**) the most effective (IC_{50} = 10 μmol L^{-1}). Experiments conducted to evaluate the influence of (**10a**-**10d**) on basal and phosporylating non-cyclic electron transport revealed that (**10c** and **10d**) at 100 μmol L^{-1} increase the basal electron flow by 60 and 25%, respectively. Thus, they act as uncouplers. On the contrary, esters (**10a**) and (**10b**) inhibited the basal electron flow. Considering phosporylating electron flow, compound (**10a**) enhanced this activity by 20% at 50 μmol L^{-1}, while 82% inhibition was observed for (**10b**) at the same concentration. As a consequence, compound (**10a**) behaves as a Hill reaction inhibitor and secondarily as uncoupler, whereas (**10b**) acts as Hill reaction inhibitor. At 50 μmol L^{-1}, compounds (**10c**) and (**10d**) inhibited phosphorylation electron flow by 75% and 39%, respectively. By using several appropriate artificial electron donors, electron acceptors, and electron transport chain inhibitors, it was determined that the esters (**10a-10d**) inhibit the electron transport chain in the span from Q_A to Q_B-D1 protein of PSII, by displacing Q_B like diuron (DCMU, Figure 1). These compounds did not affect uncoupled electron flow from water to SiMo (Figure 5). Finally, *in vivo* bioassays carried out to evaluate the effect of (**10a-10d**) on the weed species *Echinochloa crusgalli* and *Amaranthus hypochondriacus* revealed that they exert herbicidal effects and inhibit plant growth.

Another last example of triterpene endowed with inhibitory properties on the photosyhnthetic apparatus is trachyloban-19-oic acid (**11**) [48] (Figure 2). This steroid was purified from the chloroform extract of *Iostephane heterophylla*. It was found that, in addition of inhibiting ATP synthesis (IC_{50} value = 9 μmol L^{-1}), compound (**11**) also interferes with basal, uncoupled, and phosphorylating electron flow. All rates were completely suppressed at 100 μmol L^{-1}. It did not affect PSI, but it acts as Q_A-Q_B inhibitor [48].

LACTONES

The chlorinated γ-lactone cyanobacterin (**12**) (Figure 8) is a secondary metabolite isolated from the cyanobacterium (blue-green alga) *Scytonema hofmanni* [49].

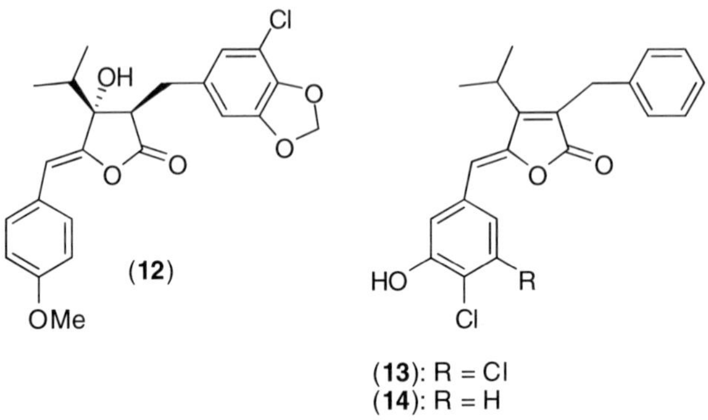

Figure 8. Structures of cyanobacterin (**12**) and nostoclides (**13,14**).

At a concentration as low as 5 μmol L^{-1}, this compound is toxic to most cyanobacteria, and also inhibits the growth of eukaryotic algae and several mono- and dicotyledons [50, 51]. It acts mainly by inhibiting the photosynthetic electron transport, with a mechanism similar to that of the herbicide 3-(3,4-dichlorophenyl)-1,1-dimethylurea [52]. In isolated *Euglena gracilis* chloroplasts, such interference leads to a cascade of events that result in the disruption of the thylakoid membranes [53]. The two natural lactones nostoclide I (**13**) and nostoclide II (**14**) (Figure 8) were isolated from a cyanobacterium, *Nostoc* sp., which is a symbiont of the common lichen *Peltigera canina*. Due to the structural similarity between cyanobacterin and nostoclides, and also considering the effect of cyanobacterin on photosynthesis [51,52], several nostoclides analogues were synthesized (Figure 9).

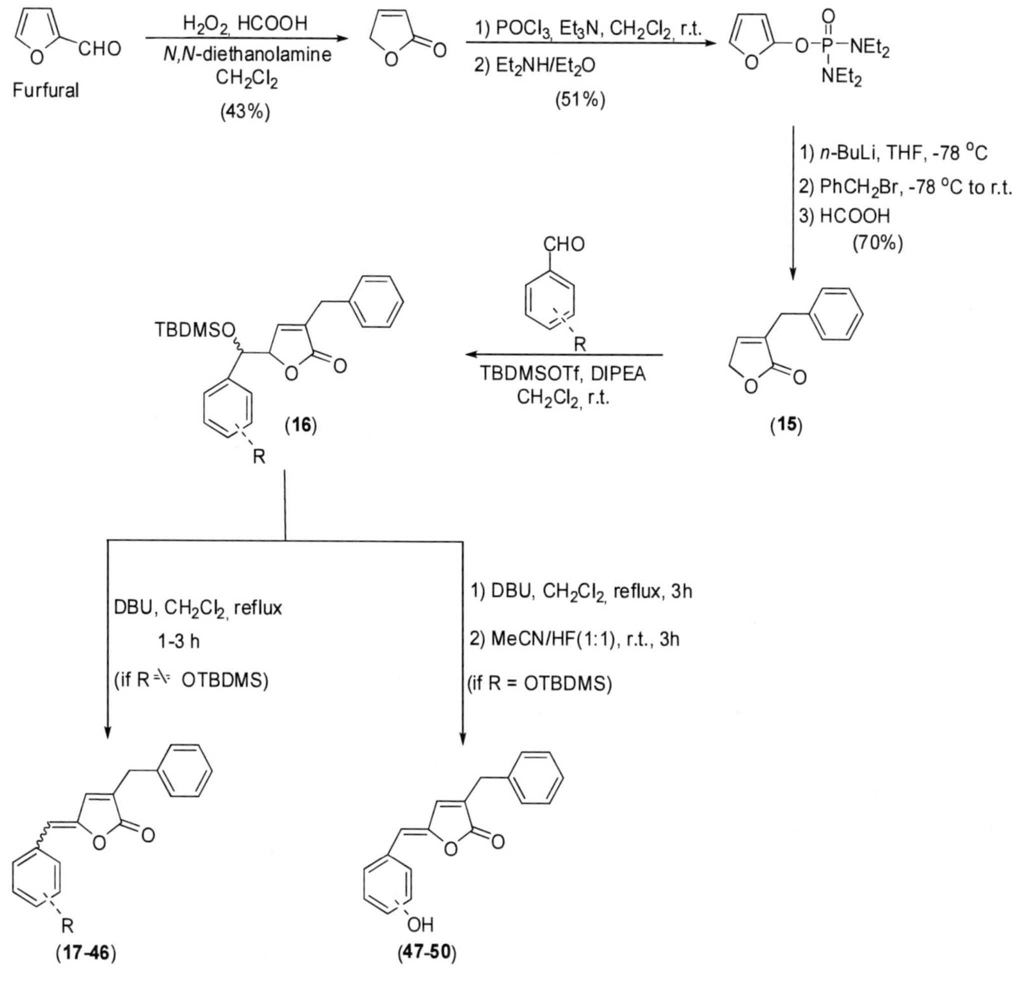

Figure 9. Synthetic route used to prepare nostoclide analogues **17-50**.

The reaction of lactone **15**, which can be prepared as shown in Figure 9, with pertinent aldehydes in the presence of *tert*-butyldimethylsilyltrifluoromethanesulfonate and diisopropylethylamine followed by treatment of the silyl ether (general structure **16**, Figure 9) *in situ* with DBU afforded compounds **17-50** (Table 1) in yields ranging from 12 to 91% [53]. For the synthesis of lactones **47-50**, *tert*-butyldimethylsilyloxy aldehydes were used. Removal of this protecting group by MeCN/HF (1:1 v/v) afforded compounds **47-50**. The majority of the synthesized nostoclide analogues exihibited *Z* stereochemistry concerning the exocyclic double bond. One exception to this generalization is compound (**19**), which was obtained as *E* stereoisomer. Rationals for these stereochemical aspects have been described [55].

Table 1. Effects of nostoclide derivatives 17-50 on ferricyanide reduction by functionally intact chloroplasts[1]

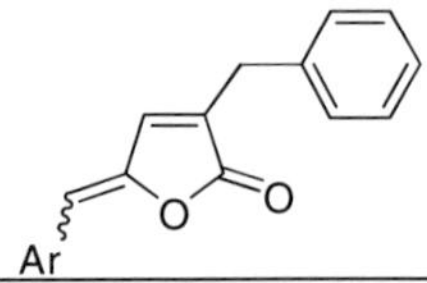

Compound	Arylidene Group	5 μmol L^{-1}	10 μmol L^{-1}
17	benzylidene	12.7 ± 3.2	29.2 ± 1.4
18	1,3-dioxalanebenzylidene	N	N
19	2,4,6-trimethoxibenzylidene	25.2 ± 2.4	43.6 ± 2.1
20	4-*N,N*-dimethylaminobenzylidene	5.7 ± 2.7	7.1 ± 1.5
21	2-chloro-4-*N,N*-dimethylaminobenzylidene	13.8 ± 1.5	25.37 ± 4.2
22	2,5-dimethoxybenzylidene	30.3 ± 0.6	43.5 ± 2.7
23	4-nitrobenzylidene	49.8 ± 1.6	57.8 ± 1.0
24	3-bromobenzylidene	13.8 ± 3.6	22.6 ± 2.9
25	4-methylbenzylidene	N	8.0 ± 1.7
26	3-nitrobenzylidene	30.0 ± 1.0	39.5 ± 0.8
27	3-methylbenzylidene	8.0 ± 2.1	16.7 ± 2.6
28	4-chlorobenzylidene	N	5.8 ± 1.4
29	3-chlorobenzylidene	5.8 ± 0.7	20.3 ± 4.8
30	4-methoxybenzylidene	8.5 ± 0.7	15.0 ± 1.4
31	4-fluorobenzylidene	9.2 ± 1.1	25.6 ± 1.8
32	4-bromobenzylidene	N	N
33	3-fluorobenzylidene	9.5 ± 2.9	28.8 ± 1.2
34	2-fluorobenzylidene	24.5 ± 3.3	49.5 ± 1.9
35	2-chlorobenzylidene	19.4 ± 2.3	30.8 ± 3.7
36	2-methylbenzylidene	16.7 ± 0.4	26.0 ± 0.1
37	4-ethylbenzylidene	40.6 ± 1.5	50.7 ± 4.1
38	2-bromobenzylidene	N	21.8 ± 2.2
39	pentafluorobenzylidene	N	6.6 ± 2.8
40	4-trifluoromethylbenzylidene	34.6 ± 2.7	55.5 ± 1.7
41	3-trifluoromethylbenzylidene	27.3 ± 4.8	49.9 ± 1.7
42	2-trifluoromethylbenzylidene	23.5 ± 4.5	44.9 ± 2.6
43	4-cyanobenzylidene	N	N
44	3-cyanobenzylidene	N	N
45	3-*N,N*-dimethylaminobenzylidene	10.9 ± 3.2	38.6 ± 5.4
46	4-phenylbenzylidene	N	N
47	4-hydroxybenzylidene	95.7 ± 5.2	93.5 ± 2.3
48	3-hydroxybenzylidene	N	6.9 ± 0.8
49	2-hydroxybenzylidene	N	N
50	3-hydroxy-4-methoxybenzylidene	N	N

[1] Activity values are expressed as percentage inhibition of the rate of ferricyanide reduction in untreated controls. N indicates no effect.

The nostoclide analogues **17-50** were evaluated as the ability to interfere with the basal photosynthetic electron transport from water to the artificial electron acceptor potassium ferricianyde at two different concentrations, and the results are presented in Table 1.

At the lower concentration, among the 34 synthesized derivatives, 21 were able to interfere with the electron transport, whereas 13 analogues were substantially ineffective. Among the active compounds, 8 analogues caused more than 40% inhibition. Lactones (**23**) and (**40**), which contain strong electron withdrawing substituents at the *para* position of the aromatic ring, were the most active compounds [54].

Another series of nostoclide analogues (compounds in Table 2), presenting a bromine attached to the *para* position of the benzyl group [56], were prepared by using a synthetic route similar to that presented in Figure 9.

Table 2. Nostoclide analogues 51-62

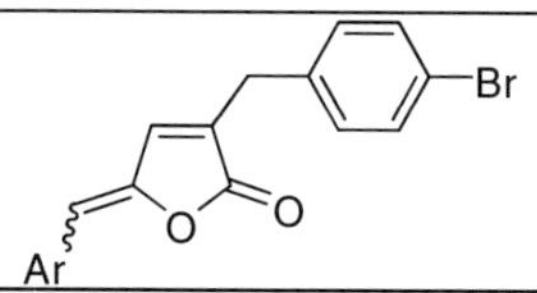

Compound	Arylidene group
51	benzylidene
52	1,3-dioxalanebenzylidene
53	4-fluorobenzylidene
54	4-bromobenzylidene
55	4-trifluoromethylbenzylidene
56	2,5-dimethoxybenzylidene
57	3-nitrobenzylidene
58	2,4,6-trimethoxibenzylidene
59	3-methylbenzylidene
60	4-chlorobenzylidene
61	4-cyanobenzylidene
62	2-chloro-4-*N,N*-dimethylaminobenzylidene

[1] Compounds presented *Z* stereochemistry concerning the exocyclic double bound except compound 58 showing E stereochemistry.

The effect of compounds **51-62** on the Hill reaction was also evaluated [56]. Thus, it was possible to establish a comparison between the effects of compounds **17-19**, **21**, **22**, **27**, **28**, **31**, **32**, **40**, **43** (Table 1) and their (*para*-bromobenzyl)-counterparts **51-62** (Table 2). As a general trend, it was found that lactones **17-19**, **21**, **22**, **27**, **28**, **31**, **32**, **40**, **43** are more effective in inhibiting the basal electron flow than compounds **51-62** [54].

A quantitative structure-activity relationship (QSAR) investigation was carried out on nostoclides analogues (**17-50**) to correlate molecular descriptors with their biological activity *in vitro*. The results suggested that the inhibitory potential of this class of compounds is intimately associated with their polarity [57]. Thus, it is likely that new nostoclide analogues with higher polarity could display improved activity against photosynthesis.

Compounds possessing a γ-lactone moiety fused to an aromatic ring are known as isobenzofuran-1(*3H*)-ones (phtalides). Natural and synthetic substances belonging to this class present various biological activities [58-64] including effects on photosynthesis.

Demuner and co-workers reported the isolation of 4-methoxy-5-methyl-6-(3-methylbut-2-enyloxy)isobenzofuran-1(*3H*)-one (**63**) from the phytopathogenic fungus *Nymbia alternantherae*, cultivated on solid media [65]. This isobenzofuranone acts as Hill reaction inhibitor and uncoupler of photosynthesis, displaying inhibitory activity on ATP synthesis (IC_{50} = 66 μmol L^{-1}). Natural product (**63**) was converted to (**64**) in quantitative yield via catalytic hydrogenation using Pd/C (Figure 10). The hydrogenated derivative (**64**) was less efficient in inhibiting ATP synthesis (IC_{50} was not determined).

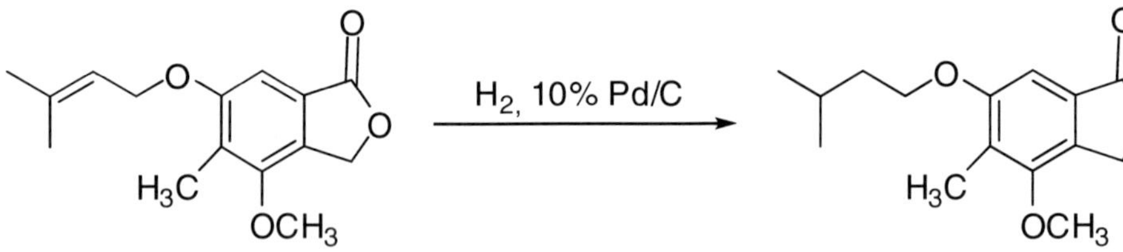

(**63**) (**64**)

Figure 10. Catalytic hydrogenation of lactone (**63**).

Inspired by the activities of isobenzofuranones (**63** and **64**) on the aforementioned photosynthetic process, and also considering the phytotoxic effects described for cryphonectric acid (Figure 11) [66], we undertook the task to synthesize and investigate the effects of C-3 functionalized isobenzofuranones on the photosynthetic machinery [67].

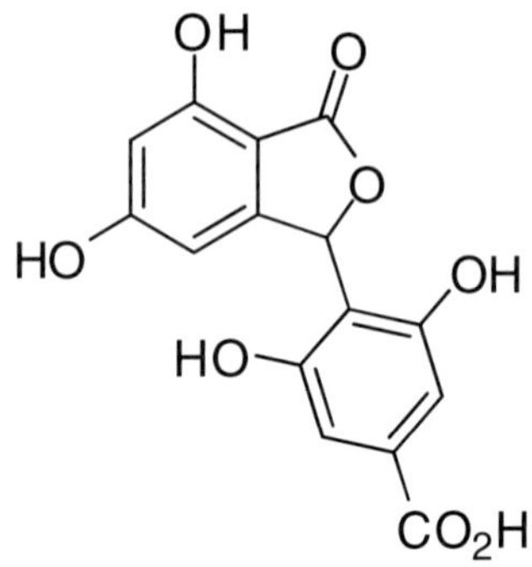

Figure 11. Structure of the secondary metabolite cryphonectric acid.

Thus, C-3 functionalized isobenzofuranones (**66-76**) were prepared employing the synthetic route outlined in Figure 12.

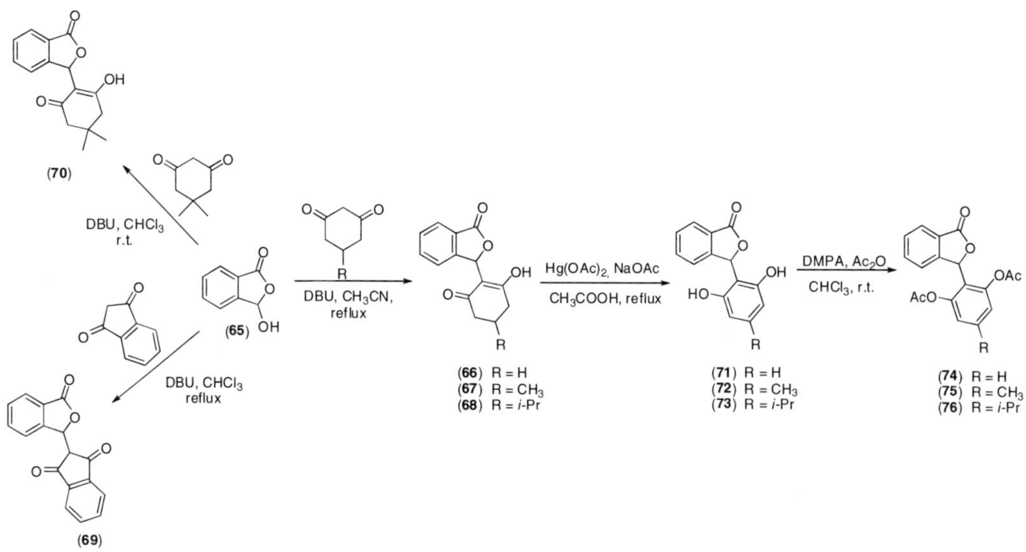

Figure 12. Preparation of C-3 functionalized isobenzofuranones (**66-76**).

The condensation reaction between commercially available phthalaldehyde (**65**) and various 1,3-dicarbonyl compounds afforded derivatives (**66-68**, **69** and **70**). Subsequent aromatization of compounds (**66-68**) gave phenolic substances (**71-73**). Finally, straightforward acetylation reactions of (**71-73**) produced the acetate esters (**74-76**).

The effects of isobenzofuranones on the chloroplastic electron transport were evaluated *in vitro* as their ability to interfere with the basal photosynthetic electron transport from water to potassium ferricyanide and the results are depicted in Table 3.

Table 3. *In vitro* effects of compounds 66-76 on ferricyanide reduction by functionally intact chloroplasts isolated from *Spinacia oleracea* leaves[a]

Compound	50 μmol L^{-1}	100 μmol L^{-1}	200 μmol L^{-1}
66	99.2 ± 1.6	114.1 ± 1.5	104.7 ± 1.3
67	102.7 ± 0.3	96.5 ± 2.0	93.8 ± 3.2
68	93.2 ± 0.5	92.0 ± 2.8	97.5 ± 2.8
69	108.6 ± 0.5	105.0 ± 3.4	78.2 ± 4.3
70	86.1 ± 0.0	68.0 ± 1.4	44.2 ± 0.2
71	90.2 ± 0.1	92.4 ± 1.3	55.7 ± 3.0
72	133.5 ± 2.0	125.6 ± 0.8	116.1 ± 4.5
73	185.7 ± 2.4	156.7 ± 4.9	87.1 ± 2.5
74	218.3 ± 4.0	222.5 ± 1.1	189.9 ± 0.9
75	201.9 ± 3.7	192.4 ± 5.9	161.1 ± 6.5
76	92.0 ± 1.5	83.1 ± 1.3	75.9 ± 1.6

[a] Basal activity was measured either in the absence or in the presence of a given compound at increasing concentration, as indicated. Values were expressed as percentage of untreated controls, and are means ± SE over at least three replicates.

Derivatives **70, 71** and **76** exerted mild inhibitory effects on basal electron transport, whereas compounds **72-75** stimulated it. Derivatives **72-75** did not interfere with uncoupled electron transport from water to ferricyanide. Thus, these isobenzofuranones act as uncouplers. Interestingly, addition of isobenzofuranones to the culture medium inhibited the growth of the cyanobacterium *Synechococcus elongatus*, with patterns consistent with the effects measured *in vitro* upon isolated chloroplasts [67].

The fractioning of the ethanolic extract of the fungus *Botryosphaeria rhodina* led to the isolation of four natural products, among which the lactone lasiodiplodin (**77**) (Figure 13).

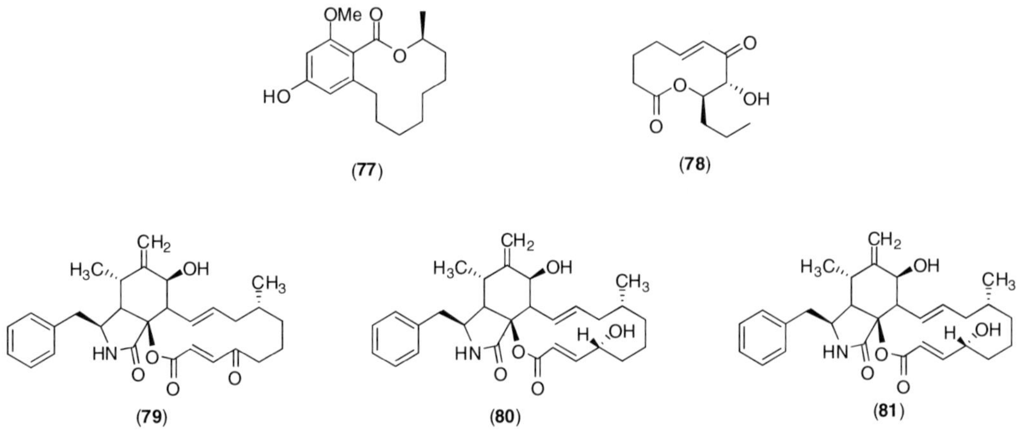

Figure 13. Structures of lactones (**77-81**).

In vitro evaluation of its influence on the photosynthetic apparatus in isolated spinach chloroplasts revealed that as the concentration of (**77**) raised, ATP synthesis is inhibited (IC_{50} = 35.6 μmol L^{-1}). Further investigations to determine the mechanism of action of lactone (**77**) revealed that it acts at three different sites: at the CF_1-ATPase complex, in the oxygen evolving complex, and at the electron transfer path between P_{680} and Q_A [68]. This mode of action is different from that displayed by synthetic herbicides targeting photosynthesis. This finding corroborates the possibility that the exploitation of the natural pool products can afford compounds presenting different molecular targets compared to well known herbicides, and provide tools to overcome weed resistance problems.

A set of fifteen substances belonging to two groups of organic compounds, nonenolides and cytochalasins, had their phytotoxic effects evaluated against the perennial weed species *Cirsium arvense* and *Sonchus arvensis* [69]. Among the nonenolides evaluated by leaf disc-puncture bioassay, stagonolide A (**78**) (Figure 13) displayed the highest phytotoxic activity (necrosis diameter > 6 mm) on *C. arvense*. Considering the cytochalasins, only cytochalasin A (**79**) exerted high phytotoxic effects on this weed species (necrosis diameter ~3 mm) [69]. The most phytotoxic compound against *S. arvensis* was the cytochalasin deoxaphomin (**80**) (necrosis diameter ~7 mmm). High phytotoxic effects were also observed for stagonolide A (**78**), cytochalasin A (**80**) and cytochalasin B (**81**) (necrosis diameter ~ 4.5, 5.5 and 4 mm, respectively) [69].

Photometric assays in the range of 450-950 nm showed that stagonolide A (**78**) and cytochalasin B (**81**) presented effects on photosynthesis. Twenty four hours after appearance of necrosis on leaf discs, both compounds caused significant decrease of the light absorption at 450 nm by *C. arvense*. This experimental observation was associated with the reduction of β-carotene or/and chlorophyll *b* content in leaf tissue of *C. arvense* since these pigments have a peak of resonant absorption near 450 nm. At the wavelengths of 530 nm and 550 nm, an increase in the light absorption for compounds (**78**) and (**81**) was noticed. However, stagonolide A (**78**) had a significantly stronger effect at 550 nm compared to cytochalasin B (**81**). Since cytochromes have a peak of absorption in the range of 530-550 nm, the substances (**78**) and (**81**) probably interfere with these proteins, and did not affect electron transport. Treatment of leaf discs with stagonolide (**78**) also led to reduction of light absorption in the wavelenghth region between 630-690 nm by *C. arvense* leaves [69]. The peaks of light absorption in this region are characteristic of chlorophyll intermediates, phytochlorophyllide and chlorophyllide [69, 70].

It has been demonstrated that some coumarins, lactones derived from *orto*-hydroxy cinnamic acid, are capable of interfering with the photosynthetic process, for instance siderin (**82**) (Figure 14) a secondary metabolite produce by *Toona ciliata* (Meliaceae). Lotina-Hennsen and co-workers showed that this coumarin inhibits ATP synthesis with an IC_{50} = 27.0 μmol L^{-1} and does not inhibit PSI electron transport. Siderin (**82**) does inhibit partial reactions of photosystem electron flow from water to DCPIP, and from water to sodium silicomolybdate, and partially inhibits electron flow from DPC to DCPIP (Figure 5). All these results support the fact that the targets of (**82**) are the donor and acceptor sites of PSII, between P_{680} and Q_A [71].

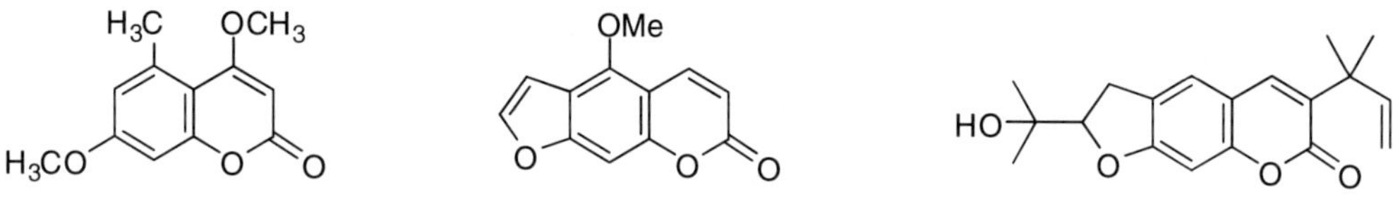

(**82**) (**83**) (**84**)

Figure 14. Structures of coumarins (**82-84**).

The furocoumarins bergapten (**83**) and chalepin (**84**) (Figure 14) were isolated from *Ruta graveolens* and chlorophyll *a* fluorescence experiments demonstrated that these compounds present a remarkable activity as photosynthetic inhibitors [72]. Because of this, the authors of this investigation suggested that these compounds are good candidates to be chemically modified in order to obtain even more effective derivatives [72].

FLAVONOIDS

The bioactivity-guided chemical analysis of *Selaginella lepidophylla* resulted in the isolation of the biflavonoids robustaflavone (**85**), 2,3-dihydrorobustaflavone (**86**), and 2,3-dihydrorobustaflavone-5-methyl ether (**87**) (Figure 15).

Inhibition of ATP synthesis coupled to electron flow from water to methyl viologen in freshly-lysed spinach chloroplasts is brought about by these compounds as their concentration increases. The determined IC_{50} values were 44, 39, and 79 μmol L^{-1} for compounds (**85**), (**86**), and (**87**), respectively. The mode of action of these compounds was further investigated.

The electron transport was partially inhibited under either basal, uncoupling or phosphorylating conditions. Compound (**85**) did not affect PSI. However, it interacts with PSII between the oxygen evolving complex and P_{680}. The target of (**86**) was located between cyt b_6f complex and plastocyanin (Figure 3). Flavonoid (**87**) had no effect on PSI or II. All three compounds act as energy transfer inhibitors since increasing concentrations of (**85**), (**86**), and (**87**) inhibited the Mg^{+2}-ATPase activity [73].

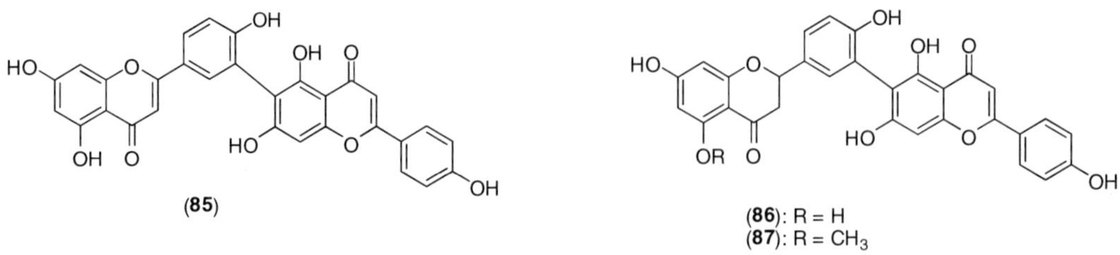

Figure 15. Structures of flavonoids (**85-87**).

Other Natural Products

Artemisinin (qinghaosu) (**88**) (Figure 16) is a secondary metabolite terpenoid produced by the Chinese medicinal herb sweet wormwood (*Artemisia annua* L.) [74]. A series of investigations have demonstrated that artemisinin (**88**) and its semisynthetic analogues have a powerful inhibitory effect on plant growth [75-77]. Some of the analogues, for instance the *aza*-compound (**89**), were more potent than artemisinin (**88**). Since deoxyartemisinin (**90**) is inactive, the phytotoxic activity of artemisinin (**88**) and its analogues is dependent upon the presence of the peroxide moiety [77]. These findings have inspired the design and synthesis of compounds presenting the endoperoxide motif (general structures **91** and **92**, Figure 16), some of which are endowed with plant growth regulatory activity [78, 79] and displayed phytotoxic effects comparable to artemisinin (**88**) [79].

Until recently, the mechanism of the phytotoxic action of artemisinin (**88**) was unknown. Sabat and co-workers [80] demonstrated that artemisinin is capable of interfering with photosynthesis. The compound inhibited *in vitro* the basal electron transport from water to potassium ferricyanide. The degree of inhibition was concentration-dependent and showed a tendency to saturation around 744 μmol L^{-1}, the highest concentration tested. At this concentration, a 65-70% inhibition was observed. Under uncoupling conditions, the electron transport was reduced by 50%. Stimulation of electron transport was noticed under phosphorylating conditions. On the other hand, when artemisinin (**88**) was applied to plants (*in vivo* experiments), it behaved as a potent inhibitor of the photosynthetic electron transport, as confirmed by fluorescence measurements [80]. The major site of action of artemisinin (**88**) was identified to be Q_B (Figure 5). It was determined that *in vivo* inhibitory effect are most likely associated with an unindentified artemisinin metabolite rather than by the interaction of the compound *per se* [80].

Sorgoleone (**93**), characterized as 2-hydroxy-5-methoxy-3-[8´Z,11´Z]-8´,11´-14´-pentadecatriene]-*p*-hydroquinone (Figure 16), is a metabolite isolated from the root exudates of *Sorghum bicolor* [81].

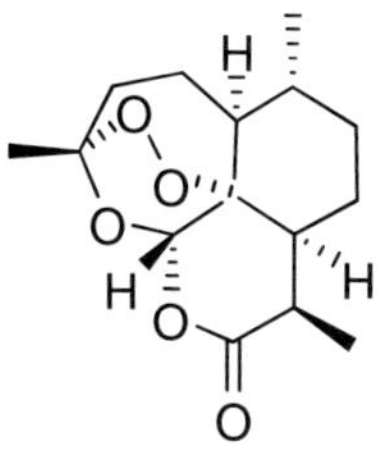

(88)

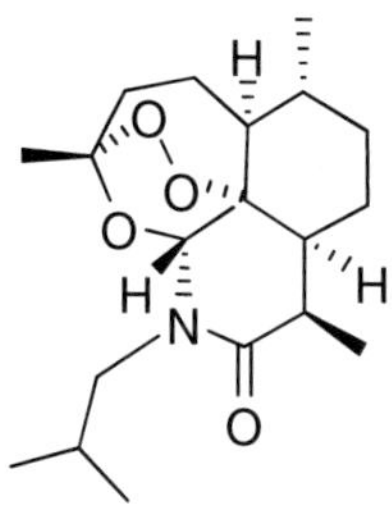

(89)

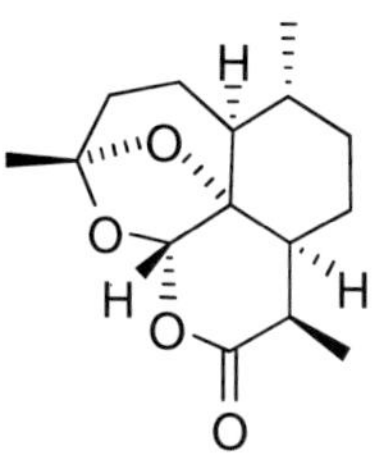

(90)

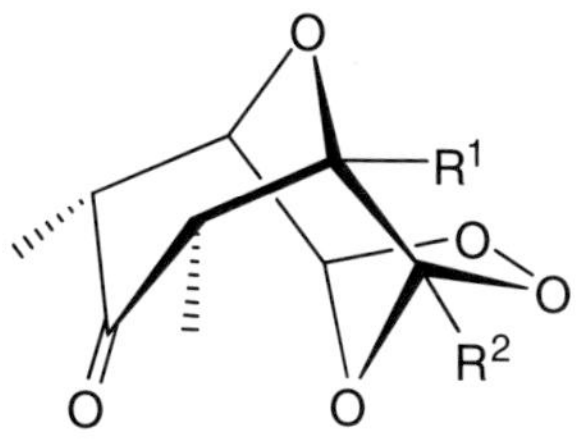

(91)

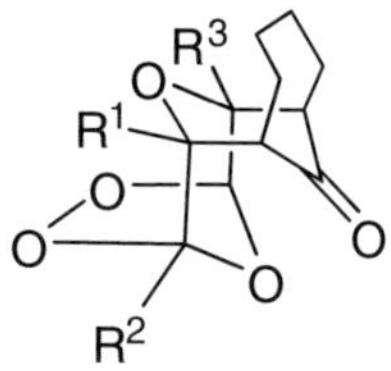

(92)

Figure 16. Structures of compounds (**88-92**).

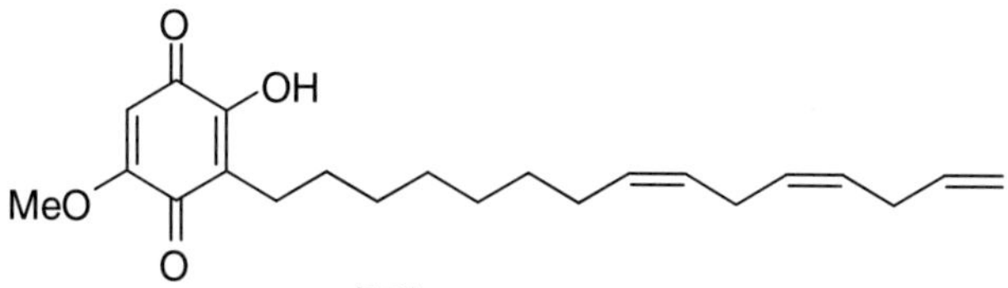

(93)

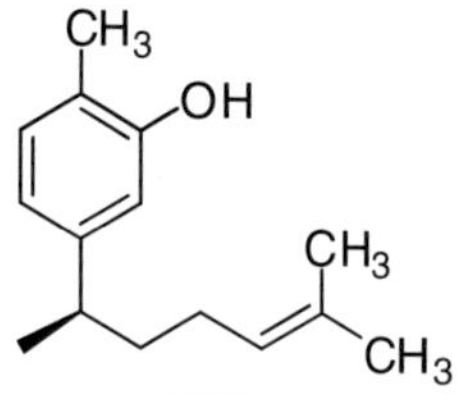

(94)

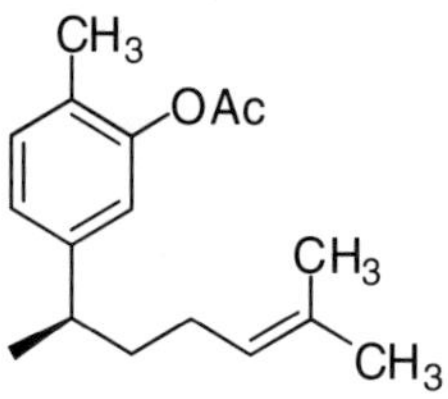

(95)

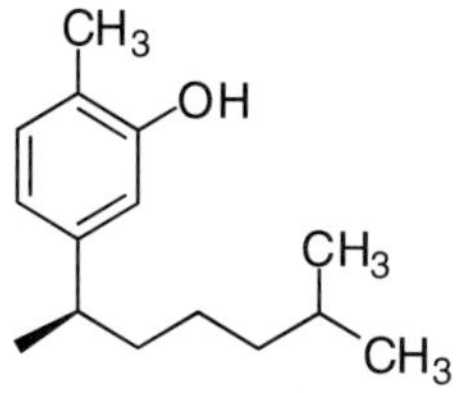

(96)

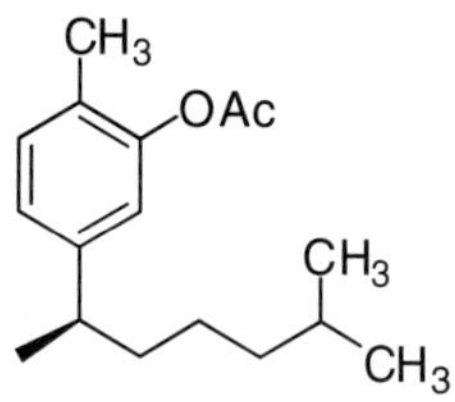

(97)

Figure 17. Structures of sorgoleone (**93**), xanthorrhizol (**94**) and xanthorrhizol analogues (**95-97**).

Since its discovery, it has been demonstrated that sorgoleone (**93**) is a potent inhibitor of chlorophyll formation in *Lemma minor* L., and it also inhibits the growth of several grass and broadleaf weeds at a concentration as low as 10 μmol L^{-1} [82]. Further investigations revealed that sorgoleone is a very effective inhibitor of electron transfer between Q_A to Q_B at the reducing side of the PSII [83], and also of the mitochondrial electron transport [84, 85]. *In vitro* experiments demonstrated that sorgoleone is as effective as diuron in inhibiting electron transport from water to potassium ferricyanide; the IC_{50} for sorgoleone (**93**) and diuron were 0.10 and 0.12 µmol L^{-1}, respectively [86]. This quinone also causes disturbance of plasma H^+-ATPase activity in root cells [87].

The phenolic compound xanthorrhizol was isolated from the methanolic extract of *Iostephane heterophylla* (**94**). The influence of (**94**), as well as that of synthetic derivatives **95**-**97** obtained from the natural metabolite via straightforward acetylation and hydrogenation reactions [88] (Figure 17), was evaluated on several photosynthetic aspects. The phenolic compounds (**94**-**97**) inhibited ATP synthesis (IC_{50} values of 68, 28, 7 and 55 μmol L^{-1} for compounds **94**, **95**, **96**, and **97**, respectively). Basal, phosphorylating, and uncoupled non-cyclic electron transport activities from water to methyl viologen were inhibited as well. For these activities, IC_{50} values ranged from 24 to 88 μmol L^{-1}. Further investigations on the mode of action on the photosynthetic apparatus revealed that **94** and **96** act by interfering with the PSII, a conclusion that was corroborated by the results of fluorescence experiments.

Conclusion

The exploitation of natural products has afforded a variety of compounds capable of interfering with the photosynthetic machinery. In some cases, the observed activity is similar to that displayed by available commercial herbicides targeting photosynthesis. Thus, some of these secondary metabolites could be considered for further development by agrochemical companies. In addition, this approach has resulted in the discovery of substances with different mechanisms of action. This is an important aspect in view of facing weed resistance problems. It is likely that several other promising inhibitors of photosynthesis will be found by further exploring the natural pool products. The consideration of photosynthesis as a target for herbicide development has resulted to date in the chemical synthesis of compounds that are endowed with low toxicity to mammals, but in several cases they are recalcitrant to biological degradation. From nature, it is also possible that active compounds with low toxicity and improved biodegradability will be found. Finally, for successful weed management one important challenge is related to selectivity: chemicals should exert their phytotoxic action only on weeds. In the case of natural products, this issue has not been properly addressed to date, and future works should also be concerned with this important matter.

Acknowledgments

We are grateful to Fundação de Amparo à Pesquisa de Minas Gerais (FAPEMIG) for financial support.

References

[1] Rutherford, A.W., Faller, P. (2004). The heart of photosynthesis in glorious 3D. *Trends Biochem. Sci.* 26, 341-344.

[2] Nelson, N., Yocum, C.F. (2006). Structure and function of photosystems I and II. *Annu. Rev. Plant Biol.* 57, 521-565.

[3] Taiz, L., Zeiger, E. (2010). Photosynthesis: The light reactions. In *Plant Physiology*, Sinauer Associates, Inc., Fifth Edition.
[4] Blankenship, R.E., Hartman, H. (1998). The origin and evolution of oxygenic photosynthesis. *Trends Biochem. Sci.* 23, 94-97.
[5] Nelson, N., Ben-Shen, A. (2004). The complex architecture of oxygenic photosynthesis. *Nat. Rev. Mol. Cell Biol.* 5, 1-12.
[6] Nelson, N. (2011). Photosystems and global effects of oxygenyic photosynthesis. *Biochim. Bophys. Acta* 1807, 856-863.
[7] Draber, W., Tietjen, K., Kluth, J.F., Trebst, A. (1991). Herbicides in photosynthesis research. *Angew. Chem. Int. Ed. Engl.* 30, 1621-1633.
[8] Heap, I.M. (1997). The occurrence of herbicide-resistant weeds worldwide. *Pestic. Sci.* 51, 235-243.
[9] Neve, P. (2007). Challenges for herbicide resistance evolution and management: 50 years after Harper. *Weed Res.* 47, 365-369.
[10] Gressel, J. (2009). Evolving understanding of the evolution of herbicide resistance. *Pest Manag. Sci.* 65, 1164-1173.
[11] Beckie, H.J. (2011). Herbicide-resistant weed management: focus on glyphosate. *Pest Manag. Sci.* 67, 1037-1048.
[12] Shaner, D.L., Lindenmeyer, R.B., Ostile, M.H. (2012). What have the mechanisms of resistance to glyphosate taught us? *Pest Manag. Sci.* 68, 3-9.
[13] Heap, I. International Survey of Herbicide Resistant Weeds [online], available at http://www.weedscience.org (accessed October 1, 2013).
[14] Pillmoor, J.B., Wright, K., Terry, A.S. (1993). Natural products as a source of agrochemicals and leads for chemical synthesis. *Pestic. Sci.* 39, 131-140.
[15] Dayan, F., Romagni, J., Tellez, M., Rimando, A., Duke, S. (1999). Managing weeds with natural products. *Pestic. Outlook* 10, 185-188.
[16] Duke, S.O., Dayan, F.E., Romagni, J.G., Rimando, A.M. (2000). Natural products as sources of herbicides: current status and future trends. *Weed Res.* 40, 99-111.
[17] Duke, S.O., Dayan, F.E., Rimando, A.M., Schrader, K.K., Aliotta, G., Oliva, A., Romagni, J.G. (2002). Chemicals from nature for weed management. *Weed Res.* 50, 138-151.
[18] Vyvyan, J.R. (2002). Allelochemicals as leads for new herbicides and agrochemicals. *Tetrahedron* 58, 1631-1646.
[19] Copping, L.G., Duke, S.O. (2007). Natural products that have been used commercially as crop protection agents. *Pest Manag. Sci.* 63, 524-554.
[20] Strange, R.N. (2007). Phytotoxins produced by microbial plant pathogens. *Nat. Prod. Rep.* 24, 127-144.
[21] Macías, F.A., Molinillo, J.M.G., Varela, R.M., Galindo, J.C.G. (2007). Allelopathy – a natural alternative for weed control. *Pest Manag. Sci.* 63, 327-348.
[22] Macías, F.A., Galindo, J.L.G., García-Díaz, M., Galindo, J.C.G. (2008). Allelopathic agents from aquatic ecosystems: potential biopesticides models. *Phytochem. Rev.* 7, 155-178.
[23] Macías, F.A., Oliveros-Bastidas, A., Marín, D., Carrera, C., Chinchilla, N., Molinillo, J.M.G. (2008). Plant biocommunicators: their phytotoxicity, degradation studies and potential use as herbicide models. *Phytochem. Rev.* 7, 179-194.

[24] Barbosa, L.C.A., Teixeira, R.R., Montanari, R.M. (2008). Phytotoxic natural products as models for the development of crop protection agents. In *Current Trends in Phytochemistry*, Ed. Francesco Epifano, Research Signpost, Kerala, India, p. 21-59.

[25] Dayan, F.E., Cantrell, C.L., Duke, S.O. (2009). Natural products in crop protection. *Bioorg. Med. Chem.* 17, 4022-4034.

[26] Beaudegnies, R., Edmunds, A.J.F., Fraser, T.E.M., Hall, R.G., Hawkes, T.R., Mitchell, G., Schaetzer, J., Wendeborn, S., Wibley, J. (2009). Herbicidal 4-hydroxyphenyl-pyruvate dioxygenase inhibitors-A review of the triketone chemistry story from a Syngenta perspective. *Bioorg. Med. Chem.* 17, 4134-4152.

[27] Hüter, O.F. (2011). Use of natural products in crop protection industry. *Phytochem. Rev.* 10, 185-194.

[28] Cantrell, C.L., Dayan, F.E., Duke, S.O. (2012). Natural products as sources for new pesticides. *J. Nat. Prod.* 75, 1231-1241.

[29] Varejão, E.V.V., Demuner, A.J., Barbosa, L.C.A., Barreto, R.W. (2013). The search for new natural herbicides – Strategic approaches for discovering fungal phytotoxins. *Crop Protection* 48, 41-50.

[30] Bicas, J.L., Dionísio, A.P., Pastore, G.M. (2009). Bioxidation of terpenes: an approach to the flavor industry. *Chem. Rev.* 109, 4518-4531.

[31] Júnior, C.V. (2003). Terpenes with insecticidal activity: an alternative to chemical crop control of insects. *Quim. Nova* 26, 390-400.

[32] Degenhardt, J., Köllner, T.G., Gershenzon, J. (2009). Monoterpene and sesquiterpene synthases and the origin of terpene skeletal diversity in plants. *Phytochemistry* 70, 1621-1637.

[33] Gershenzon, J., Dudareva, N. (2007). The function of terpene natural products in the natural worlds. *Nat. Chem. Biol.* 3, 408-414.

[34] Mann, J., Davidson, R.S., Hobbs, J.B., Banthorpe, D.V., Harborne, J.B. (1994). Terpenoids. In *Natural Products: Their chemistry and biological significance*, Longman Scientific & Technical, England, 1994, p.289-358.

[35] Morales-Flores, F., Aguilar, M.I., King-Díaz, B., Santiago-Gómez, J.-RP., Lotina-Hennsen, B. (2007). Natural diterpenes from *Croton ciliatoglanduliferus* as photosystem II and photosystem I inhibitors in spinach chloroplasts. *Photosynth. Res.* 91, 71-80.

[36] Allen, J.F. (2003). Cyclic, pseudocyclic and noncyclic photophosphorylation: new links in the chain. *Trends Plant Sci.* 8, 15-19.

[37] Mitchell, P. (1961). Coupling of phosphorylation to electron and hydrogen transfer by a chemi-osmotic type of mechanism. *Nature* 19, 144-148.

[38] Trebst, A. (2007). Inhibitors in the functional dissection of the photosynthetic electron transport system. *Photosynth. Res.* 92, 217-224.

[39] Allen, J.F., Holmes, N.G. (1986). Electron transport and redox titration. In *Photosynthesis Energy Transduction. A Practical Approach*, Hipkinns, M.F., Baker, N.R. (Eds.), IRL Press, Oxford, United Kingdom, pp. 103-141.

[40] Giaquinta, R.T., Dilley, R.A. (1975). A partial reaction in photosystem II: reduction of silicomolybdate prior to the site of dichlorophenyldimethylurea inhibition. *Biochim. Biophys. Acta* 387, 288-305.

[41] Torres-Romero, D., King-Díaz, B., Jiménez, I.A., Lotina-Hennsen, B., Bazzocchi, I.L. (2008). Sesquiterpenes from *Celastrus vulcaniola* as photosynthetic inhibitors. *J. Nat. Prod.* 71, 1331-1335.

[42] Souza e Silva, S.R., Silva, G.D.F., Barbosa, L.C.A., Duarte, L.P., King-Diaz, B., Archundia-Camacho, F., Lotina-Hennsen, B. (2007). Uncoupling and inhibition properties of 3,4-seco-friedelan-3-oic acid isolated from *Maytenus imbricata*. *Pestic. Biochem. Physiol.* 87, 109-114.

[43] Mills, J.D., Mitchell, P., Schurmann, P. (1980). Modulation of coupling factor ATPase activity in intact chloroplasts. The role of thioredoxin system. *FEBS Lett.* 112, 173-177.

[44] Torres-Romero, D., King-Díaz, B., Strasser, R.J., Jiménez, I.A., Lotina-Hennsen, B., Bazzocchi, I.L. (2010). Friedelane triterpenes from *Celastrus vulcanicola* as photosynthetic inhibitors. *J. Agric. Food Chem.* 58, 10847-10854.

[45] King-Díaz, B., Pérez-Reyes, A., Santos, F.J.L., Ferreira-Alves, D.L., Veloso, D.P., Carvajal, S.U., Lotina-Hennsen, B. (2006). Natural diterpene β-lactone derivative as photosystem II inhibitor on spinach chloroplasts. *Pestic. Biochem. Physiol.* 84, 109-115.

[46] Rubinger, M.M.M., Veloso, D.P., Stefani, G.M. (1991). Synthesis of 6α,7β-di-hydroxyvouacapan-17β-oic acid derivatives. Parte I: Hydroxamic acid and amide derivatives. *J. Braz. Chem. Soc.* 2, 124-128.

[47] King-Díaz, B., Castelo-Branco, P.A., Santos, F.J.L., Rubinger, M.M.M., Ferreira-Alvez, D.L., Piló-Veloso, D., Lotina-Hennsen, B. (2010). Furanoditerpene ester and thiocarbonyldioxy derivatives inhibit photosynthesis. *Pestic. Biochem. Physiol.* 96, 119-126.

[48] Hernández-Terrones, M.G., Aguilar, M.I., King-Diaz, B., Lotina-Hennsen, B. (2003). Inhibition of photosystem II in spinach chloroplasts by trachyloban-19-oic acid. *Pestic. Biochem. Physiol.* 77, 12-17.

[49] Pignatello, J.J., Porwoll, J., Carlson, R.E., Xavier, A., Gleason, F.K. (1983). Structure of the antibiotic cyanobacterin, a chlorine-containing γ-lactone from the freshwater cyanobacterium *Scytonema hofmanni*. *J. Org. Chem.* 48, 4035-4038.

[50] Gleason, F.K., Baxa, C.A. (1986). Activity of the natural algicide cyanobacterin on eukaryotic microorganisms. *FEMS Microbiol. Lett.* 33, 85-88.

[51] Gleason, F.K., Case, D.E. (1986). Activity of the natural algicide cyanobacterin on angiosperms. *Plant Physiol.* 80, 834-837.

[52] Gleason, F.K., Paulson, J.L. (1984). Site of action of the natural algicide cyanobacterin in the blue-green alga, *Synechococcus* sp. *Arch. Microbiol.* 138, 273-277.

[53] Gleason, F.K. The natural herbicide cyanobacterin, specifically disrupts thylakoid membrane structure in *Euglena gracilis* strain-Z. *FEMS Microbiol. Lett.* 68, 77-81.

[54] Teixeira, R.R., Barbosa, L.C.A., Forlani, G., Piló-Veloso, D., Carneiro, J.W.M. (2008). Synthesis of photosynthesis-inhibiting nostoclide analogues. *J. Agric. Food Chem.* 56, 2321-2329.

[55] Teixeira, R.R., Barbosa, L.C.A., Santana, J.O., Veloso, D.P., Ellena, J., Doriguetto, A.C., Drew, M.G.B., Ismail, F.M.D. (2007). Synthesis and structural characterization of two nostoclide analogues. *J. Molec. Struct.* 837, 197-205.

[56] Barbosa, L.C.A., Rocha, M.E., Teixeira, R.R., Maltha, C.R.A., Forlani, G. (2007). Synthesis of 3-(4-bromobenzyl)-5-(arylmethylene)-*5H*-furan-2-ones and their activity

as inhibitors of the photosynthetic electron transport chain. *J. Agric. Food Chem.* 55, 8562-8569.

[57] Teixeira, R.R., Pinheiro, P.F., Barbosa, L.C.A., Carneiro, J.W.M., Forlani, G. (2009). QSAR modeling of photosynthesis-inhibiting nostoclide derivatives. *Pest Manag. Sci.* 66, 196-202.

[58] Teixeira, R.R., Bressan, G.C., Pereira, W.L., Ferreira, J.G., Oliveira, F.M., Thomaz, D.C. (2013). Synthesis and antiproliferative activity of C-3 functionalized isobenzofuran-1(*3H*)-ones. *Molecules*, 18, 1881-1896.

[59] Logrado, L.P.L., Santos, C.O., Romeiro, L.A.S., Costa, A.M., Ferreira, J.R.O., Cavalcanti, B.C., Moraes, M.O., Costa-Lotufo, L.V., Pessoa, C. (2010). Synthesis and cytotoxicity screening of substituted isobenzofuranones designed from anacardic acids. *J. Med. Chem.* 45, 3480-3489.

[60] Yogonathan, K., Rossant, C., Ng, S., Huang, Y., Butter, M.S., Buss, A.D. (2003). 10-methoxydihydrofuscin, fuscinarin and fuscin, novel antagonists of the human CCR5 receptor from *Oidiodendron griseum*. *J. Nat. Prod.* 66, 1116-1117.

[61] Huang, X.Z., Zhu, Y., Guan, X.L., Tian, K., Guo, J.M., Wang, H.B., Fu, G.M. (2012). A novel oxidant isobenzofuranone derivative from fungus *Cephalosporium* sp. AL031. *Molecules* 17, 4219-4224.

[62] Ma, F., Gao, Y., Qiao, H., Hu, X., Chang, J. (2012). Antiplatelet activity of 3-butyl-6-bromo-1(*3H*)-isobenzofuranone on rat platelet aggregation. *J. Thromb. Thrombolysis* 33, 64-73.

[63] Cardozo, J.A., Braz-Filho, R., Rincón-Velandia, J., Guerrero-Pabón, M.F. (2005). 3-butil isobenzofuranona: un compuesto aislado de *Apium graveolens* com atividad anticonvulsivante. *Rev. Colomb. Cienc. Quim. Farm.* 34, 69-76.

[64] Lin, G., Chan, S. S.-K., Chung, H.-S., Li, S.-L. (2005). Biological action of natural occurring phtalides, In *Studies in Natural Product Chemistry*, Ed. Atta-Ur-Rahman, Elsevier, Amsterdam, Vol. 32, p. 611-669.

[65] Demuner, A.J., Barbosa, L.C.A., Veiga, T.A.M., Barreto, R.W., King-Díaz, B., Lotina-Hennsen, B. (2006). Phytotoxic constituents from *Nimbya alternantherae*. *Biochem. System. Ecol.* 34, 790-795.

[66] Arnone, A., Assante, G., Nasini, G., Strada, S., Vercesi, A. (2002). Cryphonectric acid and other minor metabolites from a hypovirulent strain of *Cryphonectria parasitica*. *J. Nat. Prod.* 65, 48-50.

[67] Teixeira, R.R., Pereira, W.L., Tomaz, D.C., Oliveira, F.M., Gilberti, S., Forlani, G. (2013). Synthetic analogues of the natural compound cryphonectric acid interfere with photosynthetic machinery through two different mechanisms. *J. Agric. Food Chem.* 61, 5540-5549.

[68] Veiga, T.A.M., Silva, S.C., Francisco, A.-C., Filho, E.R., Vieira, P.C., Fernandes, J.B., Silva, M.F.G., Müller, M.W., Lotina-Hennsen, B. (2007). Inhibition of photophosphorylation and electron transport chain in thylakoids by lasiodiplodin, a natural product from *Botryosphaeria rhodina*. *J. Agric. Food Chem.* 55, 4217-4221.

[69] Berestetskiy, A., Dmitriev, A., Mitina, G., Lisker, I., Andolfi, A., Evidente, A. (2008). Nonenolides and cytochalasins with phytotoxic activity against *Cirsium arvense* and *Sonchus arvensis*: a structure-activity relationships study. *Phytochemistry* 69, 953-960.

[70] Duke, S.O., Duke, M.V., Sherman, T.D., Nandihalli, U.B. Spectrometric and spectrofluorometric methods in weed science. *Weed Sci.* 39, 505-513.

[71] Veiga, T.A.M., González-Vázquez, R., Neto, J.O., Silva, M.F.G.F., King-Díaz, B., Lotina-Hennsen, B. (2007). Siderin from *Toona ciliata* (Meliaceae) as photosystem II inhibitor on spinach thylakoids. *Arch. Biochem. Biophys.* 465, 38-43.

[72] Sampailo, O.M., Silva, M.F.G.F., Veiga, T.A.M., King-Díaz, B., Lotina-Hennsen, B. (2012). Evaluation of furocoumarins as photosynthetic inhibitor by chlorophyll *a* fluorescence assay. *Quim. Nova* 35, 2115-2118.

[73] Aguilar, M.I., Romero, M.G., Chávez, M.I., King-Díaz, B., Lotina-Hennsen, B. (2008). Biflavonoids isolated from *Selaginella lepidophylla* inhibit photosynthesis in spinach chloroplasts. *J. Agric. Food Chem.* 56, 6994-7000.

[74] Klayman, D.L. (1985). Qinghaosu (artemisinin): an antimalarial drug from China. *Science* 228, 1049-1055.

[75] Duke, S.O., Vangh, K.C., Croom, E.M., Jr., Elsohly, H.N. (1987). Artemisinin, a constituent of annual wormwood (*Artemisia annua*), is a selective phytotoxin. *Weed Sci.* 35, 499-305.

[76] Baggchi, G.D., Jain, D.C., Kumar, S. (1998). The phytotoxic effects of artemisinin and related compounds of *Artemisia annua*. *J. Med. Arom. Plant Sci.* 20, 5-11.

[77] Dayan, F.E., Hernández, A., Allen, S.N., Moraes, R.M., Vroman, J.A., Avery, M.A., Duke, S.O. (1999). Comparative phytoxicity of artemisinin and several sesquiterpene analogues. *Phytochemistry* 50, 607-614.

[78] Barbosa, L.C.A., Pereira, U.A., Teixeira, R.R., Maltha, C.R.A., Fernandes, S.A., Forlani, G. (2008). Synthesis and phytotoxic activity of ozonides. *J. Agric. Food Chem.* 56, 9434-9440.

[79] Barbosa, L.C.A., Maltha, C.R.A., Cusati, R.C., Teixeira, R.R., Rodrigues, F.F., Silva, A.A., Drew, M.G.B., Ismail, F.M.D. (2009). Synthesis and biological evaluation of new ozonides with improved plant growth regulatory activity. *J. Agric. Food Chem.* 57, 10107-10115.

[80] Bharati, A., Kar, M., Sabat, S.C. (2012). Artemisinin inhibits chloroplast electron transport activity: mode of action. *Plos One* 7, e38942.

[81] Dayan, F.E., Rimando, A.M., Pan, Z., Baerson, S.R., Gimsing, A.L., Duke, S.O. (2010). Sorgoleone. *Phytochemistry* 71, 1032-1039.

[82] Einhellig, F.A., Souza, I.F. (1992). Phytotoxicity of sorgoleone found in grain sorghum root exudates. *J. Chem. Ecol.* 18, 1-11.

[83] Gonzales, V.M., Kazimir, J., Nimbal, C., Weston, L.A., Cheniae, G.M. (1997). Inhibition of a photosystem II electron transfer reaction by the natural product sorgoleone. *J. Agric. Food Chem.* 45, 1415-1421.

[84] Rasmussen, J.A., Hejl, A.M., Einhellig, F.A., Thomas, J.A. (1997). Sorgoleone from root exudates inhibits mitochondrial functions. *J. Chem. Ecol.* 18, 197-207

[85] Czarnota, M.A., Paul, R.N., Dayan, F.E., Nimbal, C.I., Weston, L.A. (2001). Mode of action, localization of production, chemical nature, and activity of sorgoleone: a potent PSII inhibitor in sorghum spp. root exudates. *Weed Technol.* 15, 813-825.

[86] Nimbal, C.I., Yerkes, C.N., Weston, L.A., Weller, S.C. (1996). Herbicidal activity and site of action of the natural product sorgoleone. *Pestic. Biochem. Physiol.* 54, 73-83.

[87] Hejl, A.M., Koster, K.L. (2004). The allelochemical sorgoleone inhibits root H^+-ATPase and water uptake. *J. Chem. Ecol.* 30, 2181-2191.

[88] González-Bernardo, E., Aguilar, M.I., Delgado, G., King-Díaz, B., Lotina-Hennsen, B. (2003). Photosynthetic electron transport interaction of xanthorrhizol isolated from *Iostephane heterophylla* and its derivatives. *Physiol. Plant.* 119, 598-604.

In: Photosynthesis
Editor: Nafees Khan

ISBN: 978-1-63463-304-8

Chapter 12

PHOTOSYNTHESIS IN HUMANS

Arturo Solís Herrera, **María del Carmen Arias Esparza and Martha Patricia Solís Arias***
Human Photosynthesis Study Center, Aguascalientes, México

ABSTRACT

According to established knowledge, only 0.1% of the total light energy (visible and invisible) that reaches the surface of our planet earth is transformed into free chemical energy through the chlorophyll molecule. If only the human knew he had the ability to dissociate the molecule of water,and do it irreversibly. With this small fraction the food chain is initiated. Explaining the existing biomass based only on photosynthesis in plants requires the acceptance of dogmas such as the source of human energy is glucose or the ability of the eukaryotic cell to amplify the ATP; concepts that although they are widely accepted cannot be explained rationally. Hence our discovery of the intrinsic capacity of the melanin molecule to dissociate and re-form the water molecule involves a conceptual revolution of biblical proportions. It explains and demonstrates that photosynthesis exists in humans and therefore in mammals, and in each and every living entity where melanin is present. Its primary function is the production of free chemical energy from photonic energy, a behavior unique in nature.

Keywords: Melanin, chlorophyll, Energy, water dissociation, hydrogen, photosynthesis, humans

INTRODUCTION

The word photosynthesis means building or implementing something and using it for force or energy that contains light. On the other hand no one knows what energy is; we do not know if it is a point, a line, a wave, or a particle. We only know that it is inseparable from mass and is anything that causes a change, whatever it could be. By far the best known

* Corresponding Author email: comagua2000@yahoo.com

example to date occurs in the plant leaves, where chlorophyll, in the presence of purple and red light, i.e., the ends of the visible light, and water, is capable of dissociating this and generating hydrogen and oxygen molecules. Although the idea was glimpsed by Priestley, Lavoisier and others about 350 years ago, it is not understood that what a leaf of a plant can do at room temperature would,require a laboratory process to heat water to 2000°C,.

The initial reaction in the plant leaves can be outlined as follows:

$$2H_2O \rightarrow 2H_2 + O_2$$

We use the term initial reaction because the word photosynthesis, by convention, described the entire process by which the plant merges the CO2 with water to form glucose, which is not fully understood, the molecular dissociation of water is indispensable.

The reason that the dissociation of the water molecule is so essential, is because the plant leaf is transducing light energy (photonic) in free chemical energy, which is essential to drive every one of the chemical reactions, which, as a whole, make up what we call life. The true valuable product of the dissociation of water is diatomic hydrogen, with no oxygen, since hydrogen is the quintessential carrier of energy in the entire universe. Furthermore, oxygen is toxic at any concentration, so the plant leaf expels it to the atmosphere.

The dissociation of the water molecule happens in the chlorophyll of leaves of plants. It is considered the most important chemical reaction in the world because it is the beginning of the food chain. The transduction process of light energy into chemical energy through free dissociation of water is essential, otherwise neither the plant nor the seed will hatch. Hence the expression 'without water there is no life'.

An important observation is that despite the relatively well known formula, chlorophyll has not been synthesized, so, in order to experiment with it, it is necessary to remove it from the leaf, but after 20 seconds the water molecule does not dissociate more. The explanation is not known, therefore it is said that chlorophyll can transform the light energy (photonic) end of the visible spectrum in free chemical energy, so it needs to be within a living entity. Therefore, chlorophyll cannot explain the origin of life because first there was life and then there was chlorophyll.

CAN HUMANS CREATE PHOTOSYNTHESIS?

This was a question that rarely arose; it was a dogma that seemed to admit no doubt. Since dogma tells us that the energy source of the human eukaryotic cell is glucose, it should not be thought otherwise, and cellular metabolic processes were considered perfectly studied by a significant number of scholars. But in 1990, during an observational descriptive about the three leading causes of blindness; initiated because they are the same three causes around the world for sixty years or more (glaucoma, diabetic retinopathy and age-related macular degeneration age), this indicates that available treatments are not working. Here we begin to find the first signs.

Our working hypothesis at that time (1990) was based on the fact that the metabolic needs of the tissues are filled by the blood vessels. If you were studying the tiny blood vessels in the optic nerve, whose diameter in humans is 1200 microns or about twelve human hairs together, eventually we may find early pathology data to enable us to establish early treatment (Figure 1).

In that study, from 1990-2002, we examined about 6000 patients, whose studies retain the date. To perform the study, we used fundus cameras with some modifications, in order to highlight details that were of interest to us. We also used computer programs for image analysis as well as mathematical models to collect data from patients as well as to analyze our results.

The optic nerve magnifications we required were achieved in a relatively short time (Figure 2), and anatomical details of the vascular endothelial cells were sufficiently clear for the purposes we pursued. But with these magnifications the optic nerve, the ubiquity of melanin began to attract our attention. Virtually all six thousand patients who were then studied had melanin around or near the optic nerve (Figure 3). The question emerged spontaneously: why does nature place melanin, considered a simple sunscreen, protect us (theoretically) from UV rays in an anatomical structure that is 2.5 cm deep inside the eye?

It has been known for decades, perhaps centuries, that the inside of the eye is fully upholstered in melanin. However, physiology books explained it by arguing that its purpose was to absorb excess light coming into the eye, so that it avoided reflection into the eyeball and would have better image quality.

The presence of melanin goes unnoticed because the area where the tissue is located, called the choroidal layer, takes its name from the Greek choroid, which means the color grape because of its high content of melanin, 40% more than in the skin. The choroidal layer is located below the retina. But by using high-resolution digital images in the study of the optic nerve, the omnipresence of melanin strongly demands attention (Figure 4).

By keeping a digital photographic record of each patient, we noticed some inaccuracies that appear in the books of Ophthalmology. For example it says that the melanin found within the eyeball never changes, i.e., supposedly the same amount from when we are born until we die, but pictures showed us something completely different. In some patients, the amount of melanin visible with our methods changes from one year to another, which contradicted the old concept. The digital manipulation of fundus images constituted a key part of the study that finally allowed us to discover the intrinsic property of melanin to dissociate and re-form the water molecule (Figure 5).

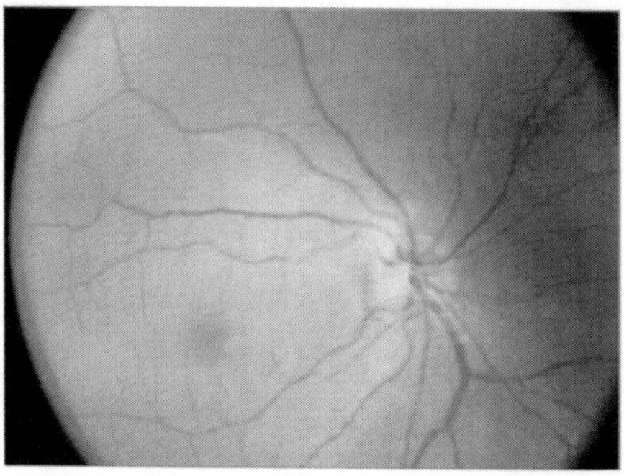

Figure 1. Photographs of right fundus of a normal patient, which can be seen clearly enough. Retinal vessels, which appear to converge in the optic nerve. To the left of the picture we can observe a small rounded area, named the macula by Leonardo Da Vinci, which is the area of maximum visual acuity.

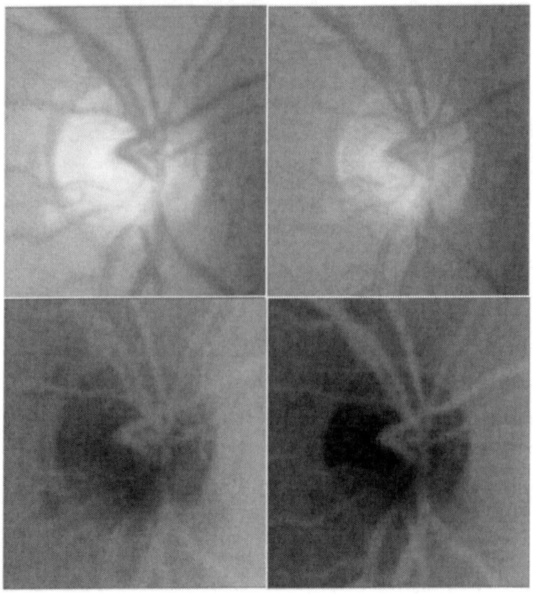

Figure 2. The analysis and digital processing of the captured images of the optic nerve in the living patient allowed us to study some important details about the anatomical characteristics of vascular endothelial cells that form the optic nerve blood vessels.

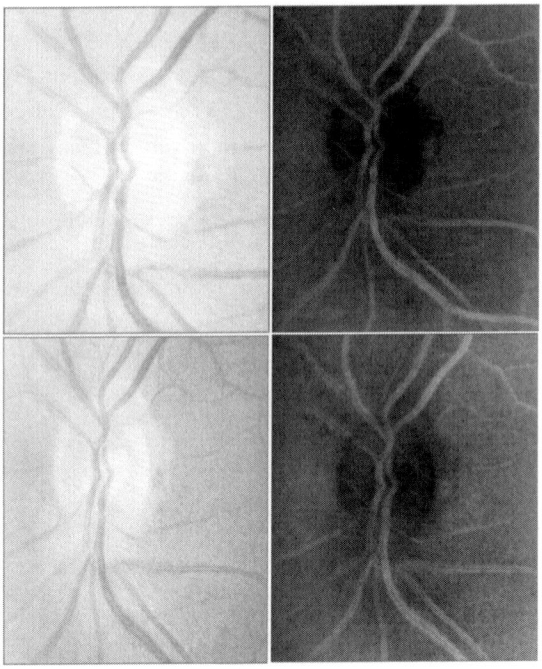

Figure 3. The uptake of fundus images in the living patient with high-resolution digital cameras allowed us to study much more clearly several interesting characteristics of the optic nerve vasculature and its surroundings. In the case of these images, note that by digitally processing them, very fine details can be observed, such as the high density of retinal capillaries. If you do not use image management programs they go unnoticed. In this patient we see a small patch of melanin in the meridian of the three. But we see a more or less extensive area of loss of melanin that is in the meridian of the twelve to six.

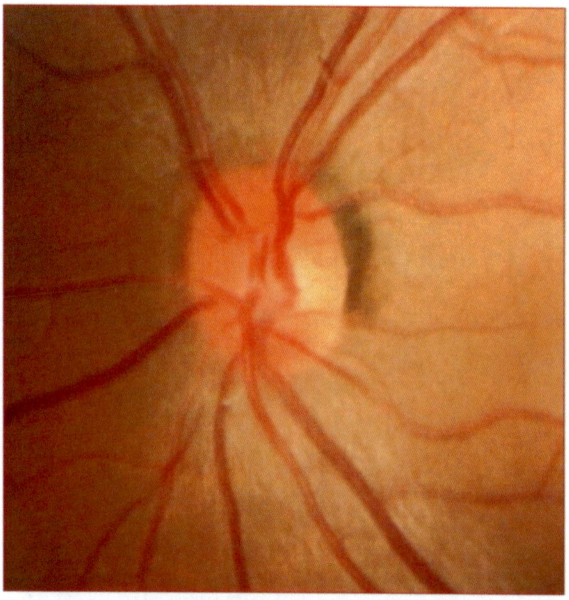

Figure 4. Left eye optic nerve of a living patient. Melanin is usually more noticeable in the temporal region of the optic nerve; such findings are considered embryonic remnants in the treaties of Ophthalmology. But we learned through the digital record that this is not so, as the amount of melanin can change, i.e., increase or decrease depending on the circumstances. We can say that the amount of melanin present in a given time is changing, is a multifactorial dynamic process.

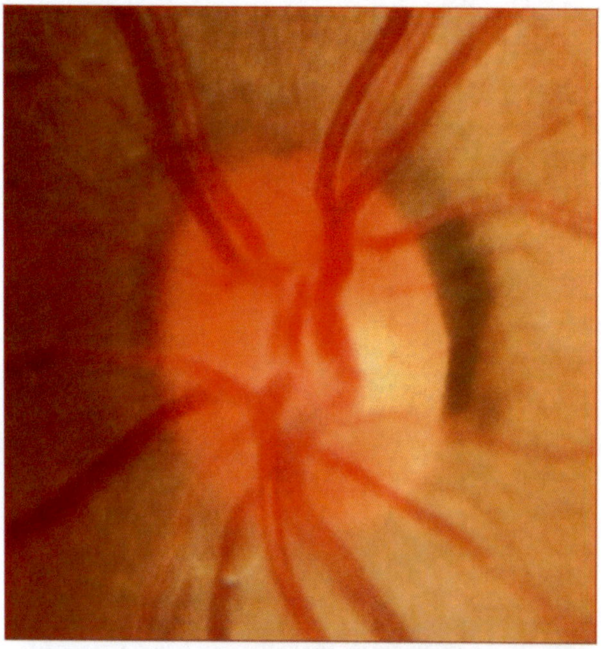

Figure 5. The image in Figure 4 at higher magnification allows us to observe in detail the presence of melanin in the edge of the optic nerve, extending the meridian from one to four. The photograph was taken using a polychromatic light source (400 to 700 nm), which is a source that emits all wavelengths of the visible spectrum.

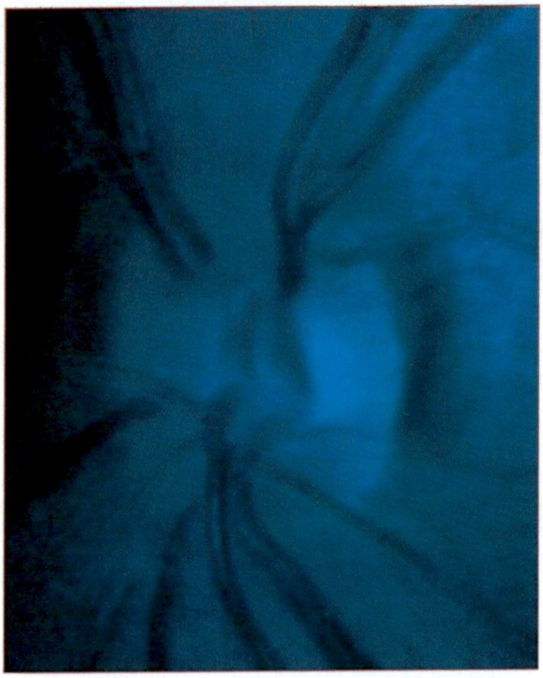

Figure 6. This photograph of the left optic nerve, the same as figure 5, was taken using a cobalt blue filter, which works in the region of 390 nm to 410 nm. As expected some details of tissue change markedly, others not so, but it maintains its dark area since it absorbs all wavelengths, which corresponds to the area where the melanin is present, namely: on the right edge of the optic nerve and meridian is one to four. The name melanin comes from the Greek melanos: dark.

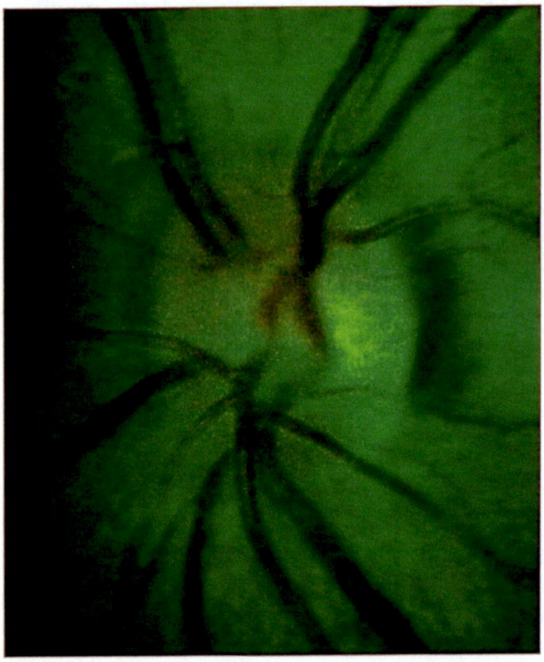

Figure 7. This photograph was taken with a light of roughly 520-570 nanometers. The details of the tissues are changed again, but the dark zone corresponding to the melanin remains unchanged.

Unlike chlorophyll, which absorbs only the ends of the visible spectrum, blue and red (400-700 nm), melanin has the amazing ability to absorb all visible and invisible wavelengths, i.e., from the waves of radio (10^6 cm) to cosmic rays (10^{-12} cm).

We can say that melanin can absorb any type of energy and is able to dissipate it in a unique way: by dissociation of water molecule. Figures 5, 6 and 7 show, in a simple manner, that melanin absorbs all wavelengths. Despite being illuminated with different wavelengths, the melanin always looks dark. Melanin is the darkest known substance.

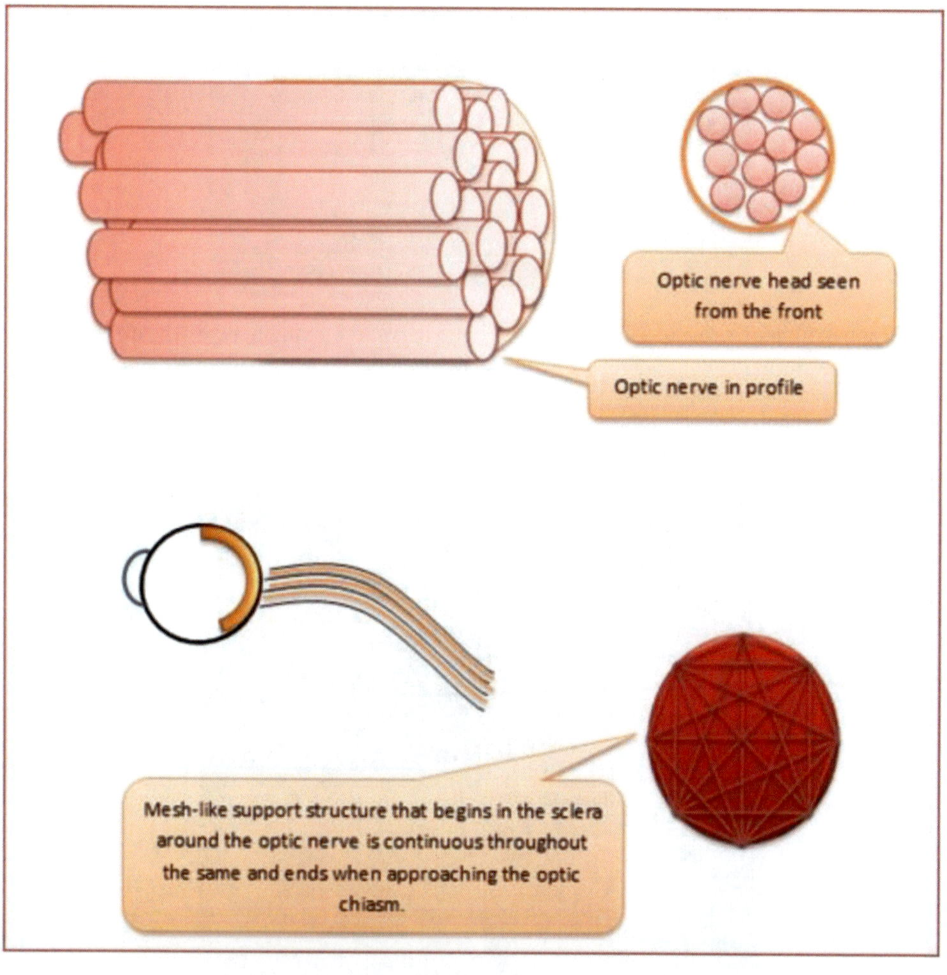

Figure 8. Axons that leave the ganglion cell layer of the retina are directed towards the optic chiasm, but they are wrapped all the way from a layer of fibrous tissue very similar to the sclera, which are indeed extensions of this, and continue along the optic nerve until it reaches the optic chiasm, where they vanish. We assume that these tubules of fibrous tissue protect the delicate optic nerve axons of intense eyeball movement as well as the damage that might cause the different movement inertia between the optic nerve and the eyeball. We could say that the lamina cribrosa sclerae is continued in the same manner starting in the eye up to the chiasm, where it disappears, providing structural support to the bundles of axons arising from the layer of retinal ganglion cells; so the concept of lamina cribrosa formation should change to sclera-chiasmatic support structure. (After the Greek letter "X" chi).

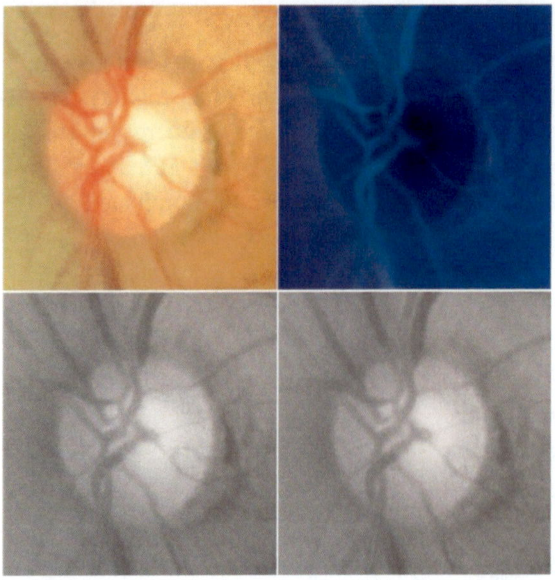

Figure 9. Composition of 4 images of the same left optic nerve of a living patient. The different wavelengths allow us to better compare the subtle anatomical differences that we are showing: the functional and anatomical status of the same optic nerve and surrounding tissues. As appears in this case where the meridian of 10 to 5 shows a zone of lower functionality, inferred by the characteristics of the tissue located between the crescent of melanin and the edge of the optic nerve.

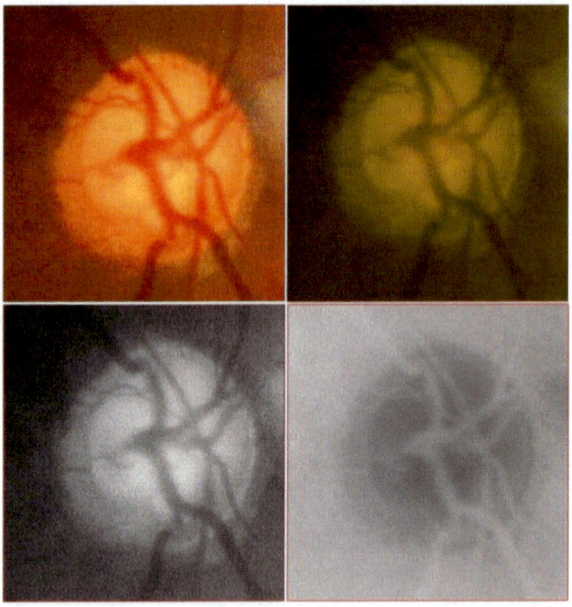

Figure 10. It seems that melanin is formed in the temporal edge as it advances in age, and after a while, it begin to move away from said edge, which in some cases means an impairment in the function of the tissue structure involved. The case illustrated in these images corresponds to glaucomatous optic neuropathy, where you can see the loss of optic nerve substance, i.e., axons and supporting tissue (glia). The digital image analysis allows determining more appropriate tissue changes, and it is observed that the tissues adjacent to the edge of the optic nerve are also affected as they demonstrate changes in coloration, which are usually more homogeneous.

Melanin

Melanin is ubiquitous in nature, being found in most organisms, in all biological kingdoms. Its role has not been understood so far. So the discovery of its intrinsic ability to dissociate and re-form the water molecule and therefore in a unique way transduce light energy (photonic) of the entire electromagnetic spectrum, transforming it into free chemical energy, which is essential as an energy supply for eukaryotic cells. This breaks new ground in several areas of knowledge, not only in biology and medicine.

The unexpected capacity of melanin to dissociate and re-form the water molecule Solis (2013) can be represented as follows:

$2H_2O \leftrightarrow 2H_2 + O_2 + 4e-$

The similarity of the reaction with the very first reaction of photosynthesis in plants was the reason the process is called human photosynthesis, so this term is used in a totally descriptive way.

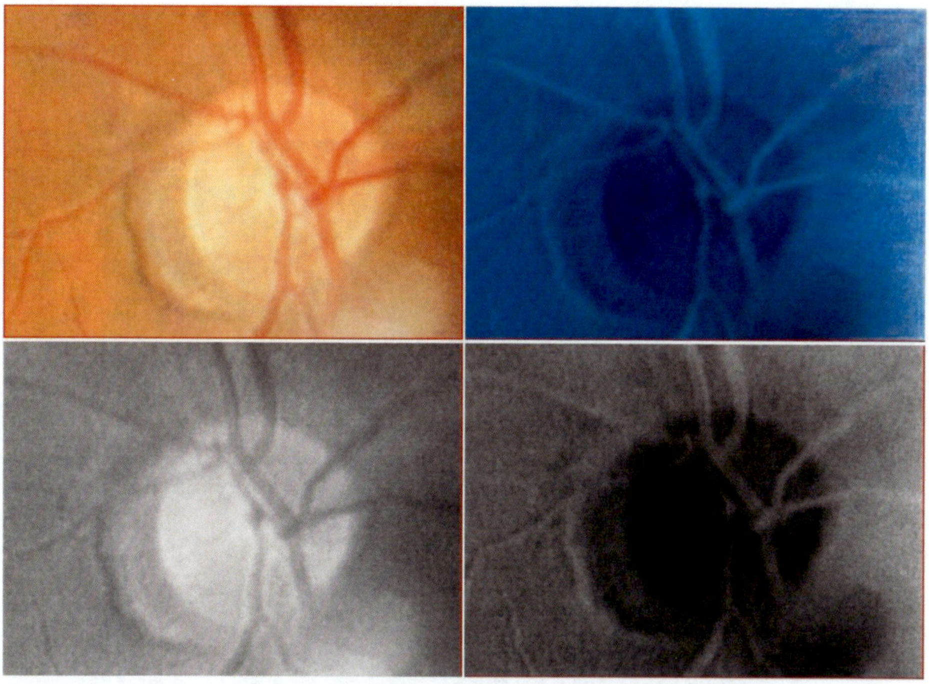

Figure 11. As we advance in age, the melanin line of figure resembling a half-moon moves away from the edge of the optic nerve, and the tissue that lies between the line of melanin and the edge of the left optic nerve is scarce of vitality and therefore decreases functionality in varying degrees.

Photosynthesis in humans is more efficient than in the leaves of plants, because in them, the chlorophyll molecule irreversibly dissociates water and only in the presence of visible light. But our photosystem, consisting of Light / Melanin / Water in order of abundance in the universe, is not only able to dissociate the water molecule in the presence of visible and invisible light, but it does so in a reversible manner; it also has the amazing ability to re-form it, producing four high-energy electrons for every two molecules of water that are reconstituted.

Therefore we can infer that nature developed a self-sustainable fuelcell, or in other words, nature is a fuel cell capable of generating its own hydrogen, so it is capable of operating for hundreds of years. Do not forget that melanin is the most stable substance known; it is found in good condition in fossilized squid ink sacks that died 180 million years ago (Goldsmith, 1981).

The formula of melanin is not known. We can say that it could be found despite the current technological advances, as any form of energy that is applied to this molecule (melanin) will be absorbed in its entirety and not return anything, which is also unique behavior. The rest of the known molecules absorb certain wavelengths of light, the melanin absorbs all alone, and then emits a characteristic signal that can be identified by some physical or chemical method.

When melanins are isolated from living tissue, they produce an insoluble and amorphous mud that defies analysis by classical techniques (Hill, 1992). In 1972 it was said that melanin was capable of transducing a type of energy into another. It was published that it transformed the light into electrical energy and heat (Crippa et.al., 1989). However the mechanism of production of electricity is explained differently and much less accurately than the concept that light melanin is also transformed into heat, as is the case with almost all known molecules.

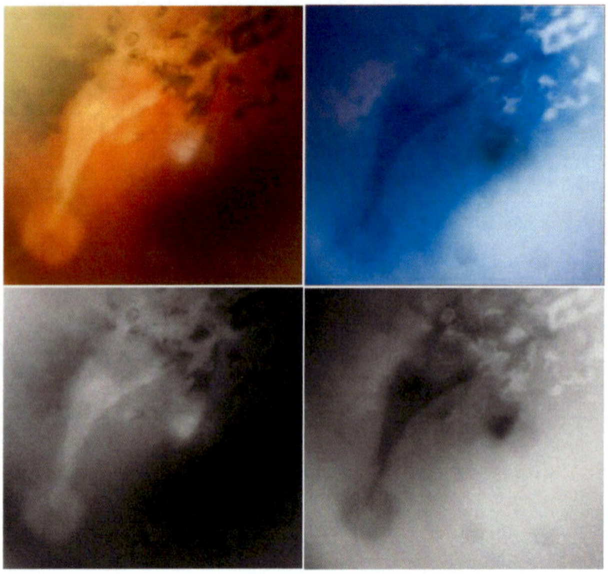

Figure 12. This composition of 4 images shows the results of the application of a laser in a patient with diabetes mellitus, the second leading cause of blindness worldwide. The dark spots are the result of the laser application, which is an example of electromagnetic radiation used therapeutically. The dark areas reflect increased melanin synthesis targeted to areas where the tissue was treated with such electromagnetic radiation. It is an observed fact that electromagnetic radiation induces the synthesis of melanin, and the explanation, previously unknown, is the increased availability of photonic energy; the cell or any living entity responds by increasing its ability to uptake this energy in order to transform it into free chemical energy, and one of the main components of such adaptation, and certainly not the only, is an increased amount of melanin.

Our finding of the previously unknown intrinsic property of melanin to dissociate and re-form the water molecule answers questions such as:

- What is the exact nature of the energy dissipation process and photodynamic?
- What causes the dramatic change in eumelanin electrical conductivity with hydration? (Meredith and Sarna, 2006).

That many different forms of energy are absorbed by melanin and dissipated as heat is a misconception about the dissipation of energy absorbed, which has remained for decades, because melanin is intractable in the laboratory.

Perhaps the difference in the findings between us and researchers who have studied melanin for years is circumstantial. We studied the melanin in one of its many natural states, which allowed us to observe the reactions of cells and tissues surrounding the presence of the pigment.

Some Examples of how the Anatomy and Function of the Optic Nerve and Surrounding Tissues Changes According to the Activity and Amount of Melanin Present

Since there was no ability to explain the main function of melanin in the human eye and therefore not in the body, pigmentary changes were always interpreted as secondary to primary changes in the other tissues, either the optic nerve, retina, the macula or choroid.

Melanin tends to approach and to surround the sources of energy, and possibly intense eyeball movement whose inertia causes an even more intense optic nerve movement, which is visible by ultrasound, in addition to other anatomical and physiological factors, which may explain the abiding presence of melanin around the optic nerve, especially in the right temporal region (Figure 8). You can describe the area of junction between the optic nerve and the eyeball as a region which converges the different inertia and traction vectors from both the optic nerve and those generated by the movement of the eyeball.

In areas where traction between tissues is intense, the synthesis of melanin is more dynamic, more mobile throughout life, as the temporal region of the optic nerve edge (Figure 9), whereas fewer retinal vessels that provide support to the structure together, reflect the higher intensity of the different inertias each converging at one point among tissues that make up the area. Two optic nerves or even equal distribution of identical melanin, even between the two eyes of the same patient, will always have at least tiny differences (Figure 10).

So far it has been considered that pigmentary changes are secondary to primary changes in other tissues such as the retina or their axons, but the discovery of the intrinsic property of melanin to dissociate and re-form the water molecule breaks the paradigm, given the critical importance of energy in cell biology, then pigmentary changes are initial and other tissue changes are secondary (Figure 11).

It is an observed fact that electromagnetic radiation activates the synthesis of melanin, and this is in all body tissues, so the retina is no exception. It empirically takes place, even without understanding why, including whether the therapeutic aims pursued are different. For example, it is thought to close choroidal vessels can be without consequences for the functioning of the eye and retina, a concept that is far from reality (Figure 12).

To date there is no explanation for the very limited beneficial effects observed in the retina of diabetic patients in whom laser photocoagulation is applied.

Initially it was to induce the formation of scars; it is a very old observation that patients with less retinopathy developed chorioretinal scars than healthy eyes, so that at first it was the end. After it was suggested that it was actually the destruction of the photoreceptors that appeared to improve the appearance of the retina and decreased oxygen demand. The closure of the abnormal vessels is another therapeutic purpose of photocoagulation, although this could be achieved without laser energy applied directly to the vessels. The application of laser to the retina has been used extensively and intensively in several countries as a treatment to try to halt the progression of diabetic retinopathy. But although it has been used since 1970 and increasingly, the statistics still show that the incidence and prevalence of diabetic retinopathy remains the second leading cause of blindness in the world to date.

Invariably pigmentary changes accompany the retinal diseases, and now we know that it precedes them (Figure 13). Since the last century, or perhaps since the invention of the direct ophthalmoscope in 1851 by Helmholzt, pigmentary changes that were observed in the eye were considered to be an entirely secondary effect, unimportant. Melanin has always been relegated to only a biological role attributed to him and the absorption of UV rays. With the discovery of its role as a source of free chemical energy, concepts changed radically thereby the pigmentary changes are initial, key and then follow the rest.

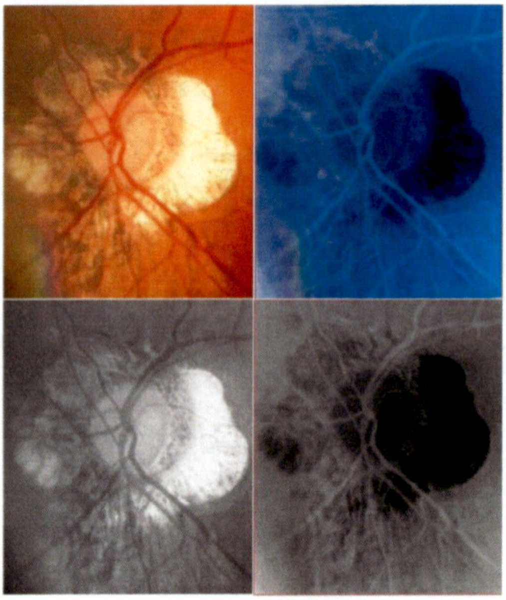

Figure 13. The composition shows changes in both the optic nerve and in the surrounding tissues. The image is very different from the other examples, because the pigment in this case seems to have disappeared entirely in the right side of the optic nerve, so the sclera, with its characteristic white color; is clearly visible and on the left side of the picture, the distribution of the pigment and choroidal blood vessels anatomical characteristics do not show the homogeneity that is a feature of a healthy tissue.

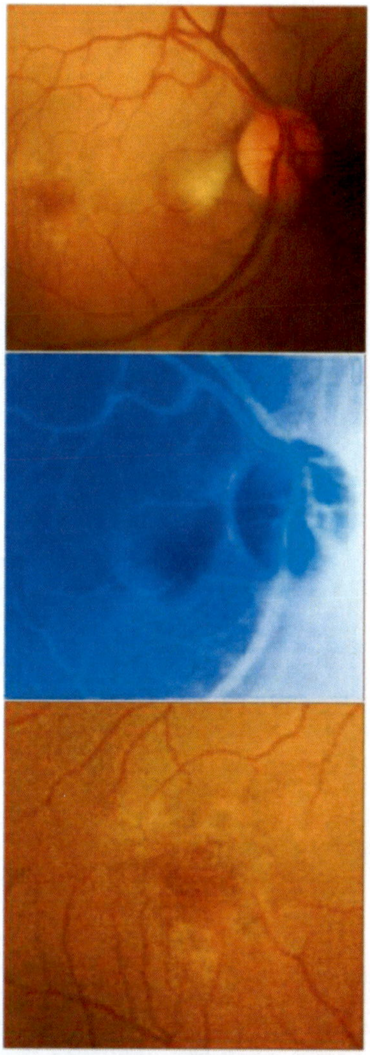

Figure 14. In this three photographs we see that pigmentary composition changes are not unique to the region of the optic nerve and therefore the macular area is no exception. In this case pigmentary changes are seen around the macula, a structure that represents the area of greatest definition of the image. The amplification can be observed in round spots, which appear to surround the bottom of the macula, called umbo.

Melanin can be found in almost all tissues of the eye and the human body in general, the amount being dependent on the location and function. Changes in its distribution, homogeneity, and relation amount changes with nearby structures in the case of disease (Figure 14).

Now we think that pigmentary changes are at initio, thereafter changes were seen in the surrounding tissues, which were the result of the alteration in the levels of free chemical energy that emanates from the melanin. If pigmentary changes, which probably reflect changes in the levels of human photosynthesis are too long or tend to increase, it will then add anatomical and functional changes of the affected tissues (Figure 15)

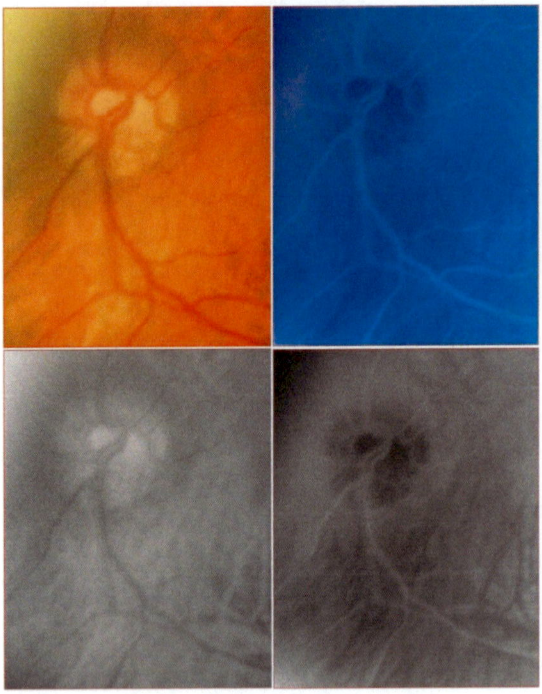

Figure 15. The composition of four photographs showing the changes that can happen to the optic nerve and surrounding tissues when the photonic energy transduction in free chemical energy has not been adequate since gestation. It can be seen that the optic nerve head shows an abnormal inclination, it seems to look down, and the usual pigment changes in the affected area are present. On the other hand the characteristics of choroidal vessels visible in these photographs are not usual. These changes are habitual present at birth, to infer low levels of free energy within the uterus.

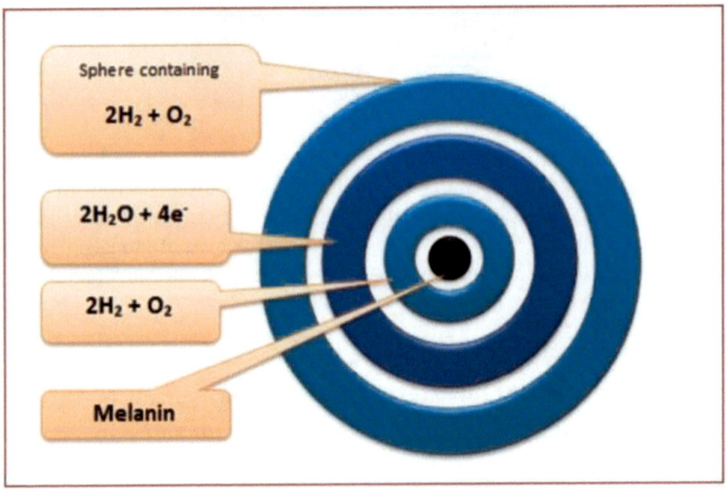

Figure 16. Melanin releases the free chemical energy symmetrically in all directions, like increasing energy fields that are gradually spreading and alternating nature, i.e., a sphere will contain diatomic Hydrogen and Oxygen, the result of the dissociation of water, and the next sphere and the previous will contain re-formed water and high-energy electrons. The reaction is not symmetrical over time, because as any chemical reaction depends on the concentration of the reactants, temperature, pressure, humidity, the amount of light, and other factors which may be known or not.

Melanin seen from Different Distances

Melanin is an amorphous substance, able to absorb any type of energy. Its formula is not known, its biological function was unexpected because it was thought that was as stable as any role could have. But the intrinsic property of melanin to dissociate and re-form the water molecule now placed in the foreground of biomolecules, fills the requirements to be considered the origin of life. Its appearance is fascinating (Figure 17).

Figure 17. Photograph of melanin taken in our laboratory of the Center for the Study of Human Photosynthesis in Aguascalientes, Mexico, at the distance of 25 cm (Solis-Herrera, 2013).

Figure 18. The appearance of melanin at a distance of less than 10 cm (Solis Herrera 2013).

The Melanin in Different Biological Kingdoms

Melanin has a remarkably steady behavior inside and outside the eukaryotic cell. We think that its function is the same wherever you are and that it is the photonic energy transduction into free chemical energy.

It is observed that melanin tends to approach energy sources, and if they reach them tends to go around. It is therefore not surprising that optimal uptake of photonic or light energy is what seems to guide its distribution in any form of life, for example in the cuticles of insects. (Figure 21)

Figure 19. Photographs of melanin taken through a light microscope is therefore a distance of less than one inch. Observe the color of melanin (Solis Herrera, 2013).

Figure 20. Melanin hundreds of light years away and the color is the same. Melanin forms large galaxy-sized mass in stellar space. (Source of photography: www.nasa.gov/).

This organization is repeated in different kingdoms described in biology, i.e., melanin is mainly found within the outer edge of the living being, but their presence is also required in each and every one of the cells that comprise it, although its amount, orientation of the molecule, granule size, combination with metals, etc., depends on its location and function (Figure 22).

Figure 21. The outer layer of the arthropod, in this case a scorpion, is completely impregnated with melanin, which is logical because of how the insect captured the photonic energy present in the environment in a more efficient way, transforming it into free chemical energy that is used within the arthropod body. The schematic representation of the scorpion bio-system is an oval-like shape, where the melanin is present mainly in the cuticle, and the interior is represented by a light brown color given hemoglobin containing copper as prosthetic group.

Figure 22. In fish we also observe the usual distribution of melanin, surrounding the whole fish, allowing for optimal absorption of electromagnetic radiation that eventually will be transformed into free chemical energy, the real source of energy for the living entities. Just think that if the energy source was glucose, diabetics then would fly. The oval represents the organization of the fish, where a significant amount of melanin is part of the outer edge and the interior is a pink color representing fish hemoglobin that contains iron as prosthetic group.

Figure 23. Photograph of the head of a common mosquito (Culexpipiens) where it is seen that the amount of melanin is higher than the rest of the body, as is usual, and in the circle that represents the bio-system we have the melanin located primarily at the interface between the environment and the biological system, in this case the eye and its interior are represented by the color blue that symbolizes water. In general terms the eyes of a living thing contain a high amount of water, in some cases close to 97%.

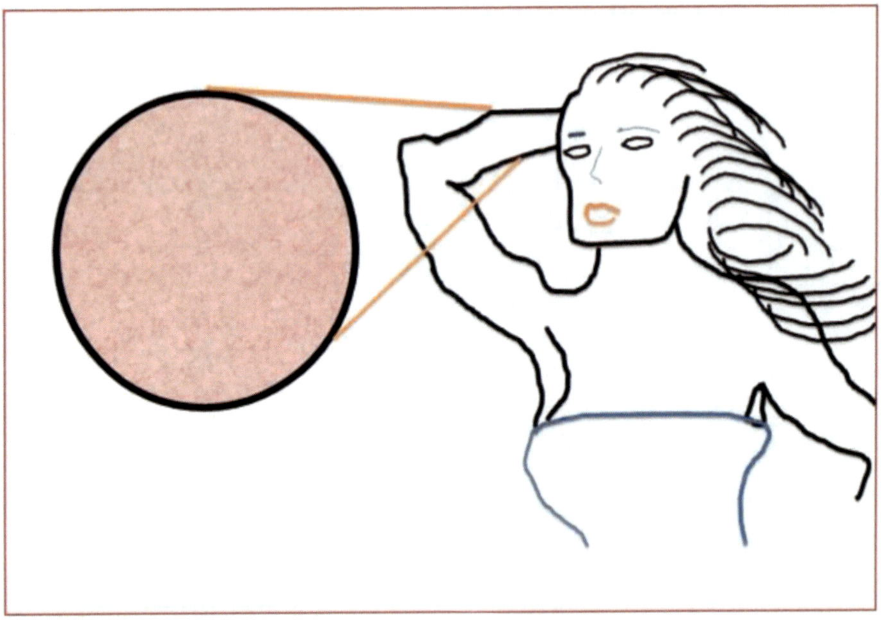

Figure 24. In man the disposition of melanin is similar to other living things, as the pigment is found mainly in the skin, but also in each and every one of the 100 trillion cells that make up the human body. So the diagram showing the organization of human tissues, we have melanin, which is an indispensable part of the interface between the environment and inside the body, much like the previous examples, and the interior is represented in pink because human hemoglobin contains iron as prosthetic group.

Since melanin is such a stable molecule (millions of years) and has a massive nature, because its molecular weight is estimated at a million Daltons; the fact that melanin is part of the interface between the environment and the living entity represents many advantages, besides the primary, which is the uptake of light energy and its subsequent transformation into free chemical energy.

The eye differs in different biological kingdoms on issues such as form, organization, maybe location, motility, including the anatomy and arrangement of the photoreceptors. But what does not change is that in each and every one of them melanin is omnipresent (Figure 23).

Hemoglobin Is to the Animal Kingdom what Chlorophyll Is to the Plant Kingdom

In 1930 Hans Fischer described heme as the compound that makes blood red and grass green. If the red blood and the green of the plants come from closely related compounds, called the heme group, we could also expect that they had similar functions within living organisms found. We could also expect that the main function of the heme group is the irreversible dissociation of the molecule of water to produce hydrogen diatomic and diatomic oxygen and therefore free chemical energy production is consistent with what we observe in nature.

The differences between the hemoglobin of mammals, plant chlorophyll and the hemolymph of arthropods mainly contain the prosthetic group and the structure that supports them (Figures 25, 26 and 27).

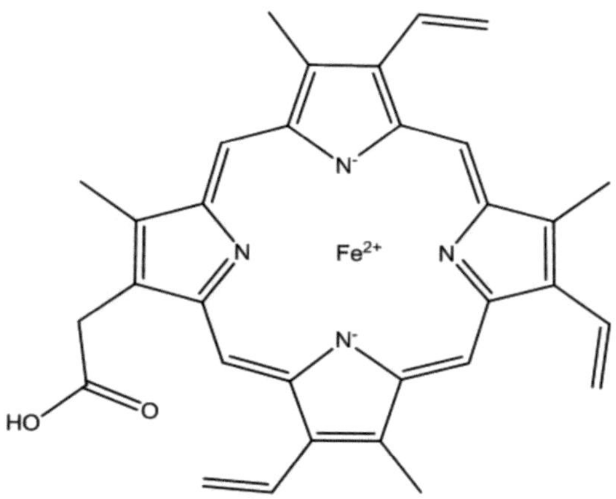

Heme group (Hemoglobin)

Chemical Formula: $C_{32}H_{28}FeN_4O_2$
Exact Mass: 556.16
Molecular Weight: 556.44
m/z: 556.16 (100.0%), 557.16 (37.3%), 558.16 (7.6%), 554.16 (6.4%), 555.16 (2.3%), 557.15 (1.5%)
Elemental Analysis: C, 69.07; H, 5.07; Fe, 10.04; N, 10.07; O, 5.75

Figure 25. The prosthetic group in hemoglobin is the Fe2 +, the structure that supports and complements its function is globin.

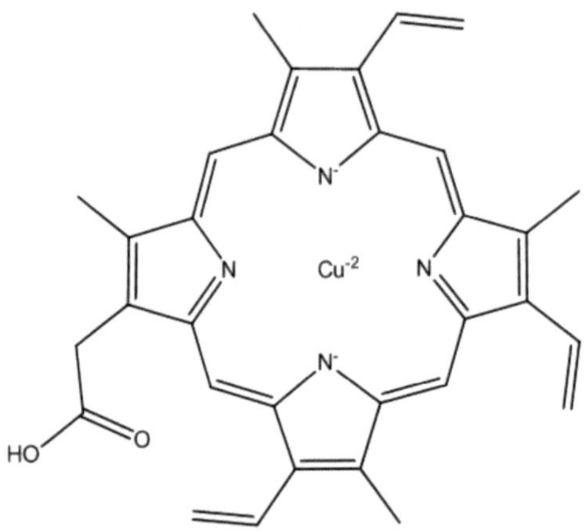

Hemacyanin in
Arthropods Blood
(Haemalymph)

Chemical Formula: $C_{32}H_{28}CuN_4O_2^{4-}$
Exact Mass: 563.15
Molecular Weight: 564.14
m/z: 140.79 (100.0%), 141.29 (45.1%), 141.04 (36.1%), 141.54 (16.2%), 141.29 (6.4%), 141.79 (2.7%)
Elemental Analysis: C, 68.13; H, 5.00; Cu, 11.26; N, 9.93; O, 5.67

Figure 26. Hemocyanin of arthropods blood containing Cu-2 as a prosthetic group, as chlorophyll and hemoglobin is capable of irreversibly dissociated in the water molecule.

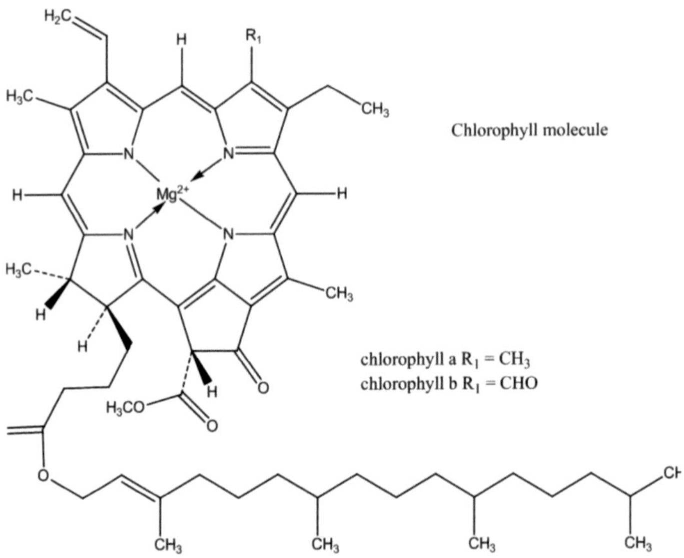

Chemical Formula: $C_{55}H_{71}MgN_4O_5^{2+}$
Exact Mass: 891.53
Molecular Weight: 892.48
445.76 (100.0%), 446.26 (73.2%), 446.76 (41.1%), 447.26 (11.2%), 447.27 (4.2%), 447.76 (2
446.26 (1.5%), 447.77 (1.2%)
Elemental Analysis: C, 74.02; H, 8.02; Mg, 2.72; N, 6.28; O, 8.96

Figure 27. The prosthetic group of chlorophyll is Mg2 +, and polar tail supports it.

Figure 28. Melanin´s distribution seems to recur in different forms of life on earth, as it forms the interface between the organism and the environment, surrounding the former in its entirety. And like in other kingdoms, melanin is not only found in the cortex, but also in the interior.

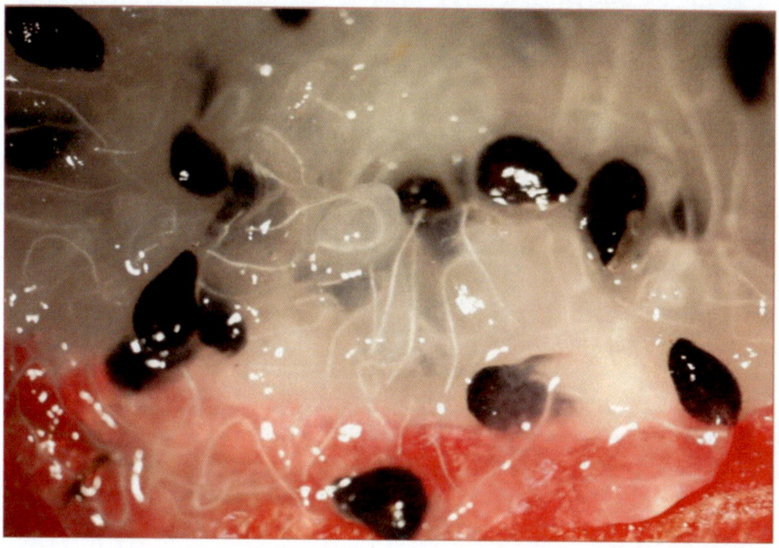

Figure 29. Dragon fruit seeds, with a stem that supports but also communicates with the rest of the internal structure. Melanin content is obvious.

Melanin Is to the Animal Kingdom that Lignin Is to the Vegetable Kingdom (Solis et al., 2012)

Melanin is also found in the plant kingdom, although it is called lignin, and is located in the trunk, but regardless of the name, it is a dark substance whose formula is also unknown, as is melanin, and also follows the distribution found in other biological entities, thus surrounding the entire structure, also forming the interface between the environment and the living entity (Figure 28).

The intrinsic property of melanin to dissociate and re-form the water molecule explains, among other things, the mystery of why seed needs water to hatch. Observation of nature finds that melanin is often contained within them (Figure 29).

The ability of melanin to dissociate the water molecule, explains in a simple way the mystery of how seeds know when there is enough rain to hatch, because when the amount of water is small, the free chemical energy produced will be small and not enough to jump-start the biochemical reactions that together produce germination (Solis, 2013).

The levels of free chemical energy are required to be present within an appropriate range so that the necessary chemical reactions can propel the way (Figure 30).

So in humans, melanin dissociates and re-forms the water molecule, and blood hemoglobin dissociates it irreversibly; whereas in plants, the leaf´s chlorophyll irreversibly dissociates the water molecule, but the melanin in the trunk, despite being called lignin, dissociates and reforms the water molecule thereby producing free chemical energy. One might say that the reason that the melanin forms the dark universal wrapper is now understood as the melanin tends to approach energy sources.

We conclude that the photosystem composed of Light/ Melanin/ Water, arranged in order of abundance in the Universe, is the source of all life, and certainly not just on planet earth.

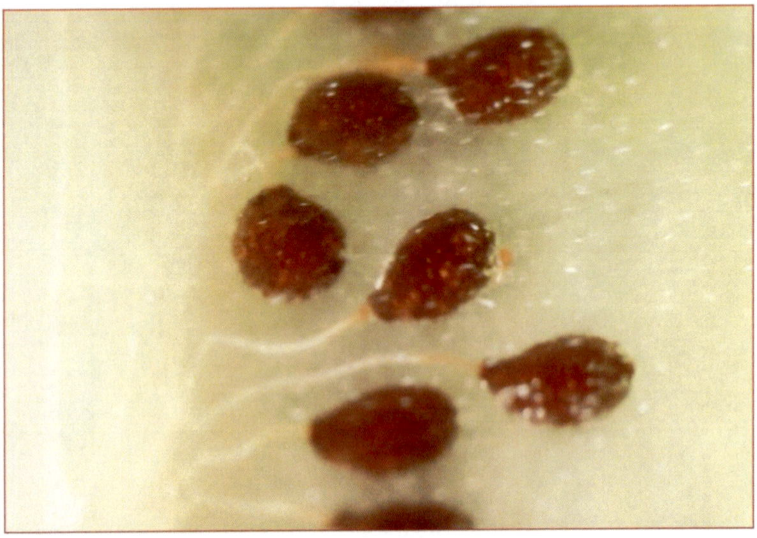

Figure 30. Kiwi seeds observed with a magnification of 20 X. The presence of melanin is unobjectionable. The free chemical energy required for flowering depends on the amount of liquid water and light present. We can say that each and every one of the chemical reactions required for germination, depend entirely of free chemical energy from the dissociation of the water molecule and its re-formation.

The Biological Event Led to a Test Tube

Transducing light energy into chemical energy freely opens the possibility of many applications. With chlorophyll this has not been achieved; in the last 20 seconds of extraction from the leaf, chlorophyll splitting water stops. Preventing the inactivation of chlorophyll is a challenge that cannot be overcome. Not so with the melanin. In 2006, it was the first time we could replicate the biological event in the test tube. The result was amazing for us although it was expected, given the extraordinary stability of the molecule of melanin, as the first LED bulb that could ignite, and continue on until date, day and night (Solis et al., 2010). The melanin to dissociate and re-form water works as a fuel cell that produces its own hydrogen, so that it breaks ground in the area of sustainable energies. In our laboratory we have continued working with energy and developing the process, and achieved solid state in 2009 (Figure 31).

Figure 31. Solid State Batteries constructed of melanin. They have shown good stability and their ability to work has been increasing based on experimentation and continuous work.

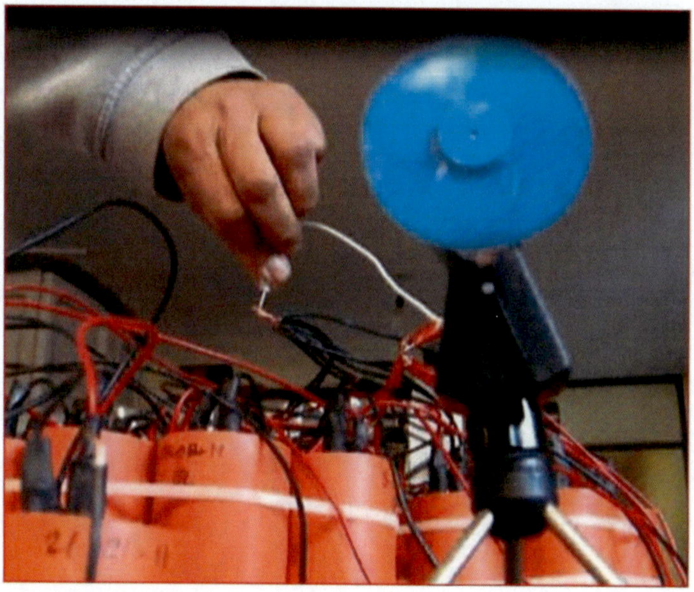

Figure 32. Photographs taken in our laboratory in 2010, where we see that the flow of energy is suitable for some small applications.

At present we can move a small engine (Figure 32), which shows that with tenacity it can move forward. When the first battery was built it could only achieve 40 millivolts of output, now, on average, each battery produces 1.2 volts on average.

The search for clean, efficient, self-sustaining energy, that can be applied on a large scale without destroying or warming the planet, seems to have come to an end with the discovery of photosynthesis in humans.

The scientific basis is already given. The following is the technological development, which of course is not without challenges, but there is no doubt that the intrinsic property of melanin to dissociate and re-form the water molecule is a totally new form of energy.

Acknowledgments

We appreciate the support of the Center for the Study of Human Photosynthesis.

References

Crippa, P.R., Cristofoletti, V., Romeo, N., 1989. A Band model for melanin deduced from optical absorption and photoconductivity experiments. *Biochim. Biophys. Acta* 538, 164-170.

Goldsmith, LA., 1981. Biochemistry and physiology of skin. New York. *Oxford University Press*.

Hill, Helene Z., 1992. The function of melanin or six blind people examine an elephant. *Bioessays*, 14(1), 49-50.

Meredith, Paul, Sarna, Tadeuz, 2006. The Physical and Chemical Properties of Eumelanin. *Pigment Cell Res*. 19, 572-594.

Solis Herrera, Arturo, Arias Esparza, M.C., Solis Arias R.I., 2012. Human Photosynthesis. In: Claudia, W., Melinda, M., (eds.), *Vegetable consumption and Health*: *New Research*. Chapter 3. *Nova Science Publishers*, New York.

Solis Herrera, Arturo. *United States Patent* No.: US 8,455, 145 B2; Date of Patent: June 4 2013.

Solis, H.A., 2013, The Human Photosynthesis, *Infinity Publishing*, 2013.

Solis, H.A., Arias Esparza, M.C., Solis Arias, R.I., 2010. The unexpected capacity of melanin to dissociate the water molecule fills the gap between the life before and after ATP. *Biomedical Research* 21(2), 224-226.

About the Editor

Nafees Khan, Ph.D.
Professor
Department of Botany
Aligarh Muslim University
Aligarh 202002, India
Tel/Fax: 91-571-2702016
E-mail: naf9@lycos.com

INDEX

A

B

C

D

E

F

G

H

I

J

K

N

O

P

S

T

U

V

W

X

Y

Z